ALLE ZEIT WACH 1842

Clinical Oxygen Pressure Measurement

Tissue Oxygen Pressure and Transcutaneous Oxygen Pressure

Edited by A. M. Ehrly, J. Hauss, R. Huch

With Contributions by
D. Abendroth, H. Baumgärtl, G. I. J. M. Beerthuizen, E. Böhm
A. Creutzig, A. M. Ehrly, B. Endrich, U. Ewald, F. Fallenstein
W. Fleckenstein, U. K. Franzeck, W. Funk, W. Hartel, J. Hauss
R. Heinrich, R. Huch, L. H. P. Keller, T. Kersting, H. M. Kohaus
H. Landgraf, R. Lemke, D. W. Lübbers, A. Ott, H. Partsch
K. H. Rühle, O. Ruland, R. Schultheiß, N. Spannbrucker
H. U. Spiegel, J. M. Steinacker, M. Thermann

Translated by Virginia Sonntag-O'Brien

With 180 Figures and 19 Tables

Springer-Verlag Berlin Heidelberg New York
London Paris Tokyo

Professor Dr. med. Albrecht M. Ehrly
Zentrum der Inneren Medizin, Klinikum der Johann Wolfgang Goethe-Universität, Theodor-Stern-Kai 7,
D-6000 Frankfurt/Main, Federal Republic of Germany

Professor Dr. Jan Hauss
Chirurgische Klinik/Poliklinik der Universität,
Jungeblodtplatz 1,
D-4400 Münster, Federal Republic of Germany

Professor Dr. Renate Huch
Universitätsfrauenklinik, Frauenklinikstraße 10,
CH-8091 Zürich, Switzerland

Translator:
Virginia Sonntag O'Brien
Belfortstraße 19,
D-7800 Freiburg, Federal Republic of Germany

Based on papers and discussions of the 2nd Symposium on Tissue Oxygen Pressure: Measurement of tissue oxygen pressure and transcutaneous oxygen pressure ($tcpO_2$) in adult patients held on June 1/2 1984 in Frankfurt, Federal Republic of Germany

---

Title of the original German edition:
Klinische Sauerstoffdruckmessung
©1985 Münchner Wissenschaftliche Publikationen, Plinganserstraße 13, D-8000 München 70, Federal Republic of Germany

ISBN-13:978-3-642-71228-9 e-ISBN-13:978-3-642-71226-5
DOI: 10.1007/978-3-642-71226-5

Library of Congress Cataloging in Publication Data
Symposium on Tissue Oxygen Pressure (2nd : 1984 : Frankfurt am Main, Germany) *Clinical oxygen pressure measurement.* "Based on papers and discussions of the 2nd Symposium on Oxygen Pressure ... held on June 1/2, 1984 in Frankfurt, Federal Republic of Germany" - T.p. verso.
Translation of: Klinische Sauerstoffdruckmessung. 1. Tissue respiration-Congresses. 2. Oxygen in the body-Measurement-Congresses. I. Ehrly, A.M. II. Hauss, J. (Jan), 1947- . III. Huch, Renate. IV. Abendroth, Dietmar, 1950- . V. Title. [DNLM: 1. Oxygen Consumption-congresses. W3 SY538F 2nd 1984k/WF 11 S989 1984k]
QP88.S96 1984 616.07'5 87-22109
ISBN-13:978-3-642-71228-9 (U.S.)

2119/3145-543210

Preface

In January 1980, the First Symposium on the Measurement of Tissue Oxygen Pressure in Patients was held in Frankfurt. After a 4-year "rest period," the organizers of the 1984 symposium, Professor R. Huch of Zürich and Dr. J. Hauss of Münster, together with myself, extended another invitation to come to Frankfurt to find out what had gone on in the field of oxygen pressure measurement and its application in clinical medicine. As the following presentations will show, the application of oxygen pressure measurements has been broadened considerably. Furthermore, technological advances have been made, particularly with the increased use of computers. For various reasons, including technical ones, these methods have not been adapted as widely as one would want. Although congresses on tissue oxygen pressure have been held in the last few years, the clinical aspect of tissue PO_2 measurement has not been dealt with in such a concentrated and comprehensive way since 1980. It therefore seemed necessary to hold such a symposium, not only for scientific reasons, but also to enable a larger group of clinicians to gain insight into the importance of the possibilities these methods offer.

We purposely did not confine the presentations given at this symposium to tissue oxygen pressure measurement in patients, but included findings from animal experiments with close relation to clinical aspects, as well as transcutaneous PO_2 measurements. Unfortunately, there has been some confusion among the interested public about the terms tissue oxygen pressure, on the one hand, and transcutaneous PO_2, on the other. Thus, it is one of the tasks of this symposium to establish a clear distinction between these methods and terms. Given the angioarchitectonic complexity of the skin, it is no wonder that it is particularly difficult to judge the skin, not only in terms of its circulation, but especially with regard to its oxygen supply. One paper, therefore, reports on initial results of measurements of actual skin tissue oxygen pressure.

The organizers of the now published symposium are proud of the fact that it was possible to bring together in Frankfurt almost all of the groups that have carried out oxygen pressure measurements in patients. We are especially grateful to Professor D. W. Lübbers, Director of the Max Planck Institute for System Physiology in Dortmund, who held the introductory presentation, as he did 4 years ago.

The organizers also wish to thank the pharmaceutical industry, and particularly the Albert-Roussell Co., whose support made this symposium possible. Furthermore, I would like to thank the Department of Human Medicine of the University of Frankfurt and, in particular, Dean H. J. Müller, who cordially welcomed the speakers and guests.

Albrecht M. Ehrly

Table of Contents

List of Contributors

Abendroth, D., Dr.
Chirurgische Klinik, Klinikum Großhadern, Marchioninistr. 15,
D-8000 München 70, Federal Republic of Germany

Baumgärtl, H., Chem.-Ing.
Max-Planck-Institut für Systemphysiologie, Rheinlanddamm 201,
D-4600 Dortmund 1, Federal Republic of Germany

Beerthuizen, G. I. J. M., Dr.
Algemeene Chirurgie, Katholieke Universiteit Nijmegen,
Postbus 9101, NL-6500 HB Nijmegen, The Netherlands

Böhm, E., Dr.
Medizinische Forschung 1, Pharmakologisches Labor, Boehringer
Mannheim GmbH, Sandhofer Str. 116, D-6800 Mannheim 31
Federal Republic of Germany

Creutzig, A., Dr.
Abt. Angiologie, Medizinische Hochschule Hannover,
Konstanty-Gutschow-Str. 8, D-3000 Hannover 61
Federal Republic of Germany

Ehrly, A. M., Prof., Dr.
Abt. für Angiologie, Johann Wolfgang Goethe-Universität,
Theodor-Stern-Kai 7, D-6000 Frankfurt/Main 70
Federal Republic of Germany

Endrich, B., Dr.
Abt. Exp. Chirurgie, Chirurgisches Zentrum, Klinikum der
Universität, Im Neuenheimer Feld 347, D-6900 Heidelberg 1
Federal Republic of Germany

Ewald, U., Dr.
Pediatric Clinic, University Hospital, S-75185 Uppsala, Sweden

Fallenstein, F., Dr.
Universitätsfrauenklinik, Frauenklinikstr. 10, CH-8091 Zürich
Switzerland

Fleckenstein, W., Dr.
Institut für Physiologie, Medizinische Hochschule Lübeck,
Ratzeburger Allee 160, D-2400 Lübeck 1
Federal Republic of Germany

Franzeck, U. K., Dr.
Medizinische Klinik, Klinikum der Universität, Bergheimer Str. 58,
D-6900 Heidelberg 1, Federal Republic of Germany

Funk, W., Dr.
Abteilung für Anästhesiologie, Stadt-Krankenhaus
München-Schwabing, Kölner Platz 1, D-8000 München 40
Federal Republic of Germany

Hartel, W., Prof., Dr., Oberstabsarzt
Abt. II - Chirurgie, Bundeswehrkrankenhaus, Oberer Eselsberg 40,
D-7900 Ulm, Federal Republic of Germany

Hauss, J., Prof., Dr.
Chirurg. Klinik/Poliklinik der Universität, Jungeblodtplatz 1,
D-4400 Münster, Federal Republic of Germany

Heinrich, R., Dr.
Abt. Innere Medizin I, Medizinische Universitätsklinik,
Otfried-Müller-Straße, D-7400 Tübingen 1
Federal Republic of Germany

Huch, R., Prof. Dr.
Universitätsfrauenklinik, Frauenklinikstr. 10, CH-8091 Zürich
Switzerland

Keller, H. P., PD Dr.
Chirurgische Universitätsklinik und -Poliklinik,
Josef-Schneider-Str. 2, D-8700 Würzburg
Federal Republic of Germany

Kersting, Th., Dr.
Universitätsklinikum Steglitz, Freie Universität,
Hindenburgdamm 30, D-1000 Berlin 45
Federal Republic of Germany

Kohaus, H., PD Dr.
Chirurgische Abt., Marienhospital, D-4650 Gelsenkirchen
Federal Republic of Germany

Landgraf, H., Dr.
Abt. für Angiologie, Johann Wolfgang Goethe-Universität,
Theodor-Stern-Kai 7, D-6000 Frankfurt/Main 70
Federal Republic of Germany

Lemke, R., Dr.
Medizinische Klinik, Städtische Kliniken Mitte, Beurhausstr. 40,
D-4600 Dortmund 1, Federal Republic of Germany

Lübbers, D. W., Prof. Dr.
Max-Planck-Institut für Systemphysiologie, Rheinlanddamm 201,
D-4600 Dortmund 1, Federal Republic of Germany

Ott, A., Dr.
Hautklinik und -Poliklinik im Rudolf-Virchow-Krankenhaus,
Universitätsklinikum Charlottenburg (FB 3), Augustenburger
Platz 1, D-1000 Berlin 65, Federal Republic of Germany

Partsch, H., Doz., Dr.
Gefäßambulanz des Hanusch-Krankenhauses,
Heinrich-Kollin-Str. 30, A-1140 Wien, Austria

Rühle, K.-H., Prof. Dr.
Abt. Pulmologie, Med. Universitätsklinik, D-7800 Freiburg
Federal Republic of Germany

Ruland, O., Dr.
Chirurgische Poliklinik der Universität, Jungeblodtplatz 1,
D-4400 Münster, Federal Republic of Germany

Schultheiß, R., Dr.
Neurochirurgische Universitätsklinik, Sigmund-Freud-Str. 25,
D-5300 Bonn 1, Federal Republic of Germany

Spannbrucker, N., Dr.
Medizinische Universitätsklinik, Sigmund-Freud-Str. 25,
D-5300 Bonn 1, Federal Republic of Germany

Spiegel, H. U., Dr. Dipl.-Ing.
Chirurg. Klinik/Poliklinik der Universität, Jungeblodtplatz 1,
D-4400 Münster, Federal Republic of Germany

Steinacker, J. M., Dr.
Sportmedizinische Untersuchungsstelle, Universität Ulm, Oberer
Eselsberg M 25-336, D-7900 Ulm, Federal Republic of Germany

Thermann, M., Prof. Dr.
Chirurgische Universitätsklinik, Hospitalstr. 40, D-2300 Kiel 1
Federal Republic of Germany

Introduction

The Relationship Between Tissue Oxygen Pressure, Skin Surface PO_2, and Transcutaneous PO_2

D. W. Lübbers

It has long been known that oxygen and carbon dioxide are exchanged through the human skin. Figure 1 shows a section through the human skin. One can see a capillary (filled with India ink) ascending towards the epidermis, where it bends and forms a capillary loop. The blood of this capillary loop supplies oxygen to the surrounding tissue and the epidermis above it. The cells of the multilayered epidermis are formed by the cells that divide in the lowermost cell layer (stratum basale), move upwards, keratinize, and then die. A viable layer consisting of approximately three to five cell layers with a thickness of 20–79 μm can be differentiated from a dead horny layer, the stratum corneum, which is about 10–20 μm thick. These are mean values, as the dimensions of the skin and the number of capillaries per millimeter can vary according to location. In atrophic skin areas, the capillaries are sparse; during hypertrophy the skin gets thicker and the capillaries are transformed. This variable anatomic picture indicates that there can be distinct local differences in the oxygen supply to the skin.

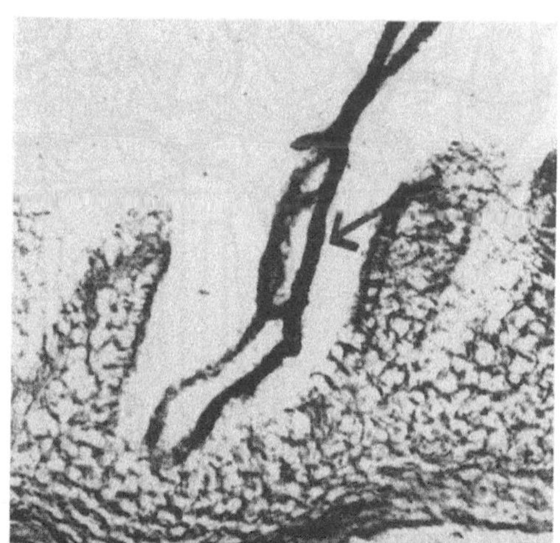

Fig. 1. Capillary blood supply of the human epidermis; capillary loop (filled with India ink) in the stratum papillare corii covered by the epidermis; capillary bridges are found between the ascending arterial and the descending venous capillary limb [3]

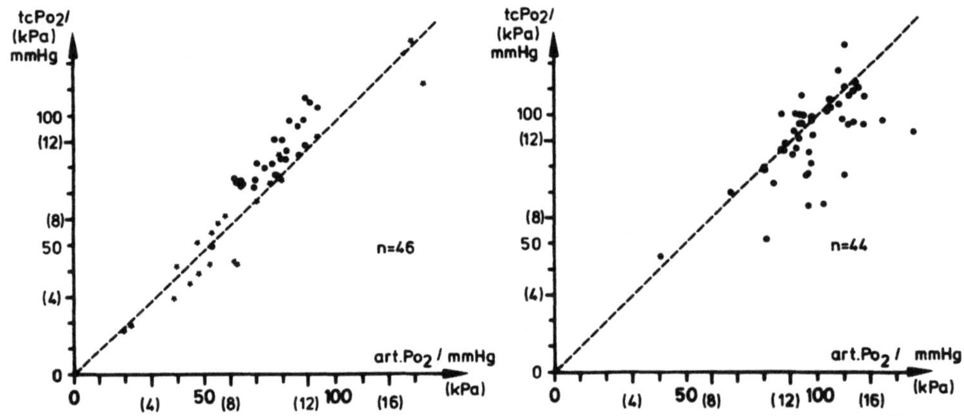

Fig. 8. Comparison of the arterial PO_2 (37 °C) and the simultaneously measured $tcPO_2$ (43 °C). On the *left*, 46 blood samples from 25 healthy (●) and 9 sick (✻) children. On the *right*, 44 blood samples from 28 women during delivery. The electrode was covered with a 12-μm Teflon membrane and a 12-μm cellophane membrane. The diameter of the platinum wire was approximately 15 μm [13]

it cannot be seen from the $tcPO_2$ value under what local flow conditions it is being measured. Unfortunately, to date there is no continuous method which allows local capillary flow to be quantitatively measured. The change in the heating energy can only provide information on changes and trends. If maximum local hyperemia cannot be safely assumed, the arterial blood PO_2 must be controlled invasively to interpret the $tcPO_2$. This interpretation can be obtained using the relation $tcPO_2/P_aO_2$ ("perfusion efficiency factor" [11] or "$tcPO_2$ index" [24]) [2, 10, 19, 22, 23, 24, 26, 27, 28, 29]. If the $tcPO_2$ index is greater than approximately 0.8, it can be assumed that the $tcPO_2$ reflects the arterial PO_2. The relationship between the $tcPO_2$ and P_aO_2 is not linear, because of the chemical bonding of oxygen to the hemoglobin and the complex diffusion conditions. Nevertheless, the $tcPO_2$ measurement under such conditions provides a sufficiently accurate, continuous, noninvasive blood gas analysis to be used in clinical practice, since continuous recording ensures that all changes in the P_aO_2 are detected. If the $tcPO_2$ index is lower than 0.8, the capillary blood flow in the skin will have an increasing effect on the $tcPO_2$. A $tcPO_2$ index of 0.6 and lower indicates - if other local influencing factors can be excluded - a blood flow disturbance accompanied by circulatory centralization or a severe peripheral vasoconstrictory tonus [16]. Clinical studies have shown that the $tcPO_2$ index in the treatment of hemorrhagic shock with liquids, for example, is a reliable indicator [19, 23, 26, 28].

The parameters influencing the $tcPO_2$ index are as follows:

1. Oxygen consumption. Skin metabolism, polarographic PO_2 measurement
2. Oxygen supply. Arterial O_2 concentration (blood): concentration of hemoglobin, O_2 saturation of hemoglobin, arterial PO_2, temperature. Capillary blood flow (skin): systemic blood pressure, vascular resistance, cardiac output, peripheral perfusion pressure, blood viscosity, number of capillaries per mm^2 of the skin surface, temperature

3. Oxygen transfer. Thickness of the epidermal layer, diffusional properties of the skin and the electrode membrane

The application of the $tcPO_2$ index can be summarized as follows:

1. $tcPO_2$ index larger than 0.8: $tcPO_2$ reflects the arterial PO_2. Suitable for noninvasive continuous blood gas analysis
2. $tcPO_2$ index smaller than 0.6: $tcPO_2$ not suitable for monitoring blood gases. $tcPO_2$ reflects the state of circulation by measuring the peripheral oxygen supply

As has already been shown (Eq. 5), the ratio of oxygen consumption to amount of oxygen available is decisive for the oxygen supply. The oxygen consumption of the skin should be constant during the measuring period (about 6 h). Not only tissue respiration, but also polarographic oxygen measurement take oxygen from the blood. This has to be taken into account when evaluating the absolute value of the measured $tcPO_2$, particularly when $tcPO_2$ values obtained with different electrodes are being compared.

The arterial oxygen supply is determined by the amount of oxygen and by the capillary flow. The parameters listed above show that the oxygen supply is dependent on several factors. Of particular importance are the hemoglobin concentration and the peripheral perfusion pressure.

The transfer of oxygen through the epidermis results in a PO_2 decrease, which in the case of membrane compensation is used to balance out the PO_2 increase caused by hyperthermia (see Fig. 7). The $tcPO_2$ index allows an interpretation of the changes in the $tcPO_2$ to be made, i.e., whether they are caused mainly by changes in blood flow or by changes in blood gas.

Unfortunately, there are some patients with normal circulation and healthy volunteers who react to local heating with maximum local vasodilatation, but not with maximum hyperemia. These persons also have a $tcPO_2$ index much lower than 0.8 although they have no pathologic circulatory condition [15]. It has been reported that in some volunteers, $tcPO_2$ slowly increased with increasing measuring time [16]. In this group, there is apparently a severe vasoconstrictor tonus of the larger feeding arteries, which cannot be broken through by local hyperthermia. It would be an important step if a noninvasive (or minimally invasive) method could be found that would provide a corresponding diagnosis. Lemke et al. [15] found – though only in a few cases – that by spraying the electrode with the local anesthetic melivacaine (Scandicaine 1%), the vasoconstrictory tonus was broken through, so that a $tcPO_2$ index corresponding to the value of a person with undisturbed blood flow was achieved. Whether the increased vasoconstrictory tonus is of diagnostic significance has yet to be proved.

In summary, it can be said that continuous, noninvasive PO_2 measurement has become an important diagnostic and therapeutic tool. It should be noted that the technique is not a substitute for blood gas analysis, but yields important information of its own.

References

1. Baumberger, J. P., R. B. Goodfriend: Determination of arterial oxygen tension in man by equilibration through the intact skin. Fed. Proc. 10 (1951) 10–11
2. Brünstler I., A. Enders, H. T. Versmold: Skin surface pCO_2 monitoring in newborn infants in shock: Effect of hypotension and electrode temperature. Pediatrics 100 (1982) 454–457
3. Ellis, R. A.: Vascular patterns of the skin. In: Montagna, W., Ellis, R. A.: Advances in biology of skin (Vol II): Blood Vessels and Circulation, (eds.) Pergamon Press, New York/Oxford/London/Paris (1961) pp. 20–37
4. Evans, N. T. S., P. F. D. Naylor: Steady state of oxygen tension in human dermis. Respir. Physiol. 2 (1966/67) 46–60
5. Evans, N. T. S., P. F. D. Naylor: The systemic oxygen supply to the surface of human skin. Respir. Physiol. 3 (1967) 21–27
6. Gottrup, F., R. Firmin, N. Chang, W. H. Goodson III, T. K. Hunt: Continuous direct tissue oxygen tension measurement by a new method using an implantable silastic tonometer and oxygen polarography. Am. J. Surg. 146 (1983) 399–403
7. Hertzmann, A. B.: Effects of heat on the cutaneous blood flow. In: Montagna, W., Ellis, R. A. (eds.): Advance Biol. of Skin (Vol. II): Blood Vessels and Circulation. Pergamon Press, Oxford (1961): pp. 98–116
8. Horstmann, E.: Blutgefäße der Haut. In: Möllendorf, V. (Hrsg.): Handbuch der mikroskopischen Anatomie des Menschen (Vol 3, 1). Springer-Verlag, Berlin (1957) S. 198–107
9. Huch, A., R. Huch, K. Meinzer, D. W. Lübbers: Eine schnelle, beheizte Pt-Oberflächenelektrode zur kontinuierlichen Überwachung des PO_2 beim Menschen. Elektrodenaufbau und -eigenschaften. Proc. Medizin-Technik Stuttgart (1972) S. 26
10. Huch, R., A. Huch: Continuous Transcutaneous Blood Gas Monitoring. M. Dekker, New York/Basel (1983)
11. Huch, R., A. Huch, D. W. Lübbers: Transcutaneous PO_2. Thieme-Stratton, Stuttgart/New York (1981)
12. Huch, R., Lübbers, D. W., A. Huch: Quantitative continuous measurement of partial oxygen pressure on the skin of adults and newborn babies. Pflüg. Arch. 337 (1972) 185–198
13. Huch, R., D. W. Lübbers, A. Huch: The transcutaneous measurement of oxygen and carbon dioxide tensions for the determination of arterial blood gas values with control of local perfusion and peripheral perfusion pressure: Theoretical analysis and practical application. In: Payne, J. P., Hill, D. W. (eds.): Oxygen Measurement in Biology and Medicine. Butterworth, London (1975) pp. 121–138
14. Hunt, T. K.: see discussion in Waxman et al. [29] S. 29
15. Lemke, R., D. Klaus, D. W. Lübbers: Experiences with the commercially available $tcPO_2$ electrode in adults. In: Huch, R., Huch, A. (eds.): Continuous Transcutaneous Blood Gas Monitoring (Vol. V). M. Dekker, New York/Basel (1983)
16. Lemke, R., D. Klaus, D. W. Lübbers: Noninvasive monitoring of $tcPO_2$ and $tcPCO_2$ in patients with normal and compromised circulation. In: R. Droh, R. Spintge (eds.): Closed circuit and other innovations in anaesthesia. Springer-Verlag, Berlin-Heidelberg-New York (1986) pp. 113–119
17. Lübbers, D. W.: Theoretical basis of transcutaneous blood gas measurements. Crit. Care Med. 9 (1981) 721–733
18. Lübbers, D. W.: Cutaneous and transcutaneous PO_2 and PCO_2 and their measuring conditions. In: A. Huch, R. Huch, J. F. Lucey (eds.): Continuous Transcutaneous Blood Gas Monitoring. The National Foundation - March of Dimes. Birth Defects: Original Article Series (Vol. XV, No. 4). A. R. Liss, New York (1979) pp. 13–32
19. Lübbers, D. W.: Transcutaneous PO_2 in shock, in print
20. Lübbers, D. W., U. Grossmann: Gas exchange through the human epidermis as a basis of $tcPO_2$ and $tcPCO_2$ measurements. In: R. Huch, A. Huch (eds.): Continuous Transcutaneous Blood Gas Monitoring. M. Dekker, New York, pp. 1–34
21. Montgomery, H.: Oxygen tension of skin and muscle. Fed. Proc. 16 (1957) 697–699
22. Nolan, L. S., W. C. Shoemaker: Transcutaneous O_2 and CO_2 monitoring of high risk surgical patients during the perioperative period. Crit. Care Med. 10 (1982) 762–764
23. Shoemaker, W. C., R. Vidyasagar: Physiological and clinical significance of $P_{tc}O_2$ and $P_{tc}CO_2$. Crit. Care Med. 10 (1981) 689

24. Shoemaker, W.C., K.K. Tremper: Transcutaneous PO_2 and PCO_2 monitoring: Experimental and clinical studies. In: Huch, R., Huch, A. (eds.): Continuous Transcutaneous Blood Gas Monitoring (Vol. V). M. Dekker, New York/Basel (1983) pp. 745-760
25. Spalteholz, W.: Die Verteilung der Blutgefäße in der Haut. Arch. Anat. Physiol. (1983) 1-54
26. Tremper, K.K., W.C. Shoemaker: Transcutaneous oxygen monitoring of critically ill adults, with and without low flow shock. Crit. Care Med. 9 (1981) pp. 706-709
27. Tremper, K.K., W.C. Shoemaker, C.R. Shippy, L.S. Nolan: Transcutaneous PCO_2 monitoring in adult patients in the ICU and the operating room. Crit. Care Med. 9 (1981) 752-755
28. Tremper, K.K., K. Waxman, R. Bowman, W.C. Shoemaker: Continuous transcutaneous oxygen monitoring during respiratory failure, cardiac decompensation, cardiac arrest and CPR. Crit Care Med. 8 (1980) 377-381
29. Waxman, K., R. Sadler, M.E. Eisner, R. Applebaum, K.K. Tremper, E.R. Mason: Transcutaneous oxygen monitoring of emergency department patients. Am. J. Surg 146 (1983) 35-38

Measuring Methods

Systematic Investigations of Needle Electrode Properties in Polarographic Measurements of Local Tissue PO_2

H. Baumgärtl

Introduction

In 1942, Davies and Brink [10], following the studies of Laitinen and Kolthoff [35], applied the polarographic method of measuring PO_2 using solid noble metal cathodes on biologic material. Since then, numerous micro- and macroneedle electrodes have been developed, which allow measurement of local oxygen partial pressure in tissue for the purpose of evaluating oxygen supply and microcirculation [9, 14, 33]. These electrode systems mainly differ in their dimensions, material properties, and the techniques used in manufacturing them. All of these systems are associated with more or less random errors, which can seriously hinder absolute PO_2 measurements [1, 6, 7, 11, 20–22, 25, 28–29, 32, 39–40, 42, 48–50].

With the objective of finding a simple way to discover such errors, we extensively examined some typical electrode properties. We developed simple calibration and control methods to be used in routine practice, which may be helpful to the user in interpreting the measurements obtained with the needle electrode.

The O_2 reduction current in the polarographic measuring circuit represents the central measured value, which is influenced by chemical reactions, by the construction of the electrode and the measuring equipment, and by the physical, chemical, and biological properties of the measuring medium. In the following, a comparison of the interaction and relationship of the measuring signal of bare and membranized platinum cathodes is presented using several examples. Included are data on measurement accuracy, selectivity, sensitivity, O_2 consumption of the cathode, linearity, catchment area, diffusion error, and response time of the measuring system. Furthermore, the influence of certain interfering ions, the effect of mechanical pressure on the PO_2 measuring signal, and possible morphological changes in the tissue caused by the electrode are discussed.

Material and Methods

The measuring set-up shown in Fig. 1 was developed to check the most important measuring properties of the PO_2 needle electrode and its calibration. The testing equipment basically consists of three measuring units, the reaction vessels of which

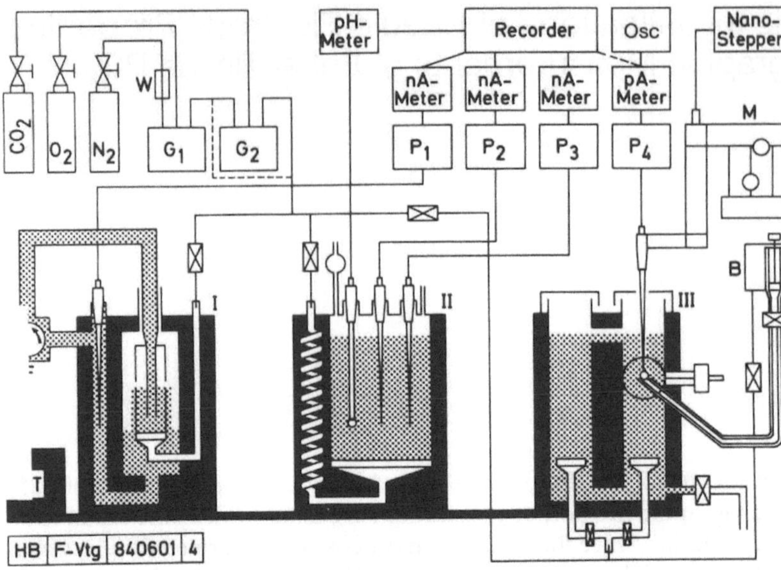

Fig. 1. Measuring set-up for function control and calibration of PO_2 microcoaxial needle electrodes. *I*, glass vessel for determining the stirring effect and the flow dependence on the current-voltage characteristics; *II*, calibration cell for calibrating several electrodes at the same time; *III*, set-up for determining response time and oscillating processes and the reduction current of small recessed electrodes; *B*, gas reservoir with dosage setter for generating a stable gas bubble; *F*, fluid pump; $G_1 + G_2$, gas mixing pump; *M*, micromanipulator; *nA*, nanoamperemeter; *Osc*, storage oscillograph; P_1, microprocessor-controlled polarographic voltage; P_2–P_4, polarographic voltages; *pA*, picoamperemeter; *T*, thermostat; *W*, Oxisorb and gas-washing bottle

Table 1. Composition of electrode test solutions (mmol $\times$ l^{-1})

Borate-phosphate-citrate buffer (pH 7.2)		Ca-Mg-bicarbonate buffer (pH 7.4)	
NaOH	68.6	NaCl	124.0
HCl	16.0	$NaHCO_3$	26.0
H_3BO_3	11.4	Glucose	10.0
H_3PO_4	6.7	KCl	4.9
CH_3COOH	6.7	$CaCl_2$	2.6
		KH_2PO_4	1.3
		$MgSO_4$	1.3

were constructed specially for the determination of the stirring effect (measuring chamber 1), for the simultaneous calibration of several electrodes (measuring chamber 2), and for the measurement of the response time (measuring chamber 3). The measuring cells, which are made of glass and which can be thermostatically controlled, are arranged on an insulating Lucite plate inside a Faraday cage.

The electrode test solutions used are borate-phosphate-citrate buffer, after Theorell and Stenhagen and Ca-Mg-bicarbonate buffer. The individual components of these solutions are listed in Table 1. The electrodes are mostly calibrated in an ethyl-

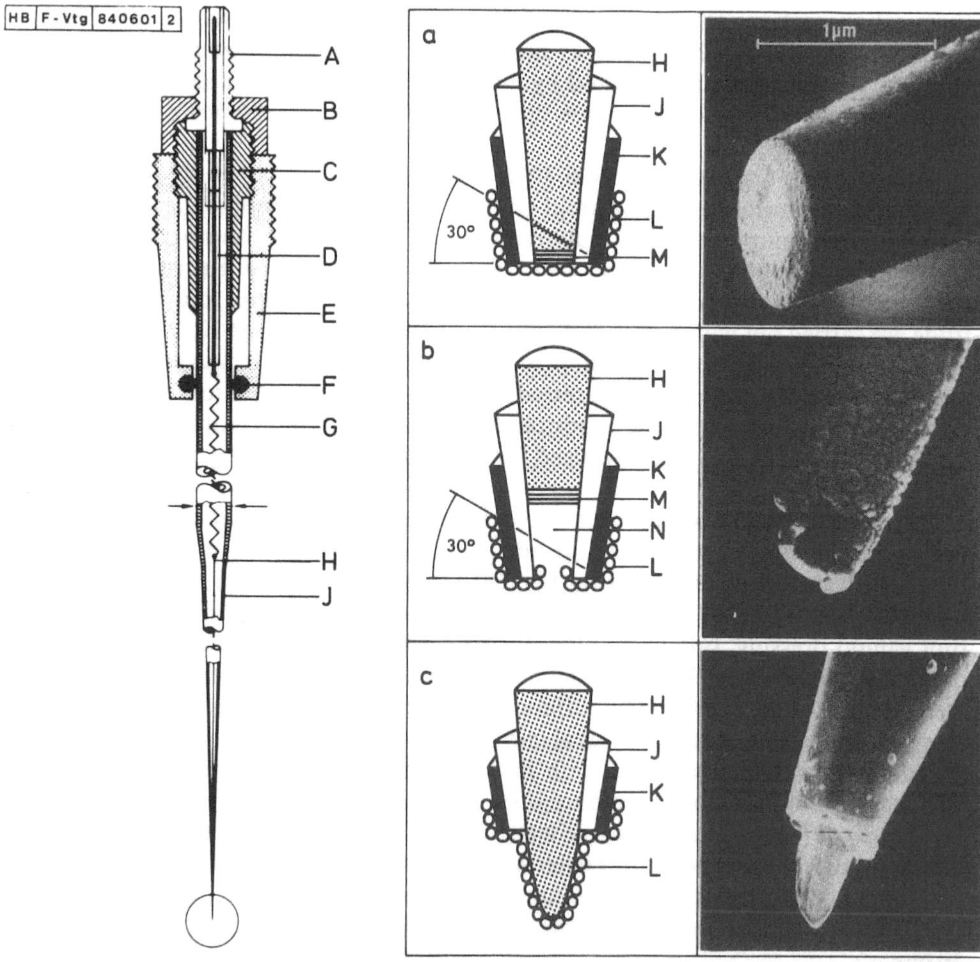

Fig. 2. Schematic drawings and scanning electron microscopic images of flat *(a)*, recessed *(b)*, and cone-shaped *(c)* PO_2 microcoaxial needle electrodes. *A*, HF chassis socket; *B*, fixing nut; *C*, metal contact to reference electrode; *D*, polyethylene coated Cu wire; *E*, quickly exchangeable Lucite housing in NS-10 form with thread; *F*, siliconized O-ring; *G*, metal coil; *H*, Pt cathode etched by alternating current electrolysis; *J*, glass insulation; *K*, sputtered three-layered reference electrode (Ta/Pt/Ag); *L*, oxygen-permeable membrane; *M*, electrochemically polished Pt surface with galvanically hardened and tempered, e.g., Au or Pd, coating; *N*, recess

ene-glycol-water mixture with a constant content of $Na_2B_4O_7$ (27 mmol $\times$ l^{-1}) and KH_2PO_4 (45 mmol $\times$ l^{-1}). The concentration of the ethylene glycol varies according to the desired O_2 conductivity ratio (see Fig. 4b). The calibration gases usually used are three-component gas mixtures of O_2, N_2 and CO_2, which are produced via two gas mixing pumps (type 1 M 100 a, Wösthoff, Bochum). The CO_2 content is maintained at 5%. The residual current is determined in "extremely pure" nitrogen (99.996%), which is additionally conducted through the gas purifier Oxisorb (Messer Griesheim, Düsseldorf) so that even the slightest traces of O_2 are bonded by

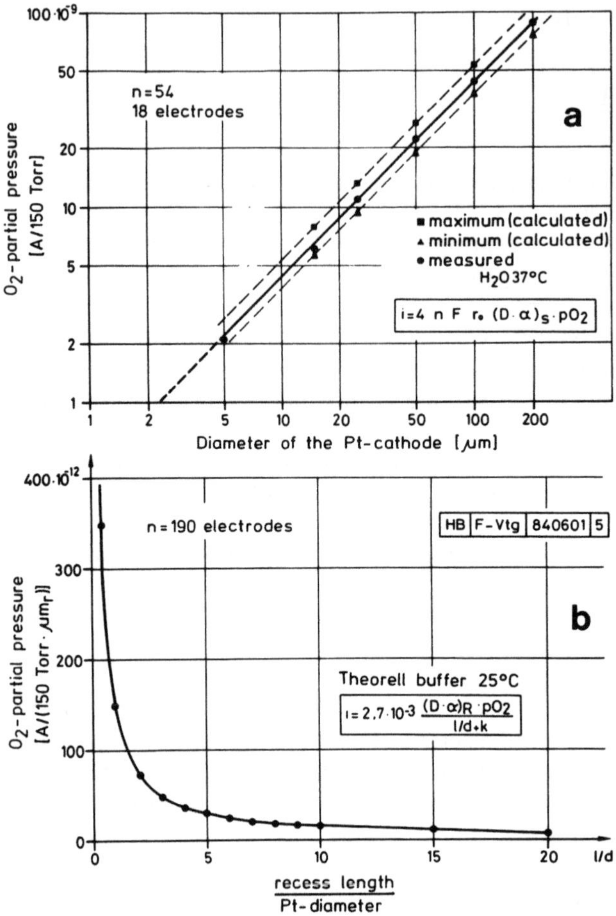

Fig. 3 a, b. Measured and theoretically expected O_2 reduction currents in water (37 °C) from flat, bare, circular Pt cathodes in dependence on cathode diameter. Measured and standardized O_2 reduction currents in Theorell buffer from bare, recessed needle electrodes in dependence on the ratio recess length/cathode diameter. The O_2 reduction current declines asymptotically with increasing recess length

Measuring Signal and O_2 Consumption of the Cathode. In our study, the measuring signal in water and diluted electrolyte solutions never exceeded the above-mentioned sensitivity range, but drifted toward lower values as the experiment went on. The reason for this may be that the peroxide and perhydroxyl anions generated are not used quickly enough in the following reactions, so that the O_2 reduction of the cathode becomes inhibited and the ideal four-electron process cannot be completed. Another complication might be the development of oxide coating or chemically absorbed oxygen on the surface of the cathode. This diminishes sensitivity, selectivity, linearity, reproducibility, and long-term stability and delays the response time of the electrode system [2, 7, 9, 23, 26, 39, 53].

In connection with this drift in the measuring signal, it should also be pointed out that the Pt cathode constantly consumes oxygen during the measurement. This consumption brings about a decrease in the O_2 content of the medium, which can be responsible for inaccurate measurements in small samples. According to the laws of electrolysis, the O_2 consumption of an electrode can be calculated using the equation

$$m = \frac{M \cdot i \cdot t}{n \cdot \mathrm{F}} \tag{8}$$

where m = amount of oxygen reduced (in g), M = molecular weight (in g), i = electrical current (A), t = time (s), n = number of electrons involved, and F = Faraday constant ($\mathrm{A} \times \mathrm{s} \times \mathrm{mol}^{-1}$). Thus, calculating the O_2 consumption of the cathode with a measuring signal of 1 nA yields 4.99×10^{-6} µg O_2/min $= 0.156 \times 10^{-6}$ µmol O_2/min $= 3.5 \times 10^{-6}$ µl O_2/min.

The oxygen pressure decrease (in Torr/min), which results from the O_2 consumption of the cathode in a measurement sample with a volume of V(cm^3), can be determined by the equation

$$\frac{\Delta \mathrm{PO}_2}{\Delta t} = 2.66\,\frac{i \times 10^{-3}}{\alpha \times V} \tag{9}$$

[18], where α = solubility coefficient and i = O_2 reduction current. The O_2 consumption values obtained, which seem low at first and which can be neglected in most cases, can be of particular importance when large, bare, or only thinly membranized cathodes are used for the PO_2 measurement in tissue with a coarse capillary network and low O_2 consumption. Under such conditions, the cathode, along with the mitochondria, becomes an additional consumer of oxygen, which also has to take its oxygen from the intercapillary O_2 supply. This can result in erroneous measurement of intercapillary PO_2 [21, 38]. This finding becomes all the more relevant when microneedle electrodes (with diameter < 5 µm) are used, since, for technical reasons, it is not always possible to see how thick the membrane is or whether there is any membrane at all.

The O_2 Reduction Current of Recessed Needle Electrodes. To be able to better judge the measuring characteristics of needle electrodes and to obtain, at the same time, a relationship to the diffusion field of a recessed electrode, we measured the O_2 reduction current of 190 bare recessed needle electrodes in an air-saturated borate-phosphate-citrate buffer solution (pH 7.2) at 25 °C. The recess length and the cathode diameter of these electrodes had been determined under light and scanning electron microscopes. The results are summarized in Fig. 3b. For better demonstration, the currents measured were normalized to a cathode radius of 1 µm. The values obtained in this manner are represented in dependence on the ratio of recess length to cathode diameter (see also Table 2).

The measuring signal drops asymptotically with the increasing length of the recess, and the course of the current roughly follows the simple mathematical relation

$$i = 2.7 \cdot 10^{-3} \cdot \frac{(D \cdot \alpha)_{\mathrm{R}} \cdot \mathrm{PO}_2}{l/d + K^*} \tag{10}$$

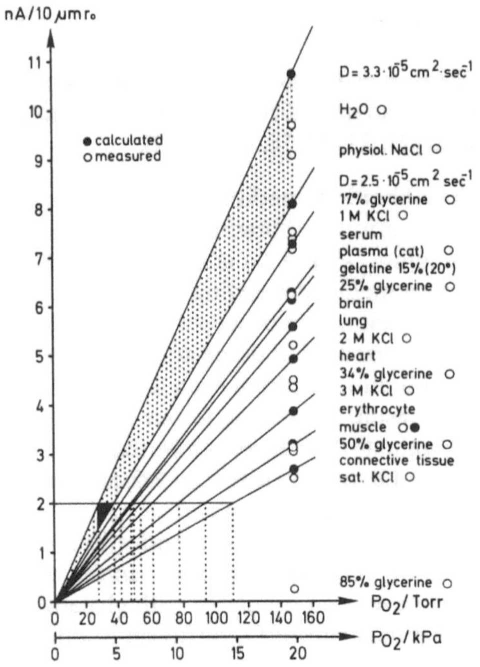

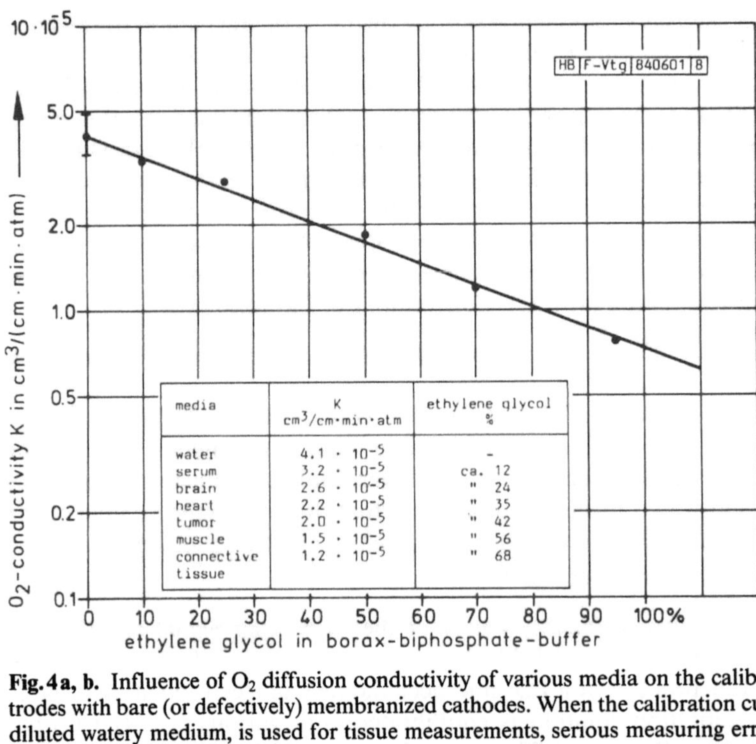

media	K cm³/cm·min·atm	ethylene glycol %
water	4.1 · 10⁻⁵	-
serum	3.2 · 10⁻⁵	ca. 12
brain	2.6 · 10⁻⁵	" 24
heart	2.2 · 10⁻⁵	" 35
tumor	2.0 · 10⁻⁵	" 42
muscle	1.5 · 10⁻⁵	" 56
connective tissue	1.2 · 10⁻⁵	" 68

Fig. 4 a, b. Influence of O_2 diffusion conductivity of various media on the calibration of PO_2 electrodes with bare (or defectively) membranized cathodes. When the calibration curve, measured in a diluted watery medium, is used for tissue measurements, serious measuring errors can result. The calibration error can be minimized by simulating the diffusion properties of the measuring medium using the corresponding calibration standard. O_2 diffusion conductivity in dependence on the concentration of ethylene glycol-water-borax-biphosphate buffer mixtures for establishing a defined calibration standard. As several examples in the diagram *(insert)* show, diffusion conditions similar to those in tissue can be simulated with various ethylene glycol concentrations

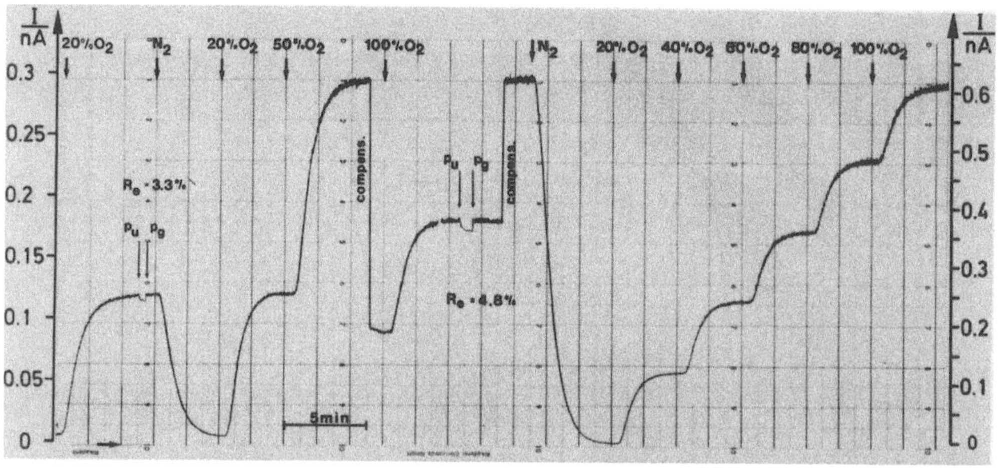

Fig. 5. Multipoint calibration of a PO_2 microcoaxial needle electrode with gas mixtures. p_u, measuring signal in unstirred medium; p_g, measuring signal in stirred medium; R_e, stirring effect in percent; *compens-1*, reduction in amplification; *compens-2*, adjusting the measuring signal to the recording width. From compens-2 on, the *right ordinate scale* applies; Table 3 shows the measured and corrected values, which yield the calibration curve

Table 3. Calibration and function test of a PO_2 microcoaxial needle electrode

Oxygen concentration (C) Volume (%)	Oxygen partial pressure (PO_2)		Oxygen reduction current		Sensitivity of electrode E_s		Calibration curve	
	kPa	Torr	i nA	$i-i_o$ nA	nA/kPa	nA/Torr	i_{korr} nA	$i_{korr}+i_o$ nA
0	–	–	0.003	–	–	–	–	–
20	19.3	145	0.118	0.115	0.00597	0.00079	0.118	0.121
40	38.5	290	0.239	0.236	0.00612	0.00081	0.236	0.239
60	57.8	435	0.354	0.351	0.00607	0.00080	0.354	0.357
80	77.1	579	0.485	0.482	0.00625	0.00083	0.472	0.475
100	96.3	729	0.622	0.619	0.00643	0.00085	0.590	0.593

To verify and to accurately determine the linearity and the proportionality constants between the measuring signal and the PO_2 (calibration curve), we prefer a multipoint calibration with a range of 0%–100% O_2. Relating the individually measured values of the reduction current to the respective O_2 pressures yields the individual electrode sensitivity and the calibration constant of each electrode by

$$E_S^T = \frac{i_i - i_o}{P_iO_2} \tag{15}$$

with

$$P_iO_2 = \frac{P_B - P_{H_2O}^T}{100} \times f_iO_2 \tag{16}$$

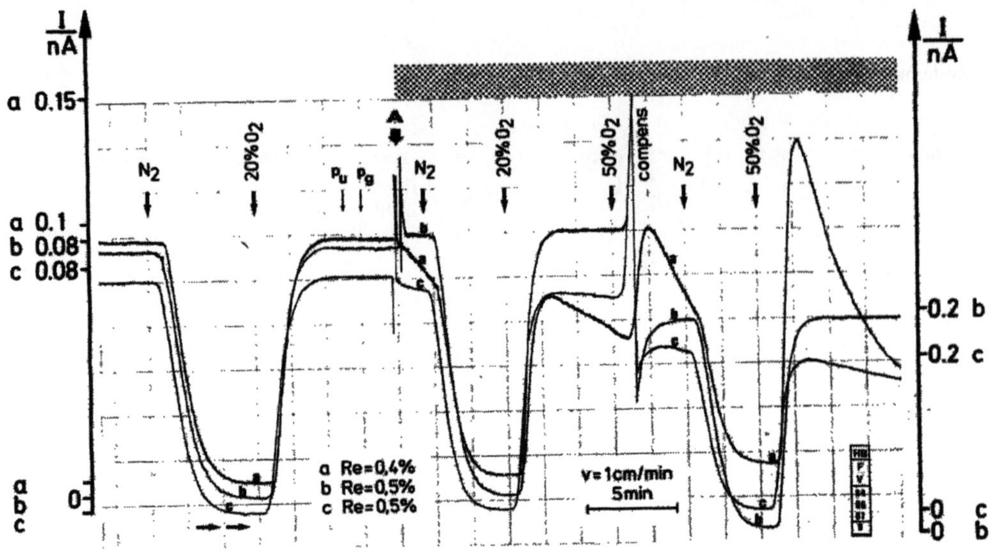

Fig. 6a. Comparison of measuring signal of PO_2 microcoaxial needle electrodes with collodion-polystyrene double membrane in Theorell and Ca-Mg-bicarbonate buffer. While all of the electrodes in the solution free of Ca^{++}-Mg^{++} ions show a reproducible measuring behavior during and after N_2 air change, the electrodes in the presence of calcium and magnesium ions react with a continuous decline in the measuring signal. The measuring signal drift is particularly pronounced when the PO_2 values are high. *A*, change in test solution; p_u, measuring signal of unstirred medium; p_g, measuring signal of stirred medium; R_e, stirring effect in percent. At *compens*, measuring signal of electrodes *b* and *c* is switched to lower amplification *(right scale)*; *a* and *c*, electrodes with membranes which are too thin; *b*, well membranized electrode

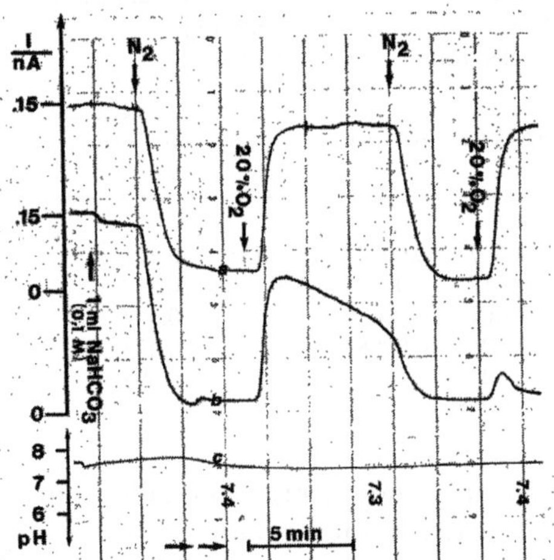

Fig. 6b. Measuring signal of a PO_2 microcoaxial needle electrode with a thick polystyrene membrane sintered at 200 °C *(a)*, in comparison with a bare recessed electrode *(b)* in $CaCl_2$ solution with sodium bicarbonate added. After addition of 1 ml $NaHCO_3$ (100 mmol × l^{-1}), a slight change in the measuring signal of a well-membranized electrode is detected. When calibration gas is changed several times, electrode *a* behaves normally; in contrast, the bare electrode *(b)*, shows a marked reduction in the measuring signal to the point where the electrode becomes insensitive to O_2

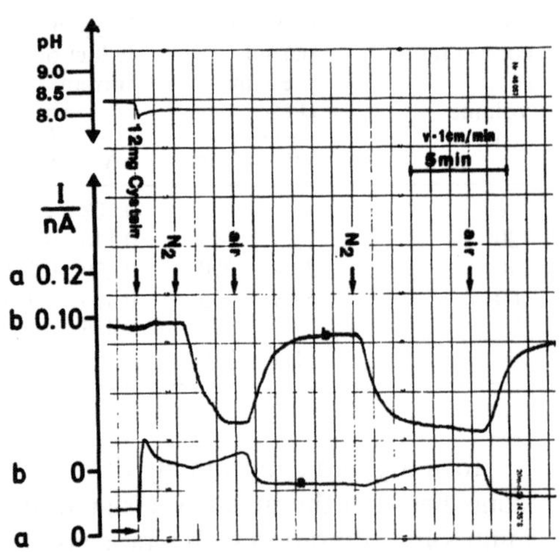

Fig. 6c. Comparison of measuring signal of a bare, recessed PO_2 microcoaxial needle electrode coated with sintered polystyrene in the presence of an SH compound. No change when a well-membranized electrode is used *(b)*. After cysteine is added in air-saturated solution, the measuring signal of the bare electrode *(a)* suddenly increases; when N_2 is added, the measuring signal increases, whereas it decreases during addition of air. *c*, continuous pH recording in the calibration cell

As the diffusion field of a membranized electrode can be compared roughly with that of a recessed electrode, the length of a recess can also be estimated from the stirring effect, as can be seen by converting Eq. 17

$$l = r_o \frac{K_z}{K_s}\left(\frac{1}{R_e/100\%} - 1\right) \tag{18}$$

l = length of recess; R_e = stirring effect; $K_z = (D \times \alpha)_R$ O_2 conductivity of the recess filling; $K_s = O_2$ conductivity of the measuring medium.

The Measuring Signal in a Steady and Nonsteady Medium; Determination of the Stirring Effect. In practice, the stirring effect of an electrode is usually determined in such a way that the flow of the calibration gas into the calibration medium is temporarily interrupted and the measuring signal is determined in gassed and nongassed states (Figs. 5, 6 a). The stirring effect R_e in percent can thus be calculated using the equation

$$R_e = \frac{i_g - i_u}{i_g} \times 100\% \quad \text{or} \quad = \left(1 - \frac{i_u}{i_g}\right) \times 100\% \tag{19}$$

where i_g is the measuring signal in the gassed, turbulent medium, and i_u the measuring signal in the nongassed, steady medium.

Numerous experiments have shown that evaluating a membrane or the length of a recess based on the stirring effect is inaccurate in large calibration vessels, because the flow velocities are mostly too low. This error can be minimized if the stirring effect is determined in special vessels in which the local flow velocities should be at least 25 cm/s. In this connection, it should be mentioned that the measurement of the stirring effect may be influenced by the flow direction. The stirring effect can be reduced by as much as 50% when the flow direction is reversed (away from the cathode). Furthermore, it should be noted that the O_2 diffusion currents measured in the presence of particularly low flow velocities or during a transition from a steady to a flowing medium are lower than in the steady medium. These phenomena, which are probably related to the diffusion layers of the electrode, have not yet been clarified and are especially relevant when needle electrodes are used for PO_2 measurements in flowing systems.

As is indicated in the literature, the stirring effects of an electrode should be smaller than 5% for absolute PO_2 measurements. It is assumed that under these conditions, the O_2 flux to the cathode is essentially limited by the diffusion resistance of the membrane or of the recess, so that disturbances such as convection or structural differences of the measuring medium only have a slight effect on the measuring signal. Unfortunately, our longstanding investigations on PO_2 microneedle electrodes have shown that a small stirring effect is no guarantee of a homogeneous, pinhole-free membrane.

The Measuring Signal of Bare and Membranized Microneedle Electrodes in the Presence of Noncomplex Bonded Calcium and Magnesium Ions. Apart from producing a defined diffusion field, the membrane has the important task of protecting the electrochemical reactions from interferences. Thus, the protective effect of the membrane has to be tested. Searching for ideal calibration standards, we found that the

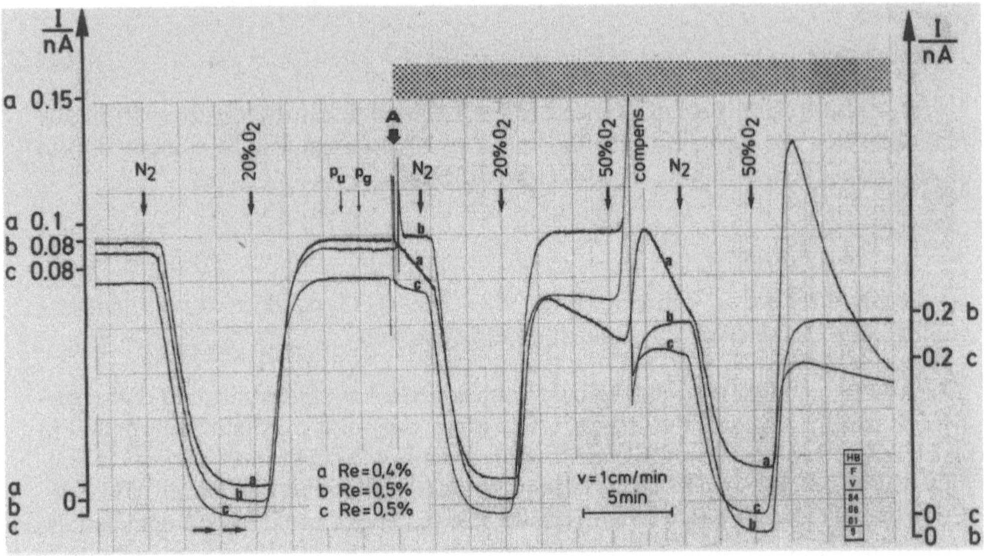

Fig. 6a. Comparison of measuring signal of PO_2 microcoaxial needle electrodes with collodion-polystyrene double membrane in Theorell and Ca-Mg-bicarbonate buffer. While all of the electrodes in the solution free of Ca^{++}-Mg^{++} ions show a reproducible measuring behavior during and after N_2 air change, the electrodes in the presence of calcium and magnesium ions react with a continuous decline in the measuring signal. The measuring signal drift is particularly pronounced when the PO_2 values are high. *A*, change in test solution; p_u, measuring signal of unstirred medium; p_g, measuring signal of stirred medium; R_e, stirring effect in percent. At *compens*, measuring signal of electrodes *b* and *c* is switched to lower amplification *(right scale)*; *a* and *c*, electrodes with membranes which are too thin; *b*, well membranized electrode

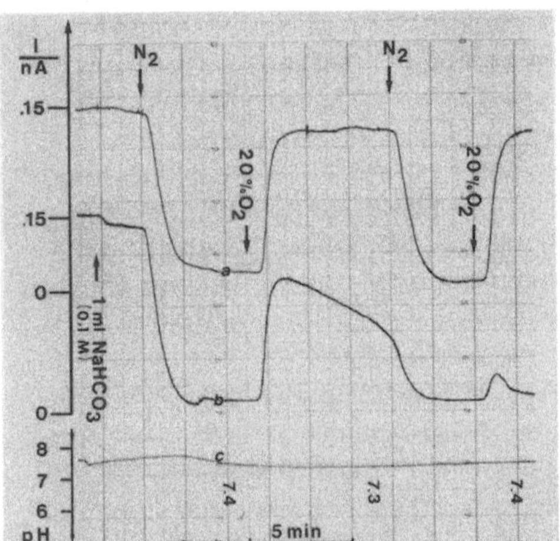

Fig. 6b. Measuring signal of a PO_2 microcoaxial needle electrode with a thick polystyrene membrane sintered at 200 °C *(a)*, in comparison with a bare recessed electrode *(b)* in $CaCl_2$ solution with sodium bicarbonate added. After addition of 1 ml $NaHCO_3$ (100 mmol × l^{-1}), a slight change in the measuring signal of a well-membranized electrode is detected. When calibration gas is changed several times, electrode *a* behaves normally; in contrast, the bare electrode *(b)*, shows a marked reduction in the measuring signal to the point where the electrode becomes insensitive to O_2

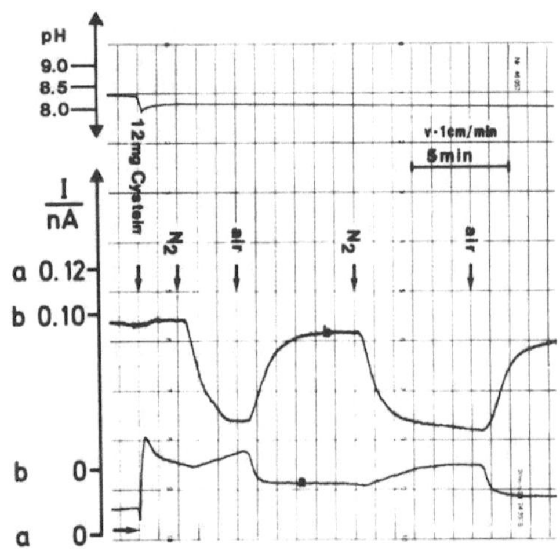

Fig. 6c. Comparison of measuring signal of a bare, recessed PO_2 microcoaxial needle electrode coated with sintered polystyrene in the presence of an SH compound. No change when a well-membranized electrode is used *(b)*. After cysteine is added in air-saturated solution, the measuring signal of the bare electrode *(a)* suddenly increases; when N_2 is added, the measuring signal increases, whereas it decreases during addition of air. *c*, continuous pH recording in the calibration cell

measuring signal in solutions with free Ca^{++} and Mg^{++} ions possesses special features [6, 7, 39]. This observation is especially elucidated in the experiment (Fig. 6a) in which three microcoaxial needle electrodes with polymer coats of collodion and polystyrene of different thicknesses were transferred in a polarized state at -700 mV from a Theorell buffer into a Ca-Mg-bicarbonate solution. In the first solution, all electrodes maintain a reproducible and stable measuring behavior, while two electrodes, despite very small stirring effects of about 0.5%, are marked by a considerable measuring signal drift after they are transferred into the second solution. This measuring signal drift is particularly pronounced in higher oxygen partial pressure ranges and, in our opinion, represents an extremely sensitive criterion for the protective effect and for the quality of the membrane of the microneedle electrode.

This variation in the behavior of the measuring signal can probably be explained by the fact that, if the membrane is too thin, the cathode cannot be evenly coated. This reduces the protection of the Pt cathode against the dissolved, positively charged Ca^{++} and Mg^{++} ions. These ions react with the OH^- ions resulting from the cathode reaction, producing calcium hydroxide, which is then transformed with CO_2 to calcium carbonate

$$2Ca^{++} + 4OH^- \rightarrow 2Ca(OH)_2 \quad (20)$$

$$\underline{2Ca(OH)_2 + 2CO_2 \rightarrow 2CaCO_3 + 2H_2O} \quad (21)$$

$$2Ca^{++} + 4OH^- + 2CO_2 \rightarrow 2CaCO_3 + 2H_2O \quad (22)$$

coating the surface of the cathode in the form of an almost insoluble deposit. This produces a kind of diffusion barrier for oxygen, so that the measuring signal decreases with the increasing calcium carbonate coating to the point where it becomes completely insensitive to O_2.

An example of this effect is demonstrated in Fig. 6b. In this experiment, two microcoaxial needle electrodes are in a 0.11 $CaCl_2$ solution, whereby the trace of the measuring signal *a* stems from the electrode with a thick sintered polystyrene membrane, while trace *b* represents the measuring signal of the bare recessed electrode. After adding a small amount of $NaHCO_3$ and changing the calibration gases several times, the measuring signal of the bare cathode drops to the point where it is completely insensitive to O_2. The measuring signal of the well membranized electrode, however, remains unaffected. Because of these significant differences, which are not detectable in other calibration media such as physiologic saline solutions or Theorell buffer, we use the bicarbonate solution containing $Ca^{++}-Mg^{++}$ ions as a testing solution to control the quality of the membrane.

The Sensitivity of the Measuring Signal to Interfering Ions. In case membrane defects or insufficiently membranized electrodes are not detected, the presence of interfering ions can cause the measuring signal to undergo certain reactions, which seriously impede the interpretation of the PO_2 measurement. Along with the influence of Ca^{++} and Mg^{++} ions, which participate in the transport of the current, sulphur compounds such as $S_2O_4^{--}$ or SH^- ions also have particular effects on bare or incompletely membranized electrodes. As can be seen in Fig. 6c, adding cysteine to an air-equilibrated solution causes a sudden increase in the current of the bare electrode (*a*). In changing equilibrated gases, the reduction current increases in nitrogen and decreases in air. On the other hand, an electrode that is well membranized (*b*) remains unaffected. The complete reversal of the measuring signal, which can also be observed in inadequately membranized electrodes in a $Na_2S_2O_4$ solution, shows that the electrochemical O_2 reduction can overlap with other current-supplying processes, which can lead to misinterpretations.

In this context, the effect of gases such as CO_2, CO, NH_3, N_2O, NO, or other substances diffusing through the membrane have to be taken into account. As has often been described in the literature, certain narcotics such as halothane, a halogenated hydrocarbon, have an effect on the polarographic microelectrode systems [3, 11, 40]. Thus, the measuring signal can be influenced by a number of factors, namely, the halothane concentration, the polarization voltage, the type and thickness of the electrode membrane, and the composition of the electrolyte solution. The current produced by the halothane reduction

$$R-CH_2Cl+H^+ +2e^- \rightarrow R-CH_3+Cl^- \tag{23}$$

adds to the measuring signal, which results from the reduction of the oxygen molecule and leads to overestimation of the PO_2 values. Such possible errors have to be given special consideration when O_2 electrodes are used in anesthetic procedures. In general, the electrode system must be specially checked when it is used to quantify the effect of drugs to clarify the influence of these substances on the polarographic reaction.

The Response Time of Needle Electrodes and the Measurement of Oscillating Processes

The response time of an O_2 electrode is of special practical importance when the electrode is used for measuring rapid processes such as rapid O_2 concentration changes on phase boundary, mass transfer coefficients [36, 37], or fast location changes when recording PO_2 profiles in tissue [12, 15].

There are a number of models for the theoretical study of response time, which is mostly calculated as the time an electrode needs to reach 90% of the new gas pressure value after a sudden gas pressure change [22, 26, 37, 47, 52, 55]. With these models, the t_{90}-response time of a membranized electrode can be roughly calculated using the equation

$$t_{90} \approx \ln 20 \frac{d_m^2}{D_m \pi^2} = 0.3 \frac{d_m^2}{D_m} \tag{24}$$

where d_m = the membrane thickness and D_m = the diffusion coefficient of the membrane. The response time of an electrode can be kept small, if the diffusibility of the membrane is large and the thickness small. Here, it must be kept in mind that the response time is quadratically dependent on the membrane thickness.

In the course of our investigations on the mass transfer mechanisms of gas bubbles, in which sudden O_2 concentration changes and flow velocities were to be determined in the region of the phase boundary, we studied, together with M. Riethues, the response time of PO_2 microcoaxial needle electrodes. For this purpose (see Fig. 7), the electrode to be tested in a concentration field of an air bubble, which is fixed in N_2-saturated water, is moved with the aid of a nanostepper from a PO_2 of zero in the water to a PO_2 of about 100 Torr in the border zone of the air bubble within 1 ms. The measuring signal is recorded with a storage oscilloscope. The evaluation of the current/time curves thus obtained yield t_{90}-values of 25–150 ms. As has often been reported in the literature, in almost every case the response time is shorter when the PO_2 increases from low to high values. When switching the electrodes from O_2 to N_2, i.e., from higher to lower oxygen partial pressures, we registered on the average a factor of 1.4 longer response times. The response time varies individually from electrode to electrode, but unexpectedly also from measurement to measurement. Usually, older electrodes have longer response times. Furthermore, our findings show that electrodes with large cathode diameters and long recesses have longer response times. "Membranizing" the microneedle electrode often has only a slight effect on the response time. This shows how difficult it is to make sufficiently thick, homogeneous, pinhole-free polymer films for microcathodes.

During these measurements, we also studied the limiting frequency of the electrode. For this purpose, a bubble swarm was produced approximately 1 cm from the electrode, and the number of PO_2 oscillations was recorded. Here we found that the limiting bubble frequency recorded in an electrode with a response time of 40 ms is approximately 40 Hz, i.e., at 40 Hz the amplitude of the electrode signal amounts to 70% of the steady-state signal.

The Effect of Mechanical Pressure of the Cathode Tip on the PO_2 Measuring Signal: Morphological Changes in Tissue Caused by the Electrode

One of the major problems associated with the local measurement of PO_2 using needle electrodes is the insertion into the tissue, since the puncturing and the possible tissue damage can affect the PO_2 measuring signal. In all punctures which we carried out with needle electrodes in brain, muscle, liver, kidney, and skin, we observed that the organ surface resists the electrode tip. Depending on the shape of the tip and the geometry of the electrode, this results in the formation of relatively large, cone-shaped dents in the tissue. In organ surfaces in vivo, a change of color is seen, indicating local ischemia. The depth of the indentation and of the impression effect associated with this dent, and thus the size of the ischemic zone, varies from experiment to experiment and from tissue to tissue. This demonstrates that this effect is closely dependent on the constitution of the electrode and its diameter, on the puncturing site, the puncturing technique, and on the experimental conditions.

Influence of the Puncturing Technique on the PO_2 Profile. Various puncturing techniques for recording PO_2 profiles are given in the literature. The following factors are discussed: (1) continuous, fast, or slow puncture; (2) continuous, fast, or slow withdrawal; (3) puncture with a vibrating electrode; (4) advancing and withdrawing in steps of different sizes; and (5) stationary position of the electrode [4, 12, 15, 31, 34]. Since all these methods have advantages and drawbacks, the best method always has to be worked out for a special application.

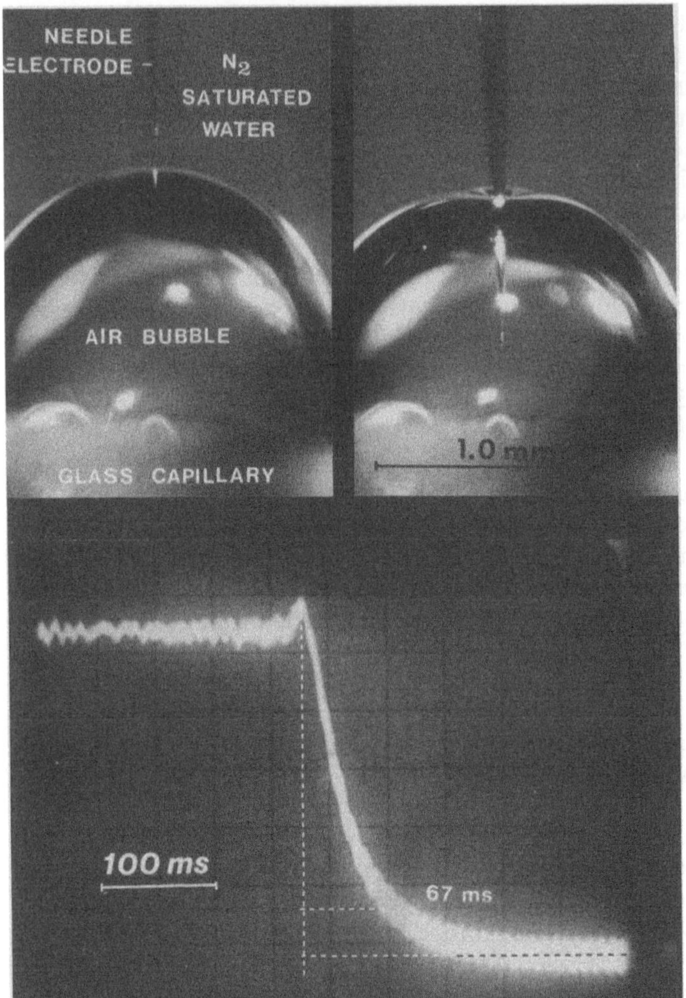

Fig. 7. Determination of the response time of microneedle electrodes. A PO_2 electrode connected to a nanostepper is transferred from the N_2-saturated calibration medium into the O_2 concentration field around an air bubble (or vice versa), which is held by a siliconized glass capillary. The current/time course yields in this case a t_{90} response time of 67 ms

To analyze the O_2 supply of the hemoglobin-free perfused guinea pig brain during hypothermia, together with U. Heinrich and D. W. Lübbers, we measured PO_2 profiles using microcoaxial needle electrodes [24] to compare the PO_2 values obtained during the vertical, stepwise advancement and withdrawal of the electrodes.

As the three examples in Fig. 8 show, along with good correlations, some severe discrepancies are observed within the same puncturing channel. The greatest differences were registered on the surface. During puncturing, in almost every case the measuring signal decreases as a result of compression of the tissue when the dura is dented too deeply. After approximately 500 μm, the PO_2 increases steeply when the electrode tip is suddenly advanced into the gray matter. The PO_2 decrease in the PO_2 profile toward the surface, which is demonstrated in the second of the three examples (solid line, out), indicates increasing formation of an edema. In several cases it was observed that, in the presence of edema, the measuring signal in the cerebral cortex can decrease to zero.

In another series of experiments, we studied how the PO_2 measuring signal can be changed as a result of pressure of the cathode tip in vitro and the effect the various puncturing techniques have on the PO_2 field. For this purpose, we affixed a thin needle electrode (cathode diameter 0.7 μm; recess length 0.7 μm) to a nanostepper and gradually guided the electrode into and out of 3% gelatine in steps of various sizes.

The most pronounced and unpredictable changes were observed during continuous puncturing (25 μm/min). Figure 9 shows the measuring signal changes in dependance on the step size of the advancement or the withdrawal. Considerable and irregular changes were observed when the electrode was moved in small steps (advanced 10×5 μm, then 20 s stop; # 10, upper trace), while there was very little change in the PO_2 when the probe was moved longer distances (advanced 2×100 μm and withdrawn 1×100 μm; # 6, upper trace). Diluting the gelatine diminished the changes in the PO_2 and raising the gelatine concentration increased them. These findings correlate with those of similar experiments performed with agar [5, 32, 48]. At present, it is not possible to determine all the factors responsible for these phenomena. An important factor is the high pressure caused by the puncturing. The pressure has been calculated to reach values of 4×10^7 Pa [25]. Such pressure is able to change the diffusion properties in front of the cathode tip, which consequently affects physicochemical processes determining the measuring signal.

Fine Structural Changes in the Tissue as a Result of Puncturing with the Needle Electrode. Apart from the sources of error associated with the polarographic measuring technique, it must be considered that each puncture in a living tissue can bring about a change in the physiologic milieu. A number of staining techniques have been worked out to make the puncturing channels visible in the tissue. Nevertheless, beyond findings on the extent of tissue changes and their effect on the O_2 diffusion field, there are surprisingly few results reported on such alterations, for example, in liver [29, 45, 46], kidney [17, 29, 42], carotid body [54], brain [29, 42], arterial wall [8, 32], muscle [42], spleen [29], skin [20], and spinal cord [51]. It is clear from these studies that the caliber of the electrode and the structure of the tissue play a decisive role in bringing about tissue changes. The smaller the electrode tip and the lower the tissue vascularization, the less severe are the morphological tissue changes.

Together with D. Schäfer (unpublished results) we carried out electron microscopic investigations to determine the fine structural changes in the tissue of the liver. We found that during vertical puncture a compression zone forms around the electrode shaft in front of the tip of the glass needle in which the lumina of the liver capillaries are almost completely compressed. Moreover, all cells and their organs are reoriented and form a ring around the glass needle. After some time, the cells in this compression zone dissolve. Some damaged and destroyed cells can also be found at other locations in the neighboring tissue areas. This could be the result of

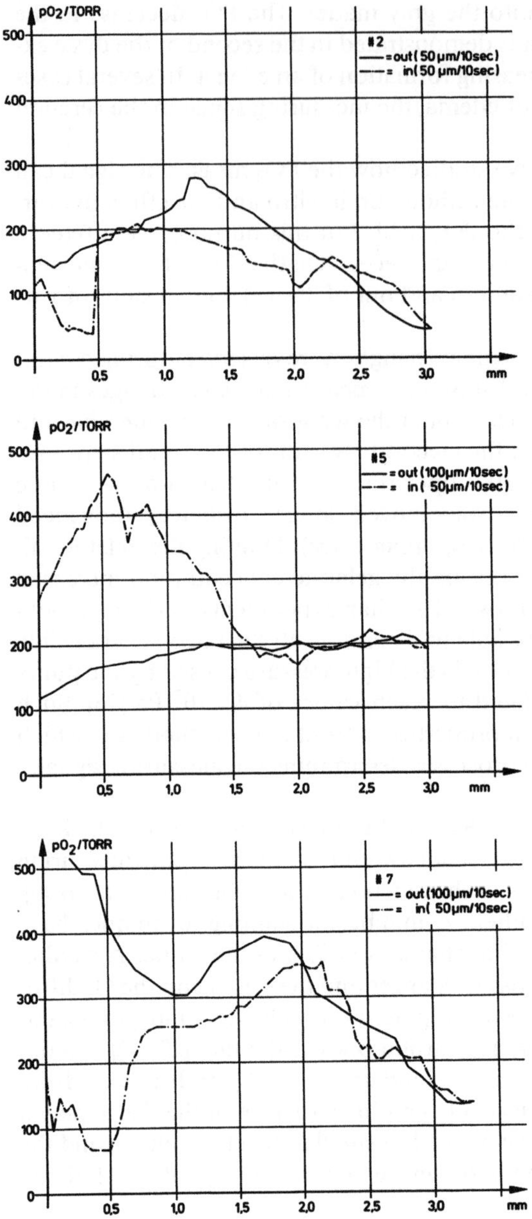

Fig. 8a–c. PO_2 profile of a hemoglobin-free perfused guinea pig brain during hypothermia (18 °C) measured by a micro Pt electrode (tip diameter 1 to 3 µm) in dependence on the puncturing technique. The PO_2 profiles are found to be different when the electrode is advanced in 50 µm steps (-----) and (——) withdrawn in the same puncturing channel. The PO_2 profiles are relatively uniform in the depths of the brain, deviations are observed in the surface region (H. Baumgärtl, U. Heinrich, D. W. Lübbers; unpublished results)

disturbed microcirculation. Systematic investigations on the thigh muscles of the rat have revealed that the fine structure is less affected here than in the liver. The structure of the connective tissue, however, is disturbed, as can be seen in Fig. 10 and a dent in the muscle fibrillae bordering the puncturing channel, as well as changes in their position, can be observed.

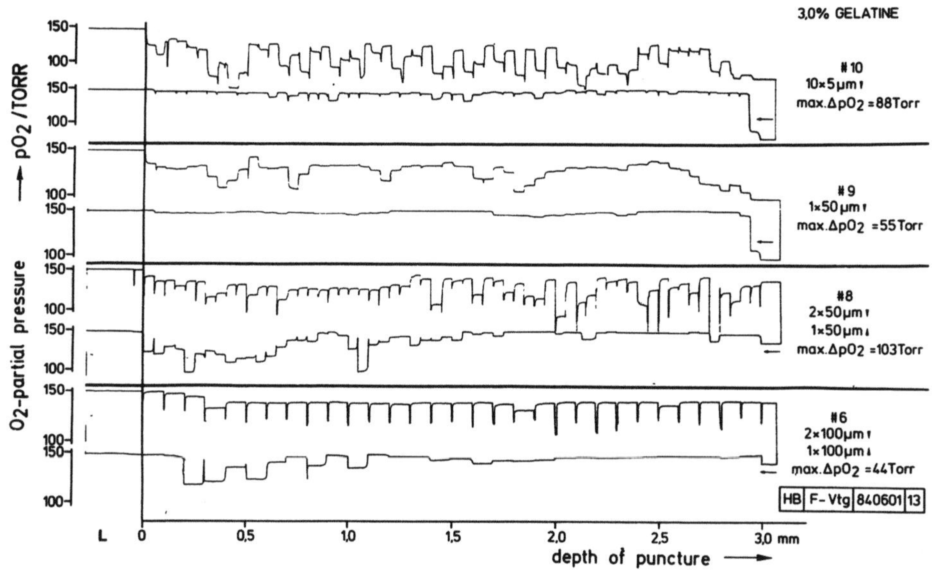

Fig. 9. The behavior of the measuring signal in air-saturated 3% gelatine (22 °C) during stepwise puncturing technique with a micro-PO_2 needle electrode (tip diameter 0.7 µm). Despite constant PO_2 of 150 Torr in the gelatine during all of the advancing punctures (*upper trace* from left to right), considerable measuring signal changes are seen resulting from effects of mechanical pressure. In contrast, uniform measuring signals are found during electrode withdrawal (*lower trace* from right to left; direction of the *small arrow*). *L*, electrolyte solution over the gelatine; *O*, gelatine surface. Protocol of the puncturing technique is shown in the right column. For example, in experiment # 8 *2 × 50 µm* ▼, *1 × 50 µm* ▲ represents two advancing steps of 50 µm followed by one withdrawing step of 50 µm. *max* ΔPO_2, maximum PO_2 change during puncture

In brain samples, E. Seidl was able to observe in many cases (unpublished results) that needle electrodes as small as 3 µm in diameter leave channels which cause vacuolization of the interstitial substance, pyknotization of the ganglion cells, and minor bleeding owing to penetration of the vessels. As is shown in Fig. 11, tissue damage is visible both in the region of the electrode tip and laterally from the shaft. The path of the electrode is usually marked by a wide trail of vacuoles in the interganglionic tissue and a zone of altered ganglions and ganglion cells.

In further studies on the relationship between PO_2 measurement data and morphological findings, we found that, apart from the mechanical effects of puncturing the tissue, there are also electrochemical reactions in the area of the electrode tip, which can cause additional damage. Such disturbances can occur when the electrode membrane is defective. The OH^- ions that form on the cathode (see Eqs. 4–6) can then no longer be held back by the membrane and can penetrate the tissue.

It is difficult to judge the magnitude of the effect of fine structural tissue changes on the PO_2 measuring signal. Since, however, it is essential to know the histologic and microtopographic conditions in the measurement area of the electrode tip in order to interprete the measurement data, this factor must be taken into consideration in every experiment.

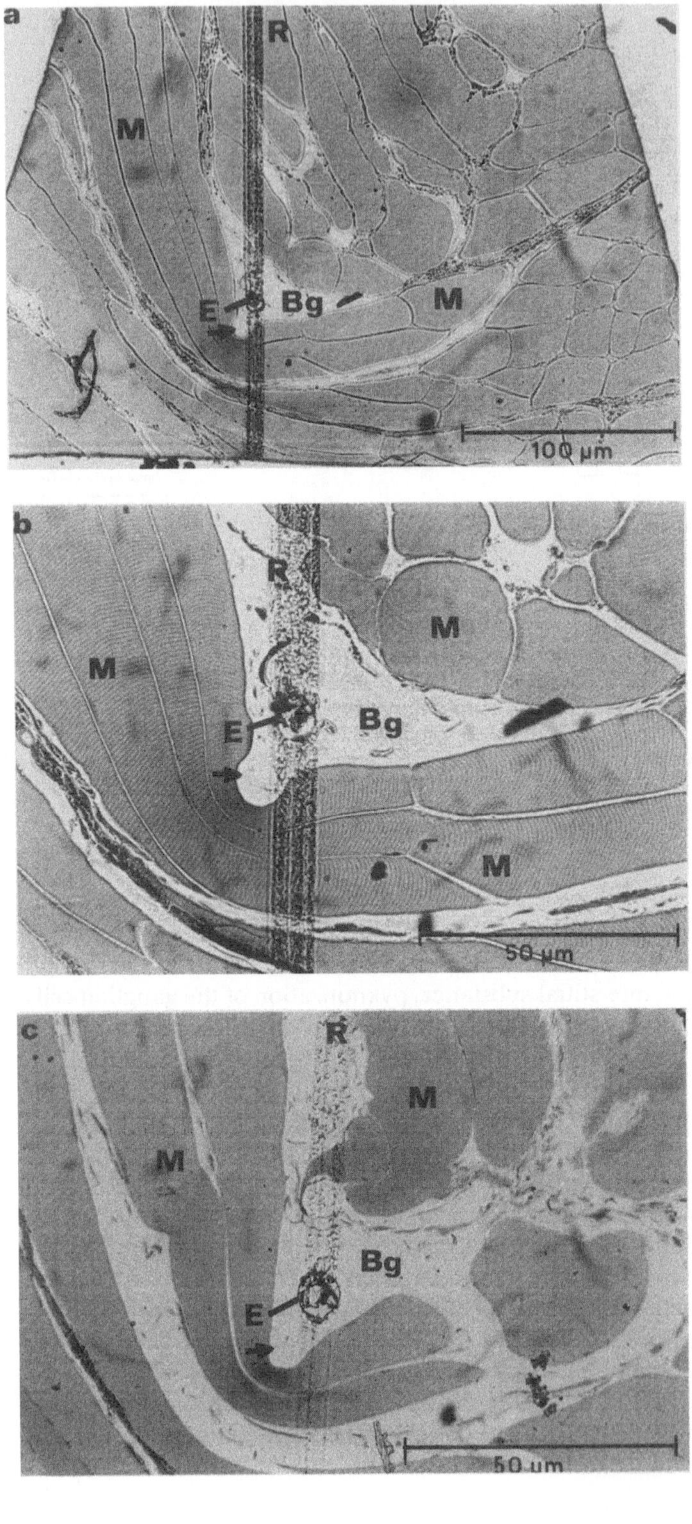
a
R
M
E
Bg
M
100 µm
b
R
M
M
E
Bg
M
50 µm
c
R
M
M
Bg
E
50 µm

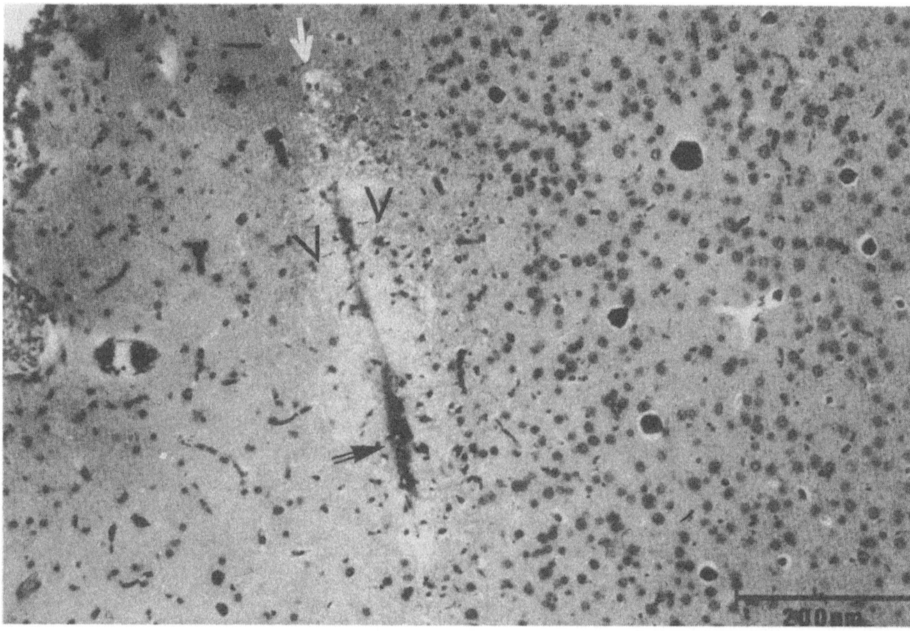

Fig. 11. Light microscopic image of an electrode channel in the cat brain. Tissue changes can be seen in the area of the electrode tip and in the neighborhood of the electrode shaft. ↓, direction of puncture; *V*, vacuolized zone; ⇓, bleeding in puncturing channel (histologic preparation and photograph from E. Seidl, unpublished results)

Summary

The O_2 reduction current, which is generated in the polarographic measuring circuit, is the central measuring value for an absolute PO_2 measurement. It is mainly determined by the electrochemical reactions, the construction of the electrode, the testing equipment used, and the properties of the measuring medium. Special function tests and calibration methods have been developed. Careful application of these tests allows certain sources of error to be detected and thus results in a more reliable interpretation of the measured data.

For the recording of PO_2 profiles in the hemoglobin-free perfused guinea pig brain, the measuring signal was more precisely quantified by a stepwise advancement and withdrawal of the electrode. Both well-agreeing PO_2 responses and severe deviations could be recorded in one and the same puncturing channel, although in

←

Fig. 10 a–c. Changes in the fine structure of the muscle tissue (rat thigh) after the needle electrode *(E)* has been in place for about 10 min. The structure of the connective tissue *(Bg)* is disturbed and the position of the muscle fibrillae *(arrow)* bordering the puncturing channel is altered. The ridges *(R)* are caused by glass, since the electrode was not removed during the dissection. **a–c** three slices of a series. (Dissection and recording from D. Schäfer, unpublished results)

our experimental model, the PO_2 profile withdrawing the electrode has to be evaluated more critically, because of the increasing edema formation.

To estimate the influence of the puncturing technique, microneedle electrodes were inserted - continuously and stepwise at various speeds and step sizes - into gelatine of different concentrations with the PO_2 kept constant. The measuring signal dropped the most during continuous and small-step movement of the electrode without withdrawal. The measuring signal was altered less when the electrode was advanced in larger steps and then withdrawn.

According to light and scanning electron microscopic investigations in liver, muscle, and brain, the electrode causes morphological changes in the tissue of all of these organs. These changes vary from organ to organ, depending on the geometry of the electrode, the puncturing technique, and the conditions and duration of the experiment.

The detailed analysis of the measuring properties of needle electrodes illustrates the limits of this measuring technique with respect to accuracy and applicability. It further shows how difficult it is to make absolute PO_2 measurements with microPO_2 electrodes in living tissue. In addition, the analysis indicates that the data obtained can only be interpreted if the interfering factors listed above are taken into consideration or have been eliminated.

References

1. Albanese, R. A.: On microelectrode distortion of tissue oxygen tension. J. Theor. Biol. 38 (1973) 143-154
2. Barr, R. E., T. E. Tang, A. W. Hahn: Variations on the response characteristics of oxygen electrodes. Adv. Exp. Med. Biol. 94 (1978) 17-23
3. Bates, M. L., A. Feingold, M. I. Gold: Effects of anesthetics on an in vivo oxygen electrode. Am. J. Clin. Pathol. 64 (1975) 448-451
4. Baumgärtl, H., A. M. Ehrly, K. Saeger-Lorenz, D. W. Lübbers: Erste Ergebnisse intracutaner Messungen von pO_2-Profilen. Verlag MWP, München (1985) S. 111-117
5. Baumgärtl, H., W. Grunewald, D. W. Lübbers: Polarographic determination of the oxygen partial pressure field by Pt microelectrodes using the O_2 field in front of a Pt macroelectrode as a model. Pflüg. Arch. 347 (1974) 49-61
6. Baumgärtl, H., D. W. Lübbers: Microcoaxial needle sensor for polarographic measurement of local O_2 pressure in the cellular range of living tissue. Its construction and properties. In: Gnaiger, E., Forstner, H. (eds.): Polarographic Oxygen Sensors, Springer-Verlag, Berlin/Heidelberg (1983) pp. 37-65
7. Baumgärtl, H., D. W. Lübbers: Die Bedeutung der Membran bei polarographisch arbeitenden O_2-Sensoren - insbesondere bei der Mikronadelelektrode - für die absolute PO_2-Mesung in Flüssigkeiten und Geweben. In: Akademie d. Wissenschaften u. d. Literatur (Hrsg.): Funktionsanalyse Biologischer Systeme, Mainz (1984) S. 155-171
8. Cole, M. A., S. Sernick, H. W. Puffer, N. E. Warner, D. W. Crawford: Methods in laboratory investigation-identification and evaluation of histopathology at microelectrode puncture sites. Lab. Inv. 49 (1983) pp. 626-631
9. Davies, P. W.: The oxygen cathode. In: V. L. Nastuk (ed.): Physical Techniques in Biological Research (Vol. IV). Academic Press, New York (1962) 137-179
10. Davies, P. W., F. Brink: Microelectrodes for measuring local oxygen tension in animal tissues. Rev. Sci. Instrum. 13 (1942) 524-533
11. Dent, J. G., K. J. Netter: Errors in oxygentension measurements caused by halothane. Br. J. Anaest. 48 (1979) 195-197
12. Ehrly, A. M.: Messungen des Gewebesauerstoffdruckes im ischämischen Muskelgewebe von Patienten mit arteriellen Verschlußkrankheiten mittels Mikro-Platin-Stichelektroden. In:

A. M. Ehrly (Hrsg.): Messung des Gewebesauerstoffdruckes bei Patienten. Witzstrock-Verlag, Baden-Baden/Köln/New York (1981) S. 36-44

13. Fabjan, Ch., M. R. Kazemi, A. Neckel: Untersuchungen über den Mechanismus und die Katalyse der kathodischen Sauerstoffreduktion an verschiedenen Metallen. Ber. Bunsenges. Phys. Chem. 84 (1980) 1026-1031
14. Fatt, I.: Polarographic Oxygen Sensors. Its Theory of Operation and its Application in Biology, Medicine, and Technology. CRC Press, Cleveland (1976)
15. Fleckenstein, W., C. Weiss: A comparison of PO_2 histograms from rabbit hind-limb muscles obtained by simultaneous measurements with hypodermic needle electrodes and with surface electrodes. Adv. Exp. Med. Biol. 169 (1984) 447-455
16. Forbes, M., S. Lynn: Oxygen reduction at an anodically activated platinum rotating disk electrode. Amer. Inst. Chem. Eng. J. 21 (1975) 763-769
17. Freund, W.: Morphologische Veränderungen bei Messungen mit Einstichsonden in der Rattenniere. Diss.-Schrift, Univ. Kiel (1983)
18. Gleichmann, U., D. W. Lübbers: Die Messung des Sauerstoffpartialdruckes in Gasen und Flüssigkeiten mit der Pt-Elektrode unter besonderer Berücksichtigung der Messung im Blut. Pflüg. Arch. 271 (1969) 431-455
19. Grote, J.: Die Sauerstoffdiffusionskonstanten im Lungengewebe und Wasser und ihre Temperaturabhängigkeit. Pflüg. Arch. 295 (1967) 245-254
20. Gruber, R. B., J. J. Amato: Errors in tissue PO_2 determinations. Mil. Med. 135 (1970) 1036-1037
21. Grunewald, W.: Diffusionsfehler und Eigenverbrauch der Pt-Elektrode bei PO_2-Messungen im steady state. Pflüg. Arch. 320 (1970) 24-44
22. Grunewald, W.: Einstellzeit der Pt-Elektrode bei Messungen nicht stationärer O_2 Partialdrucke. Pflüg. Arch. 322 (1971) 109-130
23. Hahn, C. E. W., A. H. Davis, W. J. Albery: Electrochemical improvements of the performance of PO_2 electrodes. Respir. Physiol. 25 (1975) 109-33
24. Heinrich, U., J. Hoffmann, H. Baumgärtl, B. Yu, D. W. Lübbers: Quantitative analysis of reflection spectra, oxygen supply and metabolism of the blood-free perfused guinea-pig brain. Proc. ISOTT Symp., Nijmegen (1984)
25. Himmelberg, H.: Mechanisches Verhalten von Mikrostichelektroden zur Sauerstoffdruckmessung im Muskelgewebe. Diss.-Schrift, Univ. Münster (1983)
26. Hitchman, M. L.: Measurement of dissolved oxygen. Wiley, New York/Geneva (1978)
27. Hutten, H., K. Meiners, R. Zander: Ein polarographisches Verfahren zur Bestimmung von Sauerstoff-Löslichkeitskoeffizienten in wäßrigen Elektrolytlösungen. Biomed. Techn. 27 (1982) 7-13
28. Inch, W. R.: Problems associated with the use of the exposed platinum electrode for measuring oxygen tension in vivo. Canad. J. Biochem. 36 (1958) 1009-1021
29. Jamieson, D., H. A. S. van den Brenk: Electrode size and tissue PO_2 measurement in rats exposed to air or high pressure oxygen. J. Appl. Physiol. 20 (1965) 514-518
30. Jordan, J., E. Ackermann, R. L. Berger: Polarographic diffusion coefficients of oxygen defined by activity gradients of oxygen in viscous media. J. Am. Chem. Soc. 78 (1956) 2979-2983
31. Kanabus, E. W., C. Feldstein, D. W. Crawford: Excursion of vibrating microelectrodes in tissue. J. Appl. Physiol. 48 (1980) 737-741
32. Klinowski, J., S. W. Korsner, C. P. Winlove: Problems associated with the micropolarographic measurement of the arterial wall PO_2. Cardiovasc. Res. 16 (1982) 448-456
33. Kreuzer, F., H. P. Kimmich, M. Brezina: Polarographic determination of oxygen in biological materials. In: J. Koryta (ed.): Medical and Biological Applications of Electrochemical Devices. John Wiley & Sons, New York (1980) pp 173-261
34. Kunze, K., A. Kunze: Das Sauerstoffdruckfeld in der Muskulatur-Messung mit Platin-Stichelektroden. In: A. M. Ehrly (Hrsg.): Messung des Gewebesauerstoffdruckes bei Patienten. Witzstrock-Verlag, Baden-Baden/Köln/New York (1981) S. 22-35
35. Laitinen, H. A., I. M. Kolthoff: A study of diffusion processes by electrolyses with microelectrodes. J. Am. Chem. Soc. 61 (1939) 334-3349
36. Lee, J. H., G. T. Tsao, P. C. Wankat: Hydrodynamic effect of surfactants on gas-liquid oxygen transfer. AIChE J. 26 (1980) 1008-1012
37. Linek, V., J. Sinkule, V. Vacek: Oxygen-electrode dynamics - 3-layer model chemical reaction in the liquid film. Biotech. Biol. 25 (1983) 1401-1418

38. Lübbers, D. W.: The meaning of the tissue oxygen distribution curve and its measurement by means of Pt-electrodes. In: F. Kreuzer (ed.): Oxygen Pressure Recording in Gases, Fluids, and Tissues. Prog. Res. 3 (1969) pp 112-123
39. Lübbers, D. W., H. Baumgärtl: Determination of oxygen pressure in tissue. In: Institut National de la Santé et de la Recherche Médical (ed.): New Methods in the Study of Microcirculation. Int. Comm. Microcirculation, Paris (1983) pp. 121-131
40. McHugh, R. D., R. M. Epstein, D. W. Longnecker: Halothane mimics oxygen in oxygen microelectrodes. Anesthesiol. 50 (1979) 47-49
41. Müller, L., L. Nekrassow: Zum Mechanismus der kathodischen Reduktion von Sauerstoff an einer glatten, aktiven Platinelektrode in alkalischer Lösung. J. Electroanal. Chem. 9 (1965) 282-289
42. Norton, J. M., P. W. Rand: Chronically implanted tissue oxygen electrodes in rabbits. J. Appl. Physiol. 36 (1974) 118-122
43. Rhodes, P. G., K. M. Moser: Sources of error in oxygen tension measurement. J. Appl. Physiol. 21 (1966) 729-734
44. Saito, Y.: A theoretical study on the diffusion current at the stationary electrodes of circular and narrow bond types. Rev. Polarogr. (Japan) 15 (1968) 177-187
45. Schäfer, D., J. Höper: The influence of glass needle electrodes on rat liver cells and tissue. Sixth European Congress on Electron Microscopy, Jerusalem (1976) pp. 304-306
46. Schäfer, D., J. Höper: Alterations in rat liver cells and tissue by needle electrodes. In: M. Kessler, L. C. Clark, D. W. Lübbers, I. A. Silver, W. Simon (eds.): Ion and Enzyme Electrodes in Biology and Medicine. Urban & Schwarzenberg, München/Berlin/Wien (1976) pp 217-222
47. Schneiderman, G., T. K. Goldstick: Oxygenelectrode design criteria and performance-characteristics - recessed cathode. J. Appl. Physiol. 45 (1978) 145-154
48. Schuchhardt, S., B. Lösse: Methodological problems when measuring with PO_2-needle electrodes in semisolid media. In: M. Kessler, D. F. Bruley, L. C. Clark, D. W. Lübbers, I. A. Silver, J. Strauss (eds.): Oxygen Supply. Urban & Schwarzenberg, München (1973) pp 108-109
49. Silver, I. A.: Problems in the investigation of tissue oxygen microenvironment. Adv. Chem. Ser. 118 (1973) 343-351
50. Speckmann, E. J., H. Caspers: Messung des Sauerstoffdrucks mit Platinmikroelektroden im Zentralnervensystem. Pflüg. Arch. 318 (1970) 78-84
51. Stokes, B. T., M. W. Garwood: Normal oxygen tensions in the canine spinal cord. Adv. Physiol. Sci. 25 (1981) 183-184
52. Stuck, J. D., J. A. Howell, H. T. Cullinan: Analysis of oxygen diffusion to a measuring probe in animal tissue. J. Theor. Biol. 31 (1971) 509-532
53. Tang, T. E., R. E. Barr, V. G. Murphy, A. W. Hahn: A working equation for oxygen sensing disk electrodes. Adv. Exp. Med. Biol. 94 (1978) 9-15
54. Weigelt, H., E. Seidl, H. Acker, D. W. Lübbers: The application of histological analysis for the localization of tissue PO_2 and local blood flow with the example of the glomus caroticum of the rabbit (Oryctolagus cuniculus). Adv. Exp. Med. Biol. 75 (1976) 17-23
55. Wodick, R.: Compensation of measuring errors produced by finite response time in polarographic measurements with electrodes sensitive to oxygen and hydrogen. Pflüg. Arch. 336 (1972) 327-344

Methods and Clinical Application of Local Tissue PO_2 Measurements Using the Multiwire Surface Electrode

H. U. Spiegel, J. Hauss, K. Schönleben, and H. Bünte

Introduction

In the last 50 years, a number of methods have been developed to determine the local tissue PO_2 in organs such as the skeletal muscle, the liver, and the myocardium. The local O_2 pressure provides direct information on the function of the gas exchange in tissue. The "local information" thus obtained supplements the conventional clinical measuring techniques - determination of heart rate, central venous pressure (CVP), cardiac output - which deliver only indirect information on the condition of the microcirculation. In measuring tissue PO_2, we make a distinction between direct and indirect measuring methods.

Kivisaari and Niinikoski [23] described an indirect measuring technique, the perfusion method, for which an O_2 permeable Teflon tube is inserted in the tissue and is continuously perfused with an oxygen supporting material. The measurement of the PO_2 is done polarographically or with a mass spectrometer.

Among the direct methods, we differentiate between polarographic, potentiometric, and optic measuring techniques. The historical development of the polarographic method has been described in detail by Lübbers [27].

Information Provided by Tissue PO_2 Measurement

Numerous clinical results of tissue PO_2 measurement have been reported in the last few years [9, 13, 21, 22, 26, 29, 31, 41]. The objective of our investigation was to find out whether measurement of local tissue PO_2 is able to provide diagnostic information on the oxygen supply to the organs and to alter the emphasis of therapy control [1, 10, 11, 12, 13, 14, 32, 33, 34, 38, 39, 40].

As early as 1942, Davies and Brink [5] reported on their first polarographic tissue PO_2 measurements in animal experiments. They had problems directly measuring the PO_2 in the blood or in body fluids containing protein with an uncoated platinum cathode. Since the electrode itself consumed oxygen in the area, an oxygen poor diffusion zone [30] formed, which was dependent on the convection. Moreover, the contamination of the cathode by substances, particularly proteins, induced a rapid sensitivity change so that only relative PO_2 changes could be determined with the uncoated electrode.

The introduction of an oxygen permeable membrane, which only covers the cathode or which separates the cathode and anode together with an electrolyte from the membrane (Clark Principle) [2, 3] brought about considerable improvement in the measuring properties of polarographic electrodes. The decisive technologic advancement, and thus the basis for later application of these measuring techniques for medical purposes, took place in 1966 when Kessler and Lübbers [16] developed a Clark-type multiwire surface electrode. It then took 10 years before the method was applied in patients, when in 1975, Schönleben et al. [32] began using the multiwire surface electrode in the surgical hospital of the University of Münster.

Technique of the PO_2 Measurement

Polarographic Measuring Principle

In polarographic measuring [4, 6, 15, 25], an optimum polarization voltage is applied in the range of the polarization wave, by which all of the O_2 molecules diffusing to the cathode are reduced [24]:

$$O_2 + 4e^- \rightarrow 2O^{--} \quad (1)$$

That means that in this range, the oxygen pressure on the platinum surface is zero. The reduction current is thus determined solely by the amount of oxygen coming out of the tissue. Assuming that the diffusion conditions between a medium containing oxygen and the surface of the electrode are constant, the O_2 partial pressure and the concentration of the oxygen molecules can be determined by measuring the reduction current.

The polarographic measuring circuit consists of the voltage source, the current and voltage measuring instrument, and the measuring cell (Fig. 1). With the voltage source, the polarization tension is produced, which can be taken off on the resistance. The polarization voltage is applied to the measuring cell, which consists of a platinum cathode, an Ag/AgCl anode, and a KCl electrolyte. The instrument shows the respective flowing reduction current.

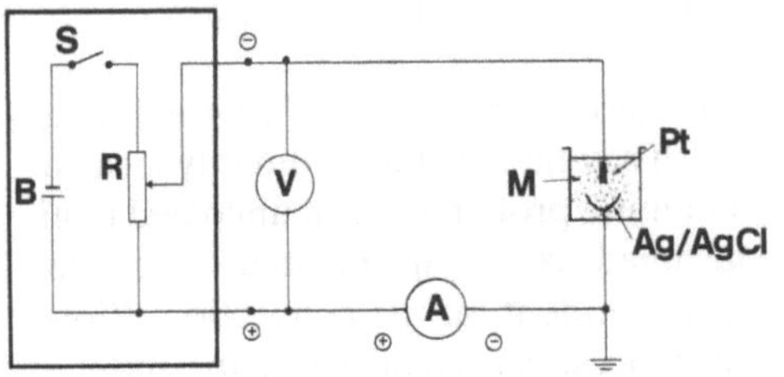

Fig. 1. Polarographic measuring circuit; *B*, voltage source; *V*, polarization voltage; *R*, resistance; *M*, measuring cell; *A*, instrument

Polarogram and Calibration Line

The relationship between the measured current and the applied voltage can be plotted in the current voltage curve, i.e., the polarogram (Fig. 2). In the presence of oxygen, a characteristic flat curve portion can be observed, the so-called polarographic wave, in which an increase in voltage results in practically no reduction in the flowing current. The current rises when the voltage shows a marked increase, since there are then other current-supplying electrochemical reactions taking place on the electrode. At 600–800 mV, the maximum diffusion current and thus the characteristic wave is reached. This voltage is applied on the electrode as the working voltage. Plotting the current intensity in relation to the oxygen partial pressure produces the calibration curve of the platinum electrode. Figure 3 illustrates the calibration degrees of a four-point calibration at different calibration temperatures. The depen-

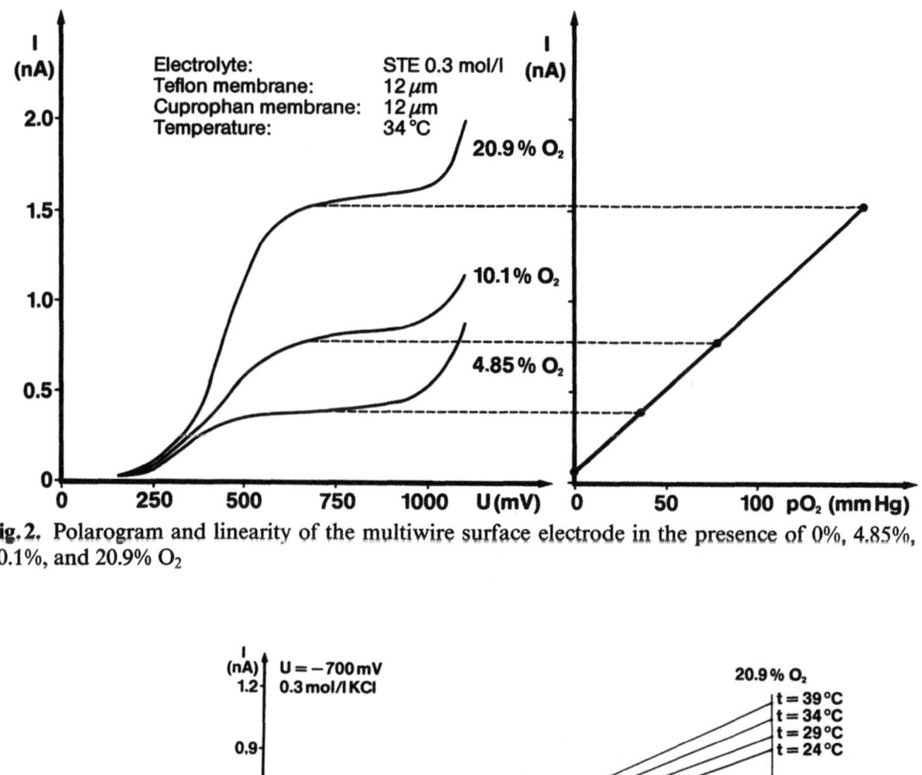

Fig. 2. Polarogram and linearity of the multiwire surface electrode in the presence of 0%, 4.85%, 10.1%, and 20.9% O_2

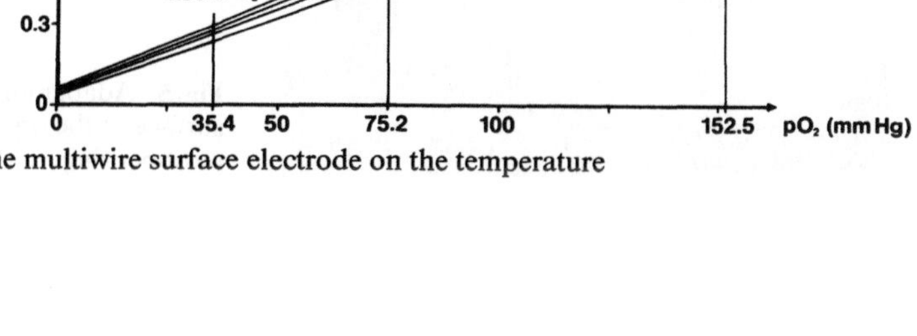

Fig. 3. Dependence of the multiwire surface electrode on the temperature

dence of the electrode on the temperature at 100 mm Hg is lower than 1% per degree Celsius.

The Multiwire Surface Electrode

The multiwire surface electrode (MDO) that is now being clinically used was specially developed for tissue PO_2 measurements by the Max Planck Institute for System Physiology in Dortmund [17, 18, 19, 32]. Figure 4 gives a diagram of the construction of this Clark-type surface electrode [2, 3]. The eight platinum cathodes are sealed in a glass and are surrounded by an annular anode consisting of Ag/AgCl. An oxygen permeable Teflon membrane separates the platinum cathodes and the anode from the measuring medium. A cuprophane membrane [8] inserted between the membrane and the platinum wires stabilizes the diffusion properties. A sterile

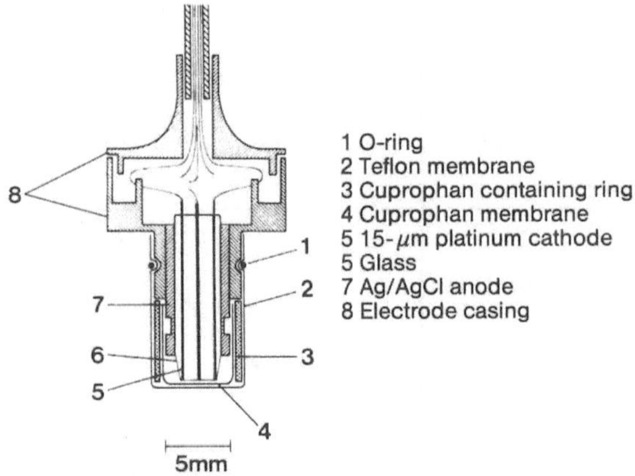

Fig. 4. Diagram of the construction of the multiwire surface electrode

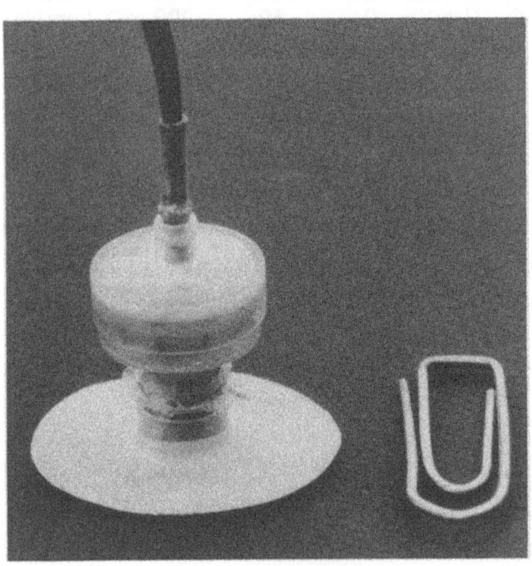

Fig. 5. Adaptation of the electrode to the surface of the organ using a Teflon cap and a rubber ring

Table 1. Specifications of the multiwire surface electrode

Cathode	15 μm platinum
Anode	Ag/AgCl
No. of wires	8
Membrane	12 μm Teflon 12 μm Cuprophan
Electrolyte	sterile 0.3 mol/l KCl solution
Electrode weight	2.1 g
Electrode diameter	5 mm

bromide-free 0.3 mol/liter KCl solution is used as electrolyte. The electrode is constructed in such a way that it can be placed directly on the organ with the aid of a small Teflon cap. A rubber ring affixes the electrode to the moist surface of the organ. No air is able to reach the electrode from the side (Fig. 5). The electrode, weighing 2.1 g, was dimensioned in such a way that pressure ischemia caused by compression of the capillaries was prevented. The technical data of the MDO electrode are listed in Table 1.

The PO_2 Measuring System

Figure 6 shows a diagram of the PO_2 measuring system including the determination, recording, and evaluation of the measuring values. With the electronic compact equipment, the local tissue PO_2 can be continuously recorded at eight different measuring points in the tissue at the same time. The PO_2 values can be calculated immediately, the drift is immediately corrected with a calculator integrated in the system. After on-line data processing, the absolute PO_2 values are plotted as function of time. To determine the pressure in the pulmonary artery, the mixed venous blood gas values, and the cardiac output (CO), a Swan-Ganz catheter is placed in the pulmonary artery under pressure and X-ray control is undertaken according to

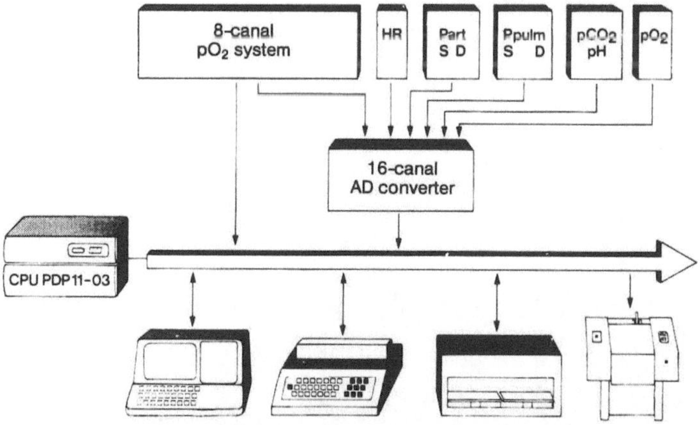

Fig. 6. Diagram of the PO_2 system with measurement determination, integrated calculator, and output equipment

the thermodilution method. To measure the arterial blood gases, intravasal PO_2, PCO_2, and pH probes were inserted in the radial artery via Teflon tubes (16 g).

Sterilization

To safeguard the patient against possible bacterial contamination, a special sterilization technique has been developed [35, 37]. The electrode and its components are sterilized separately with pure ethylene oxide at a low pressure of 680 to 700 mm Hg and at an operating temperature of 45 °C. The electrolyte is produced in sterile bottles. The material to be sterilized is packed and welded in polyethylene and polyamide foil. A gauze cuff is placed around the glass (Fig. 7), which absorbs the mixture of ethylene oxide and water vapor to prevent the chloride covering of the reference electrode from being damaged by the aggressive gas mixture. Assembly of the electrode for use on intensive care wards must be done under absolutely sterile conditions.

Calibration of the Electrode

The electrode must be calibrated before and after every measurement, since it can change during the course of a measuring period. How severe the drift is depends on the constancy of the applied polarization voltage, the purity of the platinum surface,

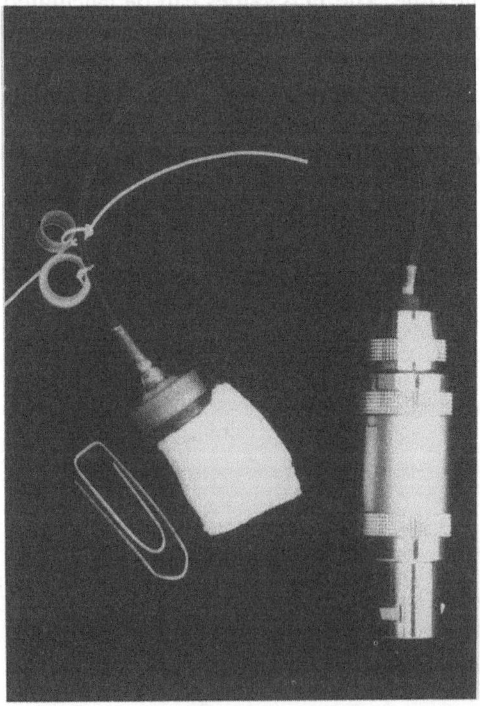

Fig. 7. Tissue electrode with gauze cuff to protect it from dirt and aggressive ethylene oxide-water vapor mixture

Table 2. Measurement parameters of the multiwire surface electrode

Polarization voltage	-700 mV
Sensitivity $\frac{\Delta I}{\Delta PO_2}$	$0.6-1.4 \times 10^{-11}$ A/mm Hg
Sensitivity/100 mm Hg in 0.9% NaCl solution at 34 °C, pH 6.0	$0.6-1.4 \times 10^{-9}$ A
N_2 value	$< 0.6 \times 10^{-10}$ A
Response time t_{90}	< 5.0 s
Electrode resistance	10^9-10^{10} ohms
Drift (sensitivity)	max. $\pm 8\%$/h
Isolation resistance	$> 1 \times 10^{12}$ ohms

and the existence of a specific boundary layer on the platinum which stabilizes the running qualities (possibly a monomolecular platinoxide film) [7].

The electrode is calibrated before and after each measurement under standardized conditions in a calibration vessel. A two-point calibration is generally carried out. The zero value and the residual current are determined with pure nitrogen. The reference gas selected for measurement in the tissue should be one that ensures that the calibration values correspond to the local oxygen tensions. For clinical use, a mixture of oxygen and nitrogen (5% O_2, 95% N_2) is chosen. Under hyperoxic conditions, when tissue PO_2 values of over 50 mm Hg can occur, a third reference gas with an adequate O_2 partial pressure is selected as calibration gas. Calibration is carried out after a response time of 2–3 min at 34 °C in the 0.9% NaCl solution equilibrated with the calibration gases. On the basis of the measured currents and taking the air and water vapor pressure into account, the sensitivity per 100 mm Hg can be calculated according to the following formula:

$$\frac{I}{100\,\text{mm Hg}} = \frac{I_{5\%}\,O_2 - I_{0\%}\,O_2}{PO_2\ 5\%\ O_2} \times 100 \qquad (2)$$

The minimum requirements listed in Table 2 must be met before the electrode can be used for measurement in the tissue.

Technique of Measurements in Patients

We chose the M. quadriceps femoris as the organ for measurement. After the skin and the subcutaneous tissue is anesthetized with 0.5% xylocaine with no vasoconstrictors, the muscle is exposed by a 1.5- to 2-cm-long incision. Care must be exercised in dissecting the fascia and in stopping any bleeding. Continuous recording of the local temperature at the point of measurement is done with a thermistor probe

Fig. 8. PO_2 multiwire surface electrode and thermistor in situ on the skeletal muscle

placed on the surface of the muscle. The leg is immobilized on a foam rubber splint. Only the measurements carried out after a period of muscle rest are taken into account. An electrode with a Teflon cap is placed on the exposed muscle. After a response time of a few minutes the measurement can begin (Fig. 8).

In recording the tissue PO_2 values, a distinction is made between two forms of measurement: continuous PO_2 measurement and the PO_2 histogram.

Forms of Measurement

Continuous PO_2 Measurement. During continuous PO_2 measurement, the electrode is left on the selected measuring site. The direct recording of the eight single measuring wires provides qualitative information on the immediate changes in the tissue PO_2 during one time interval. The individual measured values show slight fluctuations, which may possess a certain rhythm.

The investigations carried out thus far have shown that neither one single tissue PO_2 value nor the mean value from several single values is suited for evaluating the oxygen supply to an organ. One single local PO_2 value cannot provide information on the entire situation in the tissue. Knowing the amount of venous oxygen partial pressure only makes it possible to roughly determine whether there are regions with oxygen deficiency and what the PO_2 is in such regions. Judging this becomes even more difficult when at the same time there are different distances between capillaries and different capillary lengths, since in this case the venous PO_2 represents a mean value of the venous oxygen partial pressures from various supply units [20]. This mean value gives even less evidence as to whether the PO_2 has already fallen

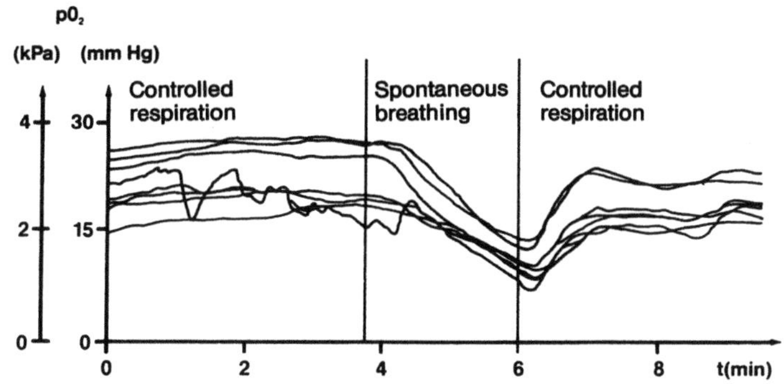

Fig. 9. Continuous PO_2 measurement in an intensive care patient during controlled ventilation and spontaneous breathing

below the critical mitochondrial oxygen pressure value. A mean tissue PO_2 of the same value can appear [28]

(1) when the arterial PO_2 is high and the circulation low (the venous and the low tissue PO_2 values drop considerably), or

(2) when the arterial PO_2 is low and the blood flow high (the venous and the low tissue PO_2 values drop only slightly).

The results of the blood gas analysis serve as a measure of the necessity of ventilation therapy and of an inspiratory supply of oxygen. They cannot, however, provide any reliable information on the O_2 supply of the organs. Low arterial PO_2 can almost always be corrected by raising the supply of oxygen. The damaging effect of too much O_2 in the microcirculation, however, has not yet been revealed.

Figure 9 shows an example of a continuous PO_2 measurement. A 69-year-old patient was examined to find out whether changes in the O_2 supply occur when the mode of ventilation is altered. The drastic decline in the tissue PO_2 showed, before the breathing insufficiency clinically appeared, that artificial ventilation had to be maintained. The PO_2 values quickly returned to normal when the ventilation was switched on again.

The Physiologic PO_2 Histogram. The second form of recording is the PO_2 histogram. When the PO_2 histogram is recorded, the patient should be in respiratory and hemodynamically steady state. In comparative studies for the exact description of PO_2 histograms, we consider the following factors necessary:

1. Information on the organ being measured
2. Mode of anesthesia
3. Mode of breathing or ventilation
4. Level of PaO_2
5. Information on the hemodynamics (HR, p_{art}, CO, p_{pulm})

The PO_2 histogram marks the statistical distribution of the local PO_2 in the organ being examined. By repositioning the electrode, at least 100 different values are measured, so that the results can be considered representative of the PO_2 in the entire organ. Since a certain time interval has to be taken into account, the PO_2 histo-

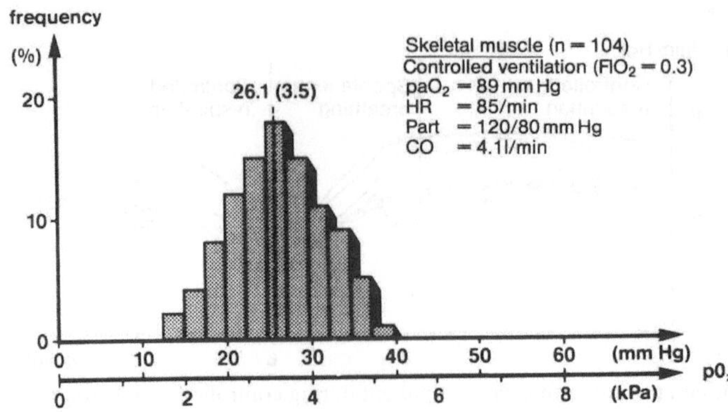

Fig. 10. Physiologic PO_2 histogram from skeletal muscle of a 43-year-old patient under basal anesthesia with piritramide

gram is not suited for gaining instantaneous information, but rather for qualitative and quantitative data on the O_2 supply in an organ. Information on the microcirculatory conditions is provided not only by the amount of local O_2 pressure, but also by the configuration of the PO_2 histogram, the absolute values, the position of the arithmetic mean and maximum values, and the existence or nonexistence of hypoxic or anoxic values.

A physiologic histogram is characterized by the following [13]: (a) There are no anoxic areas, though the lowest values can be close to zero; (b) corresponding to the spatially inhomogeneous O_2 pressure distribution in the tissue, there is an inhomogeneous statistical distribution of the measured oxygen values around a mean value; (c) the configuration of a physiologic histogram is often bell-shaped (Fig. 10).

The PO_2 histogram apparently has a normal shape, which during changing conditions is set by regulatory changes in the microcirculation. Under normal conditions, the bell-shaped configuration of the PO_2 histogram is shifted to the left. Its shape and position remains relatively constant. The working hypothesis "normal oxygen supply = normal PO_2 histogram" has proved very productive and has also been applied successfully for measurements in patients [28].

The Pathologic PO_2 Histogram. Deterioration of the oxygen supply results in an increase in low PO_2 values. In comparison with the physiologic, bell-shaped PO_2 histogram, a pathologic histogram is shifted to the left. Under severe hypoxic conditions, the low PO_2 values can increase to the point where the lowest class is the largest. The bell shape disappears when there is hypoxia - the PO_2 histogram becomes broader and irregular [28].

Following Kessler [21], we make a distinction between disturbances in the local O_2 supply (a) with anoxic or close to anoxic values and those (b) without anoxic or close to anoxic values.

The causes of disturbances in the local oxygen supply, as they are determined with the PO_2 histogram, can be divided into four phenomenologic groups [22]: (1) local or global low-flow situations, which result in partial tissue hypoxia or anoxia (e.g., hypovolemic shock); (2) disturbances in the microcirculation which

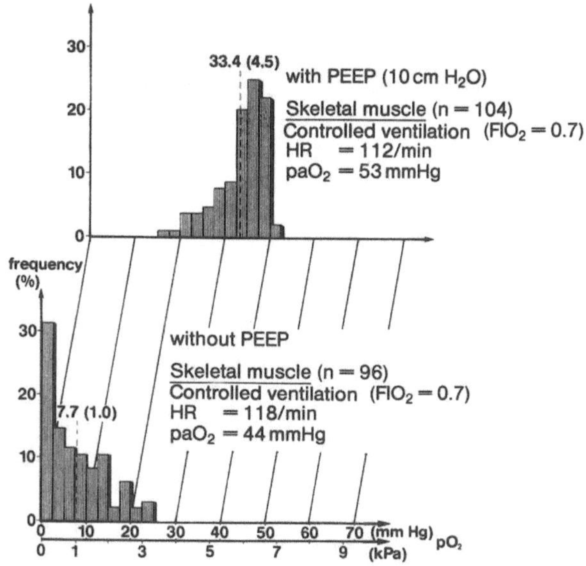

Fig. 11. Pathologic PO_2 histogram from skeletal muscle of an intensive care patient with pulmonary insufficiency

are caused by a number of pathogenetic factors, e.g., rheologic, humoral, inflammatory, immunologic factors, or by vascular sclerosis; (3) arterial hypoxia or anemia, which brings about partial tissue hypoxia or anemia; (4) distribution disturbances in the microcirculatory pattern as a result of hyperoxemia, which are characterized by a "disintegrated" PO_2 histogram without or combined with hypoxic PO_2 values.

Figure 11 shows a typical pathologic histogram recorded in a patient with pulmonary insufficiency. There is a pronounced shift to the left. Under such severe hypoxia, the hypoxic values and those close to anoxia increase to the point where the lowest class is the largest. Ventilation with a positive end expiratory pressure (PEEP) of 10 cm H_2O brings the oxygen supply of the skeletal muscle back to normal.

The Summarized PO_2 Histogram. Rather than demonstrating innumerable single findings from one measurement series, individual PO_2 histograms are compiled in a summarized PO_2 histogram under reproducible, steady-state conditions (controlled ventilation, constant inspiratory FIO_2, standardized anesthesia). Both the number of investigations and the number of measurements carried out are given. The respiratory and hemodynamic values represented are the arithmetic mean values of the individual experiments.

Figure 12 illustrates the summarized PO_2 histograms from eight single investigations before and after nitroglycerin was applied for hypotension control prior to and under the application of 8 and 16 μg/kg b.w./min nitroglycerin respectively.

The "Normal" Histogram. The similarity of the PO_2 histograms of different organs shows that the supply of oxygen to the tissue under physiologic conditions has to be regulated precisely by a highly efficient system. In comparing these PO_2 histograms, it is notable that, despite different values for flow, oxygen uptake, and consumption

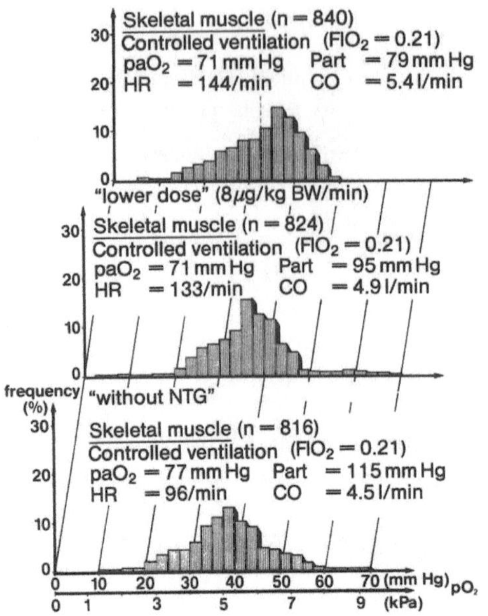

Fig. 12. Summarized tissue PO_2 histograms of the skeletal muscle before and during continuous application of nitroglycerin

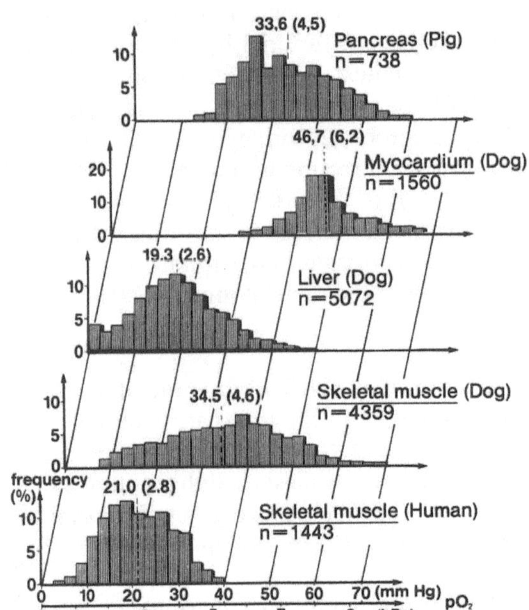

Fig. 13. Normal PO_2 histogram of the skeletal muscle of a patient *(below)*, the skeletal muscle of a dog *(above that)*, the liver of dog, the myocardium of a dog, and the pancreas of a pig *(top)*

of the organ and organ-specific capillary architecture, there is surprising agreement in the shape and position. With the exception of the heart, there was a maximum frequency between 15 and 30 mm Hg; the lowest values were in the range of the critical PO_2. The organ-specific venous PO_2 was without exception below the maximum frequency values [12]. Kessler et al. [20] explain that this obvious discrepancy between venous PO_2 and tissue PO_2 is a result of a nonhomogeneous microcirculation owing to a nonuniform capillary length distribution.

Figure 13 shows normal PO_2 histograms that were recorded on the skeletal muscle of patients, on the skeletal muscle of dogs, on the liver and myocardium of dogs, and on the pancreas of the pig. In comparison with the normal histograms obtained by other investigators, these histograms of the different species (man, dog, pig, rat) under steady-state conditions are similar.

Conclusions

Although the technique of measuring the local PO_2 in tissue (a) is always invasive, (b) is not a routine clinical method, (c) is very involved, (d) is relatively expensive, (e) and has results which are not easy to interpret, it is the only available method for directly monitoring the oxygen supply in the vascular system at several different locations at the same time on the undamaged organ. The effect of the therapy selected can thus be evaluated objectively. For the surgical treatment of intensive care patients and to answer clinical and scientific questions, the method of local tissue PO_2 measurement with the multiwire surface electrode offers a means of detecting microcirculatory disturbances at a very early stage.

References

1. Bünte, H.: Die Mikrozirkulation - eine entscheidende Kreislaufkomponente für das Überleben kritisch Kranker. In: W. Schellerer, F. W. Schildberg (Hrsg.): Aspekte moderner Chirurgie, Chirurgie aktuell (Bd. 4), Straube-Verlag, Erlangen (1977) S. 32-44
2. Clark, L. C. jr.: Monitor and control of blood and tissue oxygen tensions. Trans Am Soc Art Int Org. 2 (1956) pp 41-48
3. Clark, L. C. jr.: U. S. Patent 2 (1959) pp 913, 386
4. Danneel, H.: Zschrift Elektrochem. 4 (1897) 227
5. Davies, P. W., F. Brink: Microelectrodes for measuring local oxygen tensions in animals tissue. Rev. Sc.; Instrum. 13 (1942) 524-533
6. Fatt, I.: Polarographic oxygen sensors. CRC Press, Cleveland (1976)
7. Feldberg, S. W., C. G. Enke, C. E. Bricker: Formation and dissolution of platinum oxide film. Mechanism and kinetics, J. Electrochem Soc. 110 (1963) 826
8. Gleichmann, U., D. W. Lübbers: Die Messung des Sauerstoffdruckes in Gasen und Flüssigkeiten mit der Pt-Elektrode unter besonderer Berücksichtigung der Messung im Blut. Pflüg. Arch. 271 (1960) 431
9. Hartel, W., J. A. Schneider: Sauerstoffgewebedruck an der Magenschleimhaut - intraoperative Messung zur Kontrolle der kompletten Vagotomie. In: A. M. Ehrly (Hrsg.): Messung des Gewebesauerstoffdruckes bei Patienten, Witzstrock-Verlag, Baden-Baden (1981) S. 80-83
10. Hauss, J., K. Schönleben, U. Spiegel, M. Kessler: Measurement of local oxygen pressure in the skeletal muscle of patients suffering from disturbance of the arterial circulation. In: J. A. Siver, M. Erecińska, H. J. Bicher (eds): Oxygen transport to the tissue, Plenum Press, New York/London (1978) pp 419-422
11. Hauss, J., K. Schönleben: Klinische Beurteilung von Störungen der Mikrozirkulation. In: H. Kreuscher (Hrsg.): Praxis der Intensivpflege, Schattauer-Verlag, Stuttgart/New York (1979) S. 17-25

12. Hauss, J., K. Schönleben, H. U. Spiegel: Therapiekontrolle in der Intensivbehandlung durch kontinuierliche Gewebe-pO_2-Messung. In: A. M. Ehrly (Hrsg.): Messung des Gewebesauerstoffdruckes bei Patienten, Witzstrock-Verlag, Baden-Baden (1981) S. 45-54
13. Hauss, J., K. Schönleben, H. U. Spiegel: Therapiekontrolle durch Überwachung des Gewebe-pO_2. Huber-Verlag, Bern (1982)
14. Hauss, J., K. Schönleben, H. U. Spiegel, H. Bünte: The use of Tissue pO_2 monitoring in Determining Intensive Care Therapy. In: A. M. Ehrly (eds): Determination of Tissue Oxygen Pressure in Patients, Pergamon Press, (1983) pp 41-52
15. Heyrovsky, J.: Polarographisches Praktikum, Springer-Verlag, Berlin/Göttingen/Heidelberg (1948)
16. Kessler, M., D. W. Lübbers: Aufbau und Anwendungsmöglichkeiten verschiedener pO_2-Elektroden. Pflüg. Arch. 82 (1966) 291
17. Kessler, M.: Normale und kritische Sauerstoffversorgung der Leber bei Normo- und Hypothermie. Habil.-Schrift, Univ. Marburg (1968)
18. Kessler, M., W. Grunewald: Possibilities of measuring oxygen pressure fields in tissue by multiwire platinum electrodes. Prog. in Resp. 3 (1969) 147-152
19. Kessler, M.: Problems with the Use of Platinum Cathodes for the polarographic Measurement of oxygen. In: M. Kessler et al. (eds): Oxygen Supply, Urban & Schwarzenberg, München/Berlin/Wien (1973) pp 81-85
20. Kessler, M., J. Höper, D. Schäfer, H. Starlinger: Sauerstofftransport im Gewebe. In: Ahnefeld et al. (Hrsg.): Anästhesiologie und Intensivmedizin (Bd. V, Mikrozirkulation) (1974) S. 36-52
21. Kessler, M., J. Höper, B. Krumme: Monitoring of tissue perfusion and cellular function. Anästhesiologie 45 (1976) 184
22. Kessler, M.: Grundlegende Prinzipien der Sauerstoffversorgung des Gewebes. In: K. Messmer, B. Fagrell (Hrsg.): Mikrozirkulation und arterielle Verschlußkrankheiten. Karger-Verlag, Basel (1981) S. 45-58
23. Kivisaari, J., J. Niinikoski: Use of Siliastic Tube and Capillary Sampling Technic in the measurement of Tissue pO_2 and pCO_2. Am. J. Surg. V 125 (1973) 623-627
24. Kolthoff, I. M., J. Lingane: Polarography, 2nd Interscience, New York (1952)
25. Kolthoff, I. M., J. Jordan: J. Am. Chem. Soc. 74 (1952) 4801
26. Kopp, K. H., E. Sinagowitz, B. Kashammer, H. Weidmann: Methodische Probleme bei der Sauerstoffpartialdruckmessung im Gewebe von Intensivpatienten. In: A. M. Ehrly (Hrsg.): Messung des Gewebesauerstoffdruckes bei Patienten. Witzstrock-Verlag, Baden-Baden (1981) S. 61-66
27. Lübbers, D. W.: Methods of measuring oxygen tensions of blood and organ surfaces. In: J. P. Payne, Hill D (eds): A symposium on oxygen measurements and their significance. Churchill, London (1966) 103-127
28. Lübbers, D. W.: Die Bedeutung des lokalen Gewebesauerstoffdruckes und des pO_2-Histogramms für die Beurteilung der Sauerstoffversorgung eines Organs. Prakt. Anästh. 12 (1977) 184-193
29. Lund, N.: Skeletal muscle surface oxygen pressure fields in normal human voluntiurs and in critically ill patients. In: A. M. Ehrly (Hrsg.): Messung des Gewebesauerstoffdrucks bei Patienten. Witzstrock-Verlag, Baden-Baden (1981) S. 55-60
30. Nernst, W.: Theorie der Reaktionsgeschwindigkeit in heterogenen Systemen. Z. Physikal. Chem. 47 (1904) 52-55
31. Saborowski, F., M. Kessler, J. Höper, F. Greitschus, K. Rath, H. A. Dickmans, K. G. Thiele: Der Sauerstoffdruck im Skelettmuskel von Patienten mit chronischer Niereninsuffizienz. In: A. M. Ehrly (Hrsg.): Messung des Gewebesauerstoffdrucks bei Patienten, Witzstrock-Verlag, Baden-Baden (1981) S. 76-79
32. Schönleben, K., B. A. Krumme, H. Bünte, M. Kessler: Kontrolle der Intensivbehandlung durch Messung von Mikrozirkulation und O_2-Versorgung. In: H. Junghanns (Hrsg.): Chirurg. Forum 76 für experimentelle und klinische Forschung (1976) S. 72-76
33. Schönleben, K., J. Hauss, U. Spiegel, H. Bünte, M. Kessler: Monitoring of tissue pO_2 in patients during intensive care. In: J. A. Silver, M. Erecińska, H. J. Bicher (eds.): Oxygen transport to tissue, Plenum Press New York/London (1978) p 593
34. Schönleben, K., J. Hauss, B. Krumme, H. U. Spiegel, H. Bünte: Die Messung des Gewebesauer-

stoffpartialdrucks zur Therapiekontrolle in der Intensivmedizin. INA Bd. 17, Aktuelle Probleme der Intensivbehandlung (II). Thieme-Verlag, Stuttgart (1980) S. 115-126
35. Spiegel, H. U., J. Hauss, K. Schönleben, H. Bösenberg: Platinum multiwire surface electrodes in clinical practice. Influence of low pressure ethylenoxide sterilisation on measuring properties of the multiwire surface electrode (MDO). Arzneim.-Forsch./Drug Res. 30 12 (1980) 2204-2221
36. Spiegel, H. U., J. Hauss, K. Schönleben: Aussagekraft kombinierter Meßmethoden $tcpO_2$, lokaler Gewebe-pO_2 und intravasaler Säure-Basenstatus. Wiss. Ber. österr. Ges. Bio.-med. Techn. (1981) 279-282
37. Spiegel, H. U., J. Hauss, K. Schönleben: Einfluß der Unterdruck-Gassterilisation auf die Meßeigenschaften der Mehrdrahtoberflächenelektrode (MDO). Med. Tech. 101 (1981) 103-106
38. Spiegel, H. U., J. Hauss, K. Schönleben: Klinische Anwendung der lokalen Gewebe-pO_2-Messung. Med. Technik 101 (1981) 47-49
39. Spiegel, H. U., J. Hauss, K. Schönleben: Monitoring des lokalen Gewebe-PO_2 während Intensivtherapie. In: H. Bergmann et al. (Hrsg.): Monitoring in der Anästhesiologie und Intensivmedizin. Mandrich-Verlag, Wien/München/Bern (1983) S. 259-291
40. Spiegel, H. U., J. Hauss, K. Schönleben: Lokale Gewebe-pO_2-Messung in der experimentellen Angiologie. In: D. A. Loose (Hrsg.): Aktueller Stand der konservativen Therapie peripherer Gefäßerkrankungen, Periodica Angiologica (Bd. 5) (1984) S. 148-152
41. Sunder-Plassmann, L., A. Overkott, H. M. Becker, M. Kessler, D. W. Lübbers: Veränderungen der Mikrozirkulation bei chronisch-arterieller Verschlußkrankheit. Sauerstoffdruckmessung an der Skelettmuskulatur bei gefäßchirurgischen Eingriffen. In: A. M. Ehrly (Hrsg.): Messung des Gewebesauerstoffdrucks bei Patienten. Witzstrock-Verlag, Baden-Baden (1981) S. 67-75

Transcutaneous PO_2: Principle, Use, Clinical Experience, and Limitations of the Technique

R. Huch and A. Huch

Introduction

In the transcutaneous measurement technique, the PO_2 is polarographically determined on the intact skin after hyperemia has been induced in the local skin-capillary region. The objective of the method is to determine the arterial PO_2 level and its changes. Experimental and clinical experiences have shown that the necessary stimulation of the local skin blood flow is best achieved by local hyperthermia. The PO_2 electrode was thus combined with a heater, which both heated the skin and served as a thermostat for the electrode. This provides a possiblity for measuring, along with the skin PO_2 value, the amount of heat required to maintain a constant electrode core temperature against the cooling blood flow as a relative measure of blood flow in the skin [3]. It has been proved, that, for example, the capillary blood in the ear lobe, the fingertip, or the ankle in both adults and newborns represents arterial blood gas conditions after hyperemia has been induced. However, it is not realistic to expect transcutaneous and arterial oxygen partial pressures to be identical. This is because the factors influencing the trancutaneous PO_2 measurement are far too numerous. An explanation of each factor follows.

Hyperemia

Under normal circulatory conditions, the degree of hyperemia is primarily affected, along with the capillary density, by the heating temperature. A distinction must be made between the predetermined temperature in the electrode and the resulting temperature in the uppermost layer of the skin capillaries.

We have found that with the epidermis approximately 100 µm thick and the temperature gradients in the electrolyte and membrane layer, a drop from the electrode temperature of about 2 °C must be expected [3]. The maximum heating temperature is limited by the sensation of pain and by the danger of skin lesion in an application lasting several hours.

Temperatures above 47.5 °C on the surface of the skin cause discomfort [6]. Systematic investigations on the duration of thermal strain were able to prove that

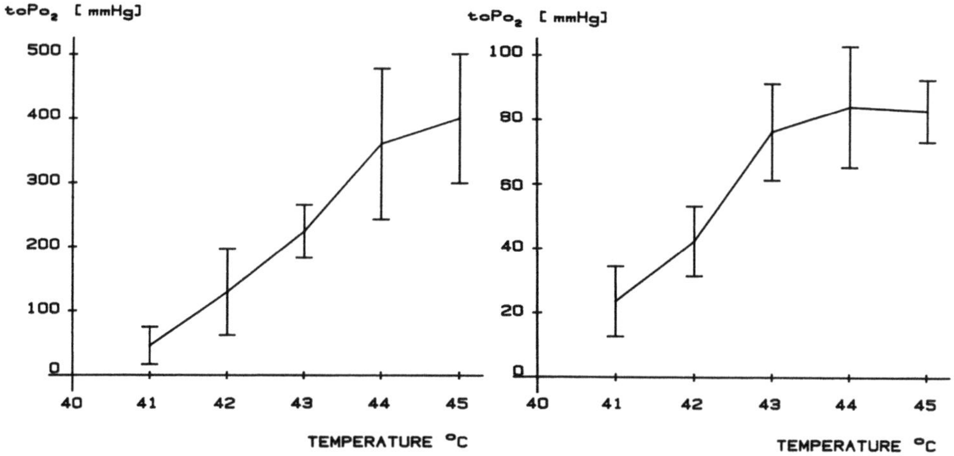

Fig. 1. Increases in transcutaneous PO_2 resulting from temperature increase in the electrode (electrode core temperature) in normoxia and hyperoxia (mean values from 10 newborns). (From [3])

after 6 h temperatures in the range of 44 °C on the skin surface caused irreversible lesions [6]. Clinical experience has shown that the temperature required to induce maximum hyperemia approximates these temperatures (e. g., electrode heating temperature of 45 °C corresponds to 43 °C on the skin).

In newborns, we raised the electrode core temperature by degrees and took the PO_2 increase as a measure of increasing circulation. While when the patients breathed under normal conditions maximum circulation appeared to have been already reached at 43 °C, our data with supplemental oxygen indicated that a temperature rise of up to 45 °C would further cause circulation to increase (Fig. 1). To ideally determine arterial values, the electrode core temperature must be high, if possible, 45 °C. Therefore, applying temperatures of 42° - 43 °C in neonatology in order to prevent skin damage is inappropriate for monitoring arterial values.

Hyperthermia

It is well known that the position of the hemoglobin binding curve is dependent on temperature. The local heating of the skin raises the capillary blood temperature and shifts the binding curve to the right. In other words, the PO_2 rises, while the O_2 content remains constant. In the PO_2 range of the binding curve, this effect is about 5% per degree Celsius. The thickness of the epidermis determines the degree of this artificial PO_2 increase in that it contributes to the determination of the temperature gradient. The thinner the epidermis, the greater the PO_2 increase caused by the binding curve. This explains why in newborns and premature babies higher values can be measured on the skin than in the blood.

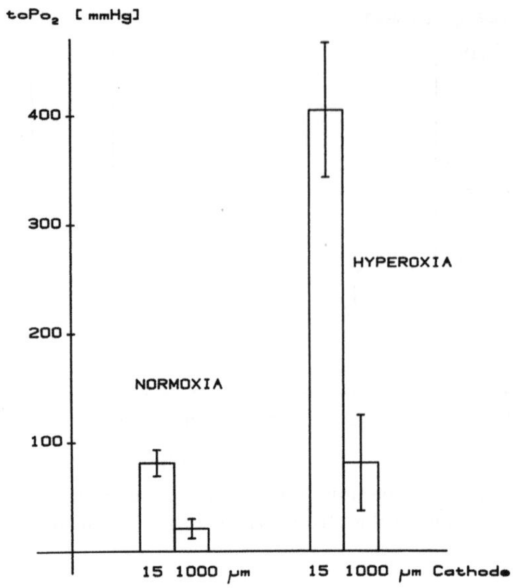

Fig. 2. Simultaneous transcutaneous PO_2 measurements in normoxia and hyperoxia with two electrodes of different cathode size in 10 infants (mean values). (From [3])

Electrode Characteristics

The O_2 consumption of the cathode is critical, particularly in the case of reduced perfusion. However, this problem can be solved by using a cathode with a small diameter. A membrane with a high diffusion resistance to oxygen can also be used. Figure 2 shows mean values on the skin during simultaneous measurement with two electrode constructions: one with 15-μm and the other with 1000-μm cathodes. The PO_2 measured on the skin is significantly lower when the larger cathode is used. This high O_2 consumption of the cathode would have to be compensated for by a corresponding membrane.

Skin Respiration

In practice, it is not realistic to try to eliminate skin respiration. On the contrary, it is even increased as a result of high local temperatures [8]. The effect of skin respiration, however, is relatively small in comparison with other factors. In our investigations, the transcutaneous PO_2 is reduced by approximately 5 mm Hg [2].

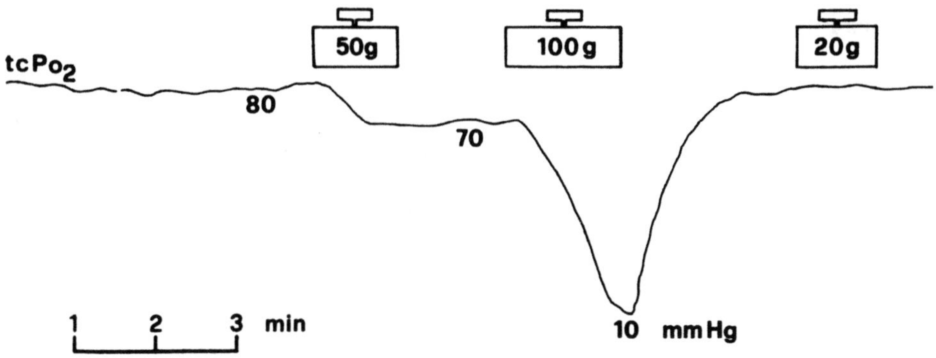

Fig. 3. Influence of added weight on transcutaneous PO_2. (From [3])

Weight of the Electrode

If skin perfusion is sufficient, the weight of the electrode is in no way critical. With an electrode weight of 4 g, which produces a pressure of about 0.01 p/mm^2, there is no capillary compression, as comparative measurements with lighter electrodes have shown. Figure 3 shows how the PO_2 recording reacts when additional weight is put on the electrode. A reduction in perfusion is not noticed up to and including a tenfold increase in pressure. It is therefore advisible to select an area of skin for electrode positioning on which the electrode can amplify skin contact with its own weight. The electrode should not be suspended from the skin, as this interferes with the continuity of the recording.

Epidermis

The last influencing factor, the skin – which may be viewed as a membrane itself – is particularly important in understanding the PO_2 measurement in adults. The polarographic measuring signal is produced by the reduction in the O_2 molecules on the cathode, which results in a so-called O_2 reduction current. Simply speaking, it can be said that with all other factors constant, the current becomes lower the less permeable the membrane is to oxygen. The avascular epidermis can be regarded as a "second Clark membrane," which cannot be taken into account during in vitro calibration.

Discussion

The transcutaneous technique for monitoring the arterial situation in adults is more problematic than in infants. Here, however, a distinction must be made between the agreement of the order of magnitude between arterial and transcutaneous values and the correlation of these two quantities. The latter can be excellent, even when

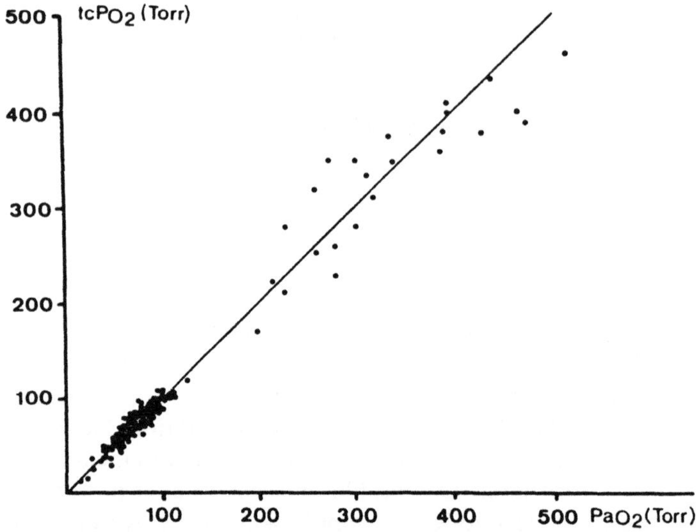

Fig. 4. Correlation between arterial and transcutaneous PO_2 in healthy and sick newborn and preterm babies. (From contribution by Schachinger in [4])

Table 1. Clinical conditions associated with poor correlation between PaO_2 and $tcPO_2$ in infants (From [3])

Condition	Characteristic
1. Shock	Defined as a blood pressure of 2–3 standard deviations below normal
2. Acidosis	Severe, pH <7.05
3. Hypothermia	Temperature of <35 °C
4. Cyanotic congenital heart disease Postcardiac surgery	Severe, $PaO_2 < 30$ mm Hg (4 kPa) Probably owing to combined effects of 1, 2, and 3
5. Anemia	Severe
6. Infants receiving high doses of tolazoline, usually for pulmonary hypertension or persistent fetal circulation syndrome	
7. Dying infants	
8. Skin edema	Severe, usually found in combination with 5, 6, and 7, above

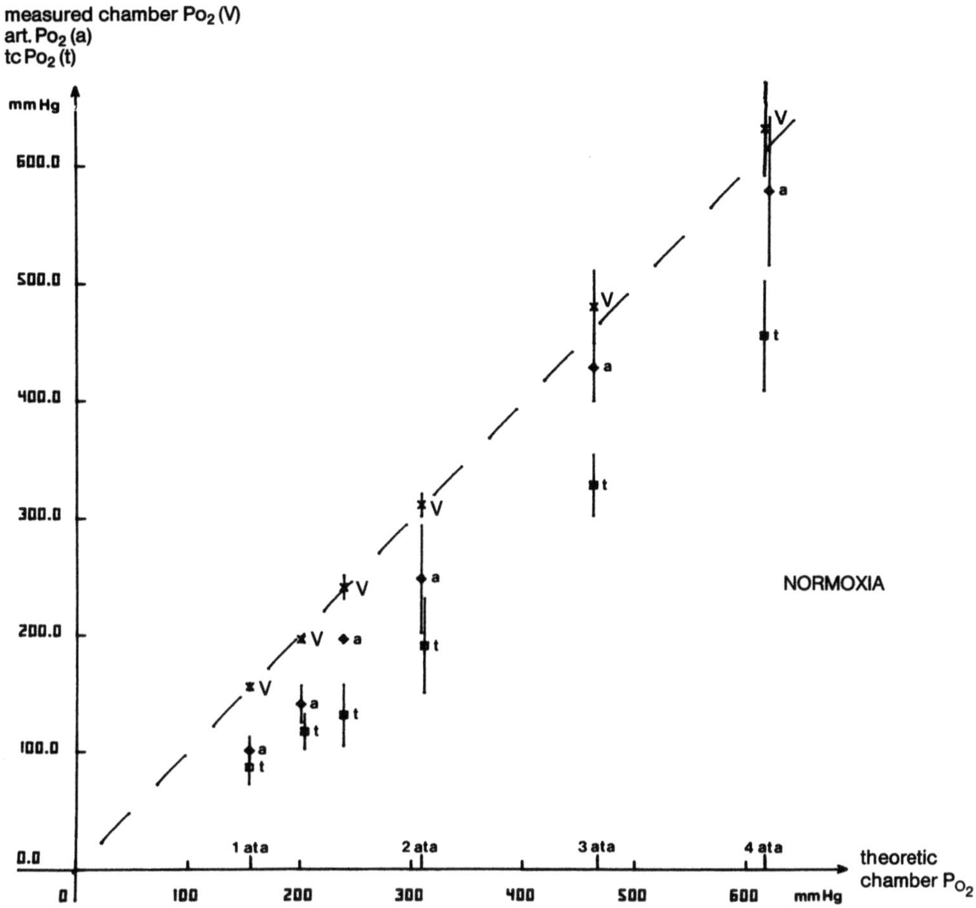

Fig. 5. Correlation between hyperbaric chamber pressure, arterial PO_2, and transcutaneous PO_2 in healthy volunteers. (From [1])

arterial and transcutaneous values are not at the same level. As we will show, this was confirmed by measurements in healthy adults. It is advisable to carry out in vivo calibration for exact measurements, because it remains valid for several hours [7].

Such an in vivo calibration, i.e., simultaneous gas measurement in the arterial blood, is not necessary for healthy preterm and newborn infants and seldom required in children with severe respiratory disease. Figure 4 shows 325 simultaneous arterial and transcutaneous PO_2 measurements performed by Schachinger [4] in healthy and sick premature and newborn infants. This correlation allows the deduction of one measured variable from the other.

Only a few clinical conditions have been identified in which a divergence between the arterial and transcutaneous PO_2 has been established. These conditions, which are shown in Table 1, are mostly pathologic and are easily diagnosed using other parameters.

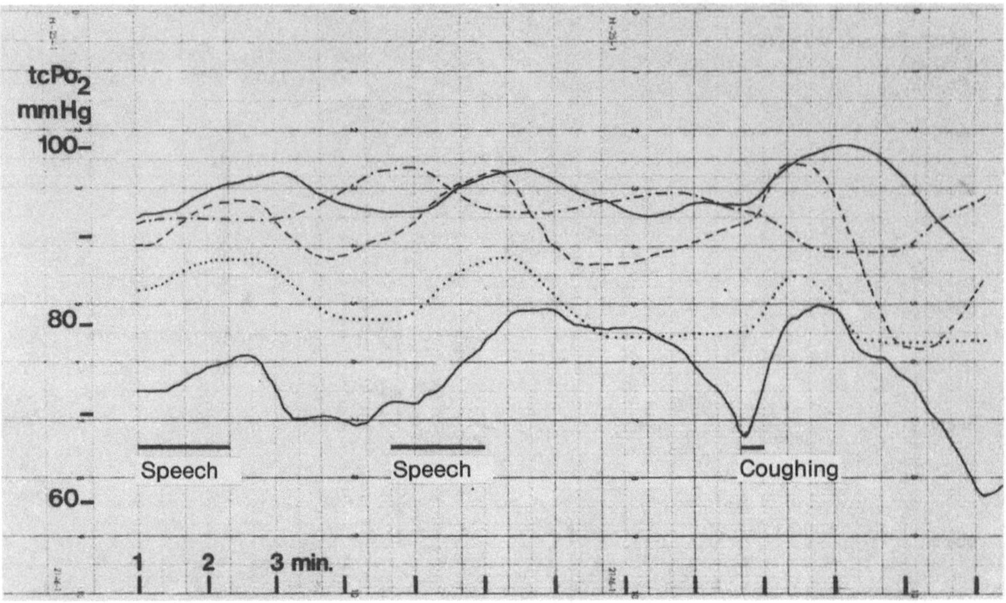

Fig. 6. Influence of speaking and coughing on transcutaneous PO_2 in five healthy volunteers

Versmold [12] studied the effect of blood pressure, hematocrit, and arterial pH on the correlation between arterial and transcutaneous PO_2. In systematic investigations, he found that the influence of such factors is not detectable until very pathologic circulatory and acid-base states exist. Experimental findings from studies in newborn animals have shown that such discrepancies have a high prognostic value [5].

Numerous investigations were carried out with simultaneous blood gas measurements in adults with normal circulation. These studies also proved a good correlation between the two values, even if there is not a 1:1 ratio.

The investigations in the hyperbaric chamber carried out together with D. W. Lübbers clearly show that the transcutaneous values follow the arterial and the chamber pressure values (Fig. 5). On the average, however, the transcutaneous values are 20% lower than the arterial values.

Measurements in adults with an undisturbed circulation have made it possible to gain an insight into the physiologic breathing behavior (Fig. 6) and breathing regulation, to study the effect of drugs on breathing, to investigate the influence of physical exercise, and to monitor the PO_2 fluctuations that result from painful labor and temporary hyperventilation.

Comprehensive investigations, however, in very sick adults in both intensive care medicine and anesthesiology - two fields in which the control of blood gases is vital - have not been able to prove the reliability of the transcutaneous technique for monitoring the arterial PO_2. Strasser et al. [11] and Stossek [10] performed particularly systematic investigations. The transcutaneous values reflect the arterial values, but in most cases of critical illnesses, the transcutaneous values correlate better with the blood pressure than with the arterial value. Furthermore, unlike in newborns, correlations in normoxia and hyperoxia differ from those in hypoxemic phases. It appears that compensatory mechanisms for the circulation begin here, which makes it necessary to differentiate whether the drops in PO_2 are pulmonary or circulatory in origin. All attempts at making this distinction using a parameter for the local blood flow have failed, at least under clinically practicable conditions.

We are convinced that we cannot tackle this limitation in the transcutaneous technique for monitoring the arterial value in adults with technical advances, for we cannot outwit physiology. We are also convinced that the transcutaneous blood gas measurement in adults provides information which reaches beyond the point of simply knowing the correct blood gas values. In this connection, Shoemaker's very impressive investigations [9] should be mentioned. The reports included in the present book discuss further aspects and possibilities of transcutaneous PO_2 measurement.

References

1. Huch, A., Huch, R., Hollmann, G., Hockerts, T., Keller, H.P., Seiler, D., Sadzek, J., Lübbers, D.W.: Transcutaneous PO_2 of volunteers during hyperbaric oxygenation. Biotelemetry 4 (1977) 88-100
2. Huch, R., Lübbers, D.W., Huch, A.: Quantitative continuous measurement of partial oxygen pressure on the skin of adults and newborn babies. Pflügers Arch. 337 (1972) 185-198
3. Huch, R., Huch, A., Lübbers, D.W.: Transcutaneous PO_2. Thieme, Stuttgart (1981)
4. Huch, R., Huch, A., Rooth, G.: Atlas der Sauerstoff-Cardiorespirographie beim Neugeborenen. Georg Fischer Verlag, Stuttgart (1984)
5. Johnson, P., Wilkinson, A.R., Sloper, J., Whyte, P.L.: Continuous transcutaneous and intraarterial oxygen measurement during experimental hypoxemia in infant monkeys. In: Huch, Huch, Lucey (Eds.): Continuous transcutaneous blood gas monitoring. Original Article Series: Birth Defects. The National Foundation March of Dimes, Liss, New York, vol. 15/4 (1979) 607-614
6. Moritz, A.R., Henriques, F.C.: Studies of Thermal Injury, II. The Relative Importance of Time and Surface Temperature in the Causation of Continuous Burns. Am. J. Pathology (1947) 695-720
7. Rooth, G., Hedstrand, U., Tyden, H., Oegren, C.: The validity of the transcutaneous oxygen tension method in adults. Crit. Care Med. 4 (1976) 162-165
8. Severinghaus, J.W., Stafford, M., Thunstrom, A.M.: Estimation of Skin Metabolism and Blood Flow with $tcPO_2$ and $tcPCO_2$ electrodes by cuff occlusion of the circulation. Acta Anaesthesiol. Scand. Suppl. 68 (1978) 9-15
9. Shoemaker, W.C.: Tremper, K.K.: Transcutaneous PO_2 and PCO_2 Monitoring: Experimental and Clinical Studies. In: Huch, Huch (Eds.): Continuous Transcutaneous Blood Gas Monitoring. Marcel Dekker, New York (1983) 745-760
10. Stosseck, K.: Transcutane Sauerstoffdruckmessung. In: Frey, Kern, Mayrhofer (Eds.): Anaesthesiologie und Wiederbelebung. Springer Verlag, Berlin (1977) 1-68
11. Strasser, K., Goeckenjan, G.: The monitoring of adult intensive care patients by transcutaneous PO_2 measurements. In: Huch, Huch, Lucey (Eds.): Continuous transcutaneous blood gas monitoring. Original Article Series: Birth Defects. The National Foundation March of Dimes, Liss, New York, vol. 15/4 (1979) pp. 525-530
12. Versmold, H.T., Linderkamp, O., Holzmann, M., Strohhacker, I., Riegel, K.: Transcutaneous monitoring of PO_2 in newborn infants: Where are the limits? Influence of blood pressure, blood volume, blood flow, viscosity, and acid base state. In: Huch, Huch, Lucey (Eds.): Continuous transcutaneous blood gas monitoring. Original Article Series: Birth Defects. The National Foundation March of Dimes, Liss, New York, vol. 15/4 (1979) 285-294

Tissue Oxygen Pressure Measurements

Oxygen Pressure in the Muscle Tissue of Patients with Intermittent Claudication After Placebo and Saline Infusions

H. Landgraf and A. M. Ehrly

Introduction

The oxygen pressure of the muscle tissue is an objective and quantitative measure for the oxygen supply of the tissue in patients with intermittent claudication [1, 2, 3, 4, 8, 18]. The oxygen pressure of the muscle tissue, which in these patients is considerably reduced, can be significantly improved by therapeutic measures, such as reconstructive vascular surgery, transluminal angioplasty, or the administration of hemorrheologically active drugs [4, 5, 7, 10]. Using certain hemorrheologically active drugs, the total blood flow under resting conditions does not necessarily need to be increased [5, 7]. It can thus be assumed that the therapeutic improvement of the maldistribution of capillary perfusion [6, 9] in arterial occlusive disease leads to a better supply of the tissue with oxygen [11, 16, 17]. Measurements of tissue oxygen pressure in claudicants were only performed when a steady state of the cardiovascular system could be assumed, a state which particularly implies that in the resting muscle no change occurs in either oxygen consumption or the O_2 tissue diffusion coefficient.

In a crossover investigation to evaluate the oxygen delivery to the tissue in patients with intermittent claudication following hypervolemic hemodilution with various plasma substitutes, the influence of an infusion of 500 ml physiologic (0.9%) saline solution was tested. It turned out, however, that the PO_2 values changed after the short-term isotonic saline infusion, which was different from the expected placebo effect.

We therefore set up a 30-min "pseudoinfusion" to find out its influence on the oxygen pressure in the tissue of the lower leg in a group of patients with peripheral arterial occlusion. The entire procedure of an infusion into the cubital vein was simulated without any fluid actually being infused.

Patients and Methods

The patients investigated suffered from arterial occlusive disease stage II b according to Fontaine, with occlusions or stenoses of the pelvic or leg arteries. The investigations were carried out with the patients resting in a horizontal position.

The tissue PO_2 in the anterior tibial muscle of the affected extremity was measured before, immediately after, and 15, 30, 60, 90, and 120 min after termination of the infusion. Measurements were carried out according to the method described by Ehrly and Schroeder [1, 18] with coated microplatinum needle electrodes developed by Kunze and Lübbers [13, 14, 15].

Parallel to the tissue PO_2 measurements, several hemorrheologic parameters (hematocrit, apparent whole blood viscosity, plasma viscosity, erythrocyte aggregation), as well as blood pressure and pulse rate were determined. A total of eight patients received infusions of the isotonic saline solution; the placebo group consisted of five patients.

Results

The pseudoinfusion into the cubital vein of the five patients, i.e., without any actual infusion of fluid, revealed no marked changes in the oxygen tension in the five patients (Fig. 1). These findings correlate with those of earlier investigations, in which measurements in patients with arterial occlusive disease over a period of 2 h showed no significant changes [4]. Apart from a slight diminution of the hematocrit values, blood pressure, pulse rate, and the rheologic parameters remained unchanged.

The infusion of 500 ml isotonic saline solution within a period of 30 min, however, brought about a gradual but marked improvement in the tissue oxygen pres-

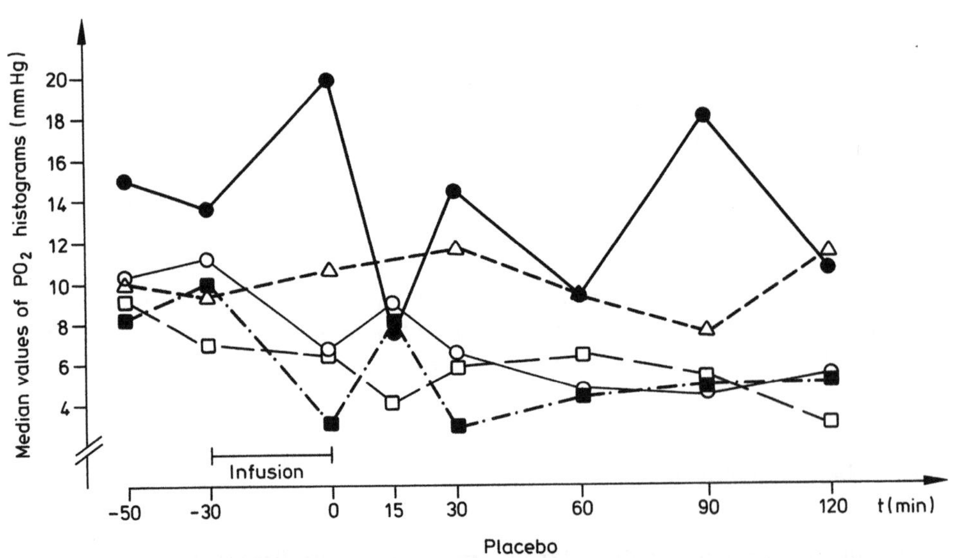

Fig. 1. Tissue PO_2 in the anterior tibial muscle represented by the median values of the PO_2 histograms during and after a pseudoinfusion in five patients with peripheral arterial occlusions. The times given in Figs. 1-4 refer to the end of the infusion procedure

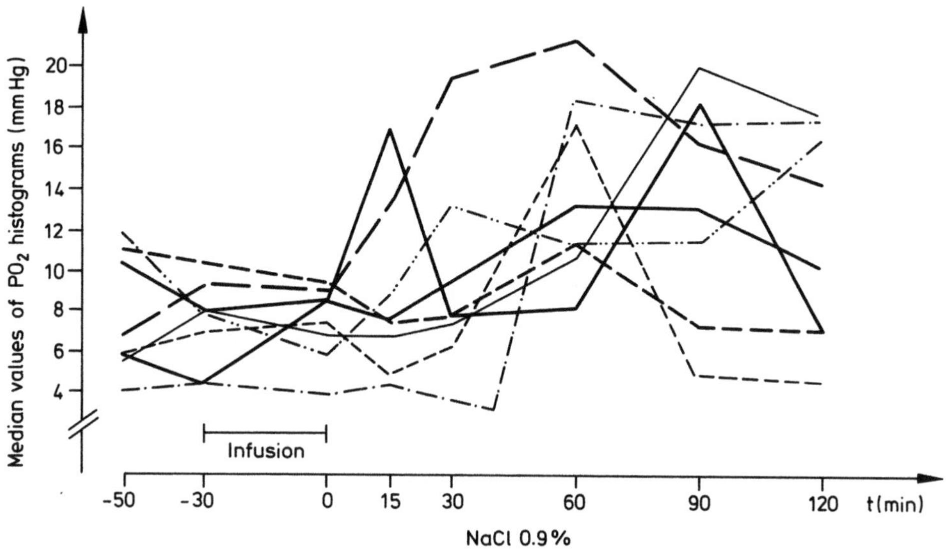

Fig. 2. Tissue PO_2 in the anterior tibial muscle represented by the median values of the individual PO_2 histograms during and after an infusion of 500 ml 0.9% saline solution within 30 min in eight patients with peripheral arterial occlusions

sure, which was manifested not only by the behavior of the individual median values, but also by the pooled histograms and their median values (Figs. 2, 3). Statistical calculations using the Wilcoxon test for pair differences showed that the median values 60 and 90 min after the infusion were significantly higher than the initial value.

The rheologic parameters determined simultaneously (Fig. 4) revealed marked changes immediately following of the infusion. The hematocrit, the apparent whole blood and plasma viscosity, and the erythrocyte aggregation were significantly reduced. The reduction in the apparent whole blood viscosity could be established up to 2 h after infusion was stopped. The hemodynamic parameters blood pressure and pulse rate, on the other hand, remained unchanged. At the time of the maximum increase in tissue PO_2, the changes in the hemorrheologic parameters had already regressed.

Discussion

Two factors are to be considered in the explanation of these results. First, it seems feasible that the transport of oxygen to the relatively hypoxic tissue is improved by the reduction in blood viscosity, but that it takes a certain amount of time before the

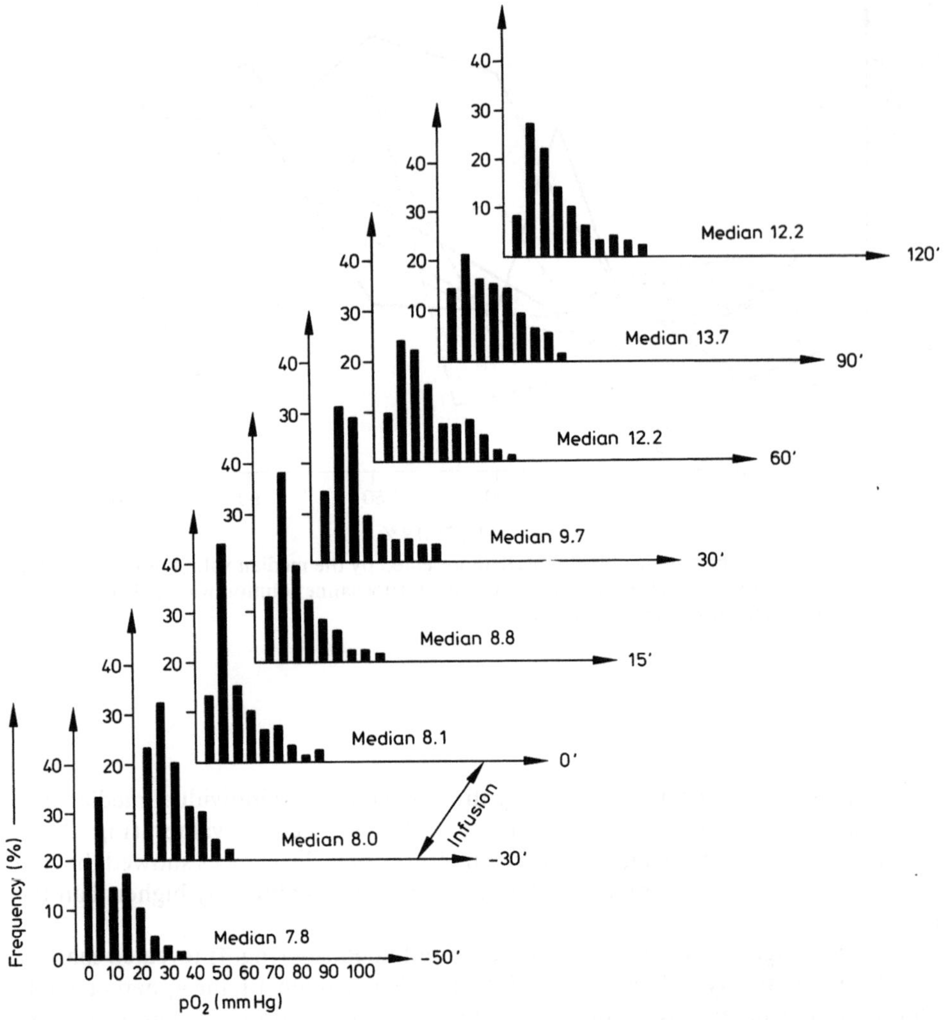

Fig. 3. Tissue PO_2 in the anterior tibial muscle represented by the pooled PO_2 histograms and their median values during and after an infusion of 500 ml 0.9% saline solution within 30 min in patients with peripheral arterial occlusions

partial oxygen pressure in the tissue shows a change from the initial values. In other words, local equilibrium occurs later than the systemic changes. The surprisingly marked increase in tissue PO_2, however, does not quite correlate to the relatively slight hemorrheologic changes.

A second possible explanation for our findings is that the diffusion conditions for O_2 in the tissue are improved by the short-term hypervolemic infusion of isotonic saline solution. This theory is corroborated not only by the fact that isotonic saline solution is distributed relatively quickly in the entire extracellular space (according to Halmagyi [12], only approximately one-fifth of the infused volume remains in the vessels) and thus changes the hydration of the tissue, but also by the

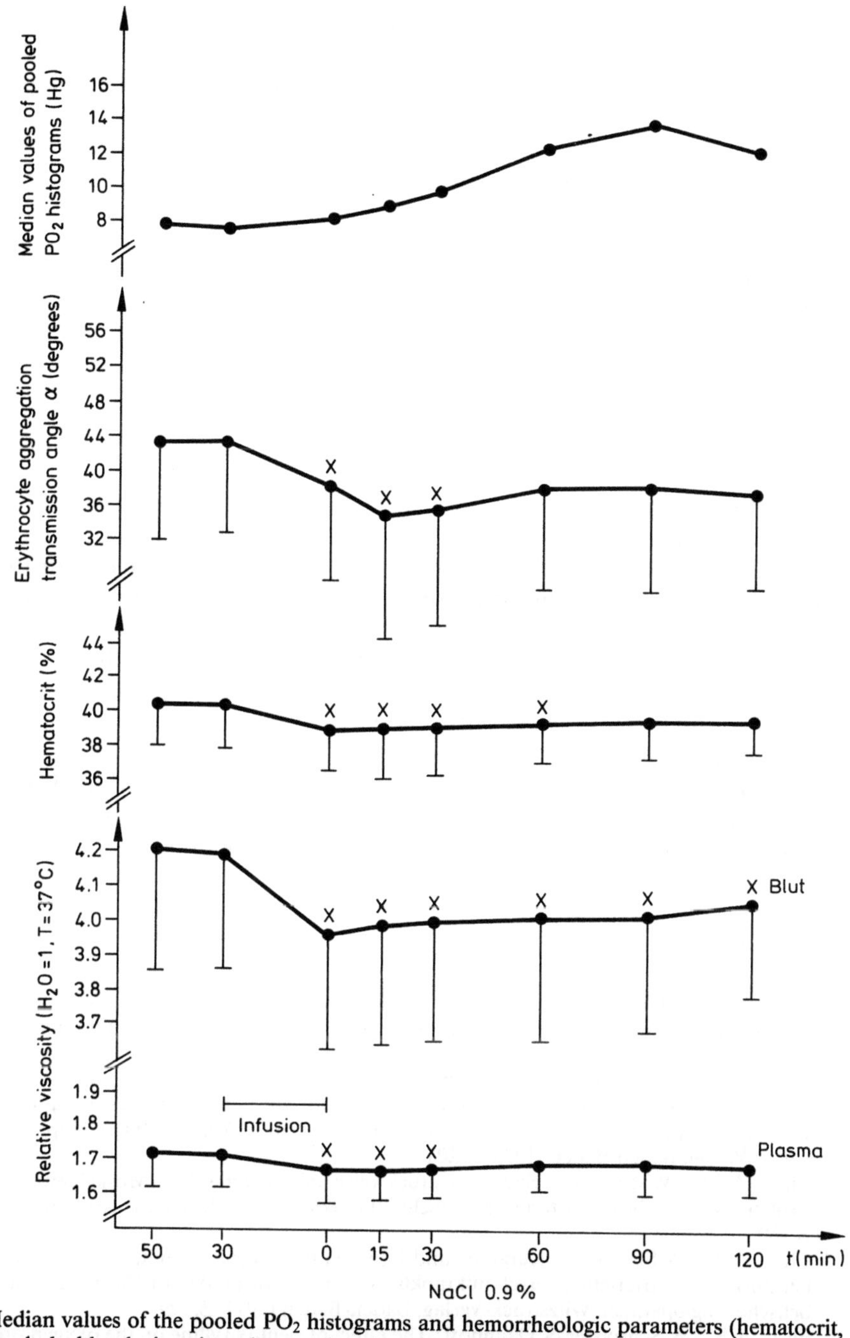

Fig. 4. Median values of the pooled PO_2 histograms and hemorrheologic parameters (hematocrit, apparent whole blood viscosity, plasma viscosity, erythrocyte aggregation) during and after a 30-min infusion of 500 ml 0.9% saline solution in eight patients with peripheral arterial occlusive disease (*x*, statistically significant difference from initial value at 30 min; Wilcoxon test for combined random tests, significant regarded as $2\alpha \leq 0.05$)

age dependent reduction of the total water content in the body [19]. It is possible that the older patients in the investigation - the average age was 65 - had latent tissue dehydration, which was compensated for by the infusion of the saline solution so that the O_2 delivered could diffuse more easily into the tissue [20].

To sum up the results of the investigation, it can be said that the short-term infusion of a certain amount of noncolloidal fluid in patients with chronic arterial occlusions can improve the oxygen pressure of the tissue. This is important to know, not only in connection with comparative studies on the effect of plasma substitutes on oxygen supply, but also for the investigation of so-called vasoactive drugs using electrolyte or glucose solutions as vehicle solutes.

Summary

The short-term hypervolemic infusion of 500 ml physiologic saline solution resulted in a significant increase in the tissue oxygen pressure in the anterior tibial muscle in patients with intermittent claudication. These findings contrasted with those obtained after pseudo infusion using the same procedure and the same measuring times.

One reason for these findings could be the brief but statistically significant improvement of the flow properties of the blood after the saline infusion. Another possibility is an increase in the diffusion coefficient of oxygen as a result of better hydration of the tissue.

The fact that the tissue PO_2 in ischemic disease can be changed solely by the hypervolemic infusion of a saline solution must be taken into consideration when, in pharmacotherapeutic studies, large amounts of fluid are infused.

References

1. Ehrly, A. M., H. J. Köhler, W. Schroeder, R. Müller: Sauerstoffdruckwerte im ischämischen Muskelgewebe von Patienten mit chronischen peripheren arteriellen Verschlußkrankheiten. Klin. Wschr. 53 (1975) S. 687
2. Ehrly, A. M., H. J. Köhler, W. Schroeder, R. Müller: Messung des lokalen Sauerstoffdrucks im ischämischen Muskelgewebe: Eine objektive Methode zur Beurteilung des Ischämiegrades bei Patienten mit peripherer arterieller Verschlußkrankheit. In: E. Zeitler (Hrsg.): Hypertonie, Witzstrock-Verlag, Baden-Baden (1975) S. 201
3. Ehrly, A. M., W. Schroeder: Sauerstoffdruckwerte im ischämischen Muskelgewebe von Patienten mit chronischen arteriellen Verschlußkrankheiten. Verh. Dtsch. Kreislauff. 42 (1976) S. 380
4. Ehrly, A. M., W. Schroeder: Kurzzeit- und Langzeitmessungen des mittleren Sauerstoffdrucks bei chronischen arteriellen Verschlußkrankheiten. In: K. Alexander, M. Cachovan (Hrsg.): Diabetische Angiopathien. Witzstrock-Verlag, Baden-Baden (1977) S. 229
5. Ehrly, A. M., W. Schroeder, S. Dannhoff: The effect of pentoxifylline on the oxygen pressure of ischemic muscle tissue of patients with chronic arterial occlusions. IRCS Med. Sciences 5 (1977) 411
6. Ehrly, A. M., W. Schroeder: Zur Pathophysiologie der chronischen arteriellen Verschlußerkrankung. I. Mikrozirkulatorische Blutverteilungsstörung in der Skelettmuskulatur. Herz/Kreislauf 11 (1979) 275

7. Ehrly, A.M., H.J.Köhler, W.Schroeder: Zeitliche Beziehungen zwischen der Erhöhung des muskulären Gewebesauerstoffdrucks und der Zunahme der schmerzfreien Gehstrecke bei der Therapie der schweren Claudicatio intermittens mit Ancrod. VASA 8 (1979) 28
8. Ehrly, A.M., H.Landgraf, K.Saeger-Lorenz: Quantification of the effects of rheologically active drugs in patients with vascular disease. In: J.F.Stoltz, P.Drouin (eds.): Hemorheology and diseases. Doin, Paris (1980) p 665
9. Ehrly, A.M.: New pathophysiological concept of ischemic diseases: Microcirculatory Blood Maldistribution (MBM). Bibl. Anat. 20 (1980) 456
10. Ehrly, A.M., H.Landgraf, K.Saeger-Lorenz: Einfluß therapeutischer Maßnahmen bei chronischen arteriellen Verschlußkrankheiten auf den Gewebesauerstoffdruck (Mikro-Pt-Stichelektroden) in der Unterschenkelmuskulatur. In: A.M.Ehrly (Hrsg.): Messung des Gewebesauerstoffdrucks bei Patienten. Witzstrock-Verlag, Baden-Baden (1981) S.84
11. Ehrly, A.M., H.Landgraf: Therapeutische Beeinflussung der gestörten Sauerstoffversorgung des Gewebes bei Mikrozirkulationsstörungen. Therapiewoche 33 (1983) 824
12. Halmagyi, M.: Zur Verweildauer kolloidosmotischer Infusionen. In: H.Reisigl, W.Fekl (Hrsg.): Infusionstherapie bei Volumenmangel und bei rheologischen Indikationen. Karger-Verlag, Basel/München/Paris/London/New York/Sydney (1979) S.51
13. Kunze, K.: Die lokale kontinuierliche Sauerstoffdruckmessung in der menschlichen Muskulatur. Pflüg. Arch. 292 (1966) 151
14. Kunze, K.: Das Sauerstoffdruckfeld im normalen und pathologisch veränderten Muskel. Schriftenreihe Neurologie. Springer-Verlag, Berlin/Heidelberg/New York (1969)
15. Lübbers, D.W., H.Baumgärtl: Herstellungstechnik von palladinierten Pt-Stichelelektroden (1-5 μm Außendurchmesser) zur polarographischen Messung des Wasserstoffdruckes für die Bestimmung der Mikrozirkulation. Pflüg. Arch. 294 (1967) 39
16. Schroeder, W.: Nutritive und nicht-nutritive Skelettmuskeldurchblutung. Arch. Kreislauff. 49 (1966) 36
17. Schroeder, W.: Wirkungsgrad der Skelettmuskeldurchblutung und nicht-nutritive Muskeldurchblutung. In: Die arterio-venösen Anastomosen. Aktuelle Probleme in der Angiologie (Bd.2). Huber-Verlag, Bern/Stuttgart (1968) S.82
18. Schroeder, W.: Die Messung des Sauerstoffdruckes in der Skelettmuskulatur - eine quantitative Methode zur Kontrolle der Sauerstoffversorgung und der Funktion der terminalen Muskelstrombahn. Herz/Kreisl 10 (1978) 146
19. Schwab, M., W.T.Dissmann, W.Schubert: Der Einfluß des Alters auf die Flüssigkeitsräume des Körpers. Klin. Wschr. 41 (1963) 1173
20. Vaupel, P.: Effect of percentual water content in tissues and liquids on the diffusion coefficients of O_2, CO_2, N_2 and H_2, Pflüg. Arch. 361 (1976) 201

Normoxia and Hypoxia in Patients with Severe Burns

G. I. J. M. Beerthuizen, R. J. A. Goris, H. P. Kimmich, A. J. van der Kley, and F. Kreuzer

Introduction

During the first 48 h after a severe burn, the clinical picture is characterized by the development of shock. The patient is treated with saline and colloids. With respect to the infusion volume to be applied, various treatment plans are used. There is still no satisfactory parameter today for determining the tissue perfusion in the individual patient.

Diuresis has been determined as a parameter of tissue perfusion. Reduced diuresis, however, is a result of a decrease in tissue circulation and therefore appears after a change in the tissue perfusion. We examined muscle PO_2 during the first 48 h after a burn as a direct parameter of tissue perfusion. Animal experiments on a hypovolemic shock model revealed an early decline in muscle PO_2 [1]. The objective of our investigation was to find out whether muscle PO_2 in patients with severe burns is already diminished before blood pressure decreases.

Method and Patients

From September 1982 to January 1984, we carried out measurements in patients with second- or third-degree burns of more than 25% of the body surface. Following an established protocol, the following parameters were registered within the first 48 h after the burn (6, 8, 10, 14, 18, 24, 30, 36, and 48 h): pulse rate, systolic and diastolic arterial blood pressure, central venous pressure, arterial PO_2, muscle PO_2, arterial pH, amount of infusion, temperature, and diuresis per hour.

The muscle PO_2 measurement in the quadriceps femoris was performed with a polarographic needle electrode [2] and with a measuring instrument controlled by a microcomputer. After iodine was applied to the skin and local anesthesia was administered, an Abbocath needle (20 gauge) was inserted 3 cm into the muscle. After the needle electrode was calibrated, it was placed in the m. quadriceps femoris using the Abbocath as a guiding cannula, and the measurement was started. With a micromanipulator, the electrode was withdrawn in steps of 200 μm, and the PO_2 was measured at each point. A total of 100 PO_2 values were obtained. The median value of these 100 measurements is characteristic of the muscle PO_2. The measurements took a total of 5 min.

The investigations were carried out on 13 patients between the ages of 18 and 71 years (mean age 42). The total area of burned skin varied between 25% and 90% (mean 49%) of the body surface. One patient had a pre-existing muscle atrophy, which made adequate PO_2 measurement impossible and forced us to disregard the results obtained from this patient in the further course of the investigation. One patient died within 48 h after the accident as a result of irreversible burn shock. Eight patients had not suffered any traumatic damage apart from the burn and could be treated according to the established protocol. Shock was assumed when the systolic blood pressure was below 80 mm Hg.

Results

In the eight patients mentioned above, the hemodynamic parameters were significantly changed within the first 48 h after the burn. Pulse rate and diastolic blood pressure were significantly higher ($P < 0.05$). The arterial PO_2 was normal during the first 48 h. Urine production was at its lowest in the period between 18 and 24 h after the burn ($P < 0.05$). Muscle PO_2 was lower during the first 24 h than 24–48 h after the burn.

There was no relationship between muscle PO_2 and urine production. The muscle PO_2 in patients with subsequent shock was significantly lower than in those with no subsequent shock. When the muscle PO_2 was higher than 3 kPa, no drop in blood pressure followed. Figure 1 shows all the PO_2 values for all the patients. The

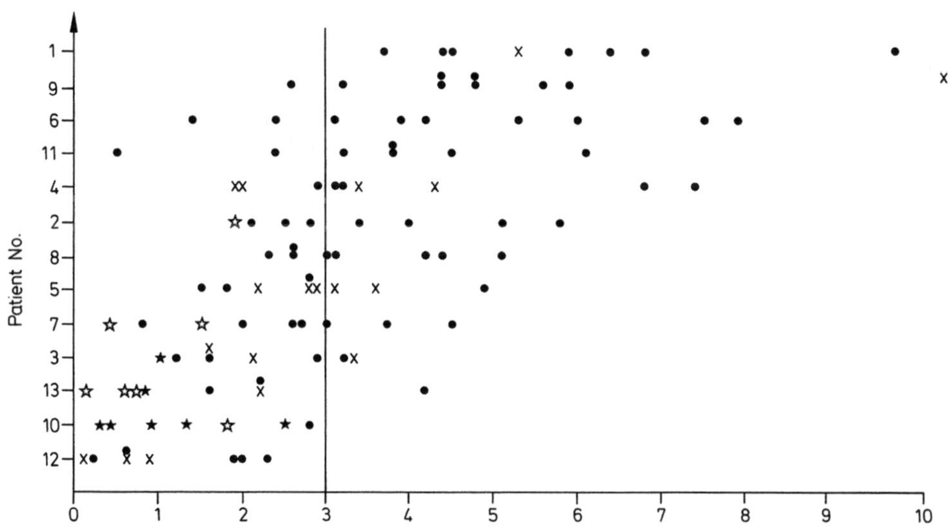

Fig. 1. Muscle PO_2 in 13 patients with severe burns, ·, no problems; x, diuresis < 0.5 ml/kg/h; ✲, systolic BP < 80 mm Hg; ★, systolic BP < 80 mm Hg and diuresis < 0.5 ml/kg/h. Patient 12 had pre-existing muscular atrophy

patients were arranged according to the level of the muscle PO_2 value throughout the investigation. For each patient, the median value was determined from all of the measurements. The patients were then arranged according to the increasing median value. The relation of muscle PO_2 to blood pressure and diuresis during the subsequent period were also included for each measuring point.

Conclusion

A decrease in diuresis in patients with severe burns is not dependent on muscle PO_2 and occurs in the presence of both lower and higher PO_2 values. The connection between muscle PO_2 and diuresis in these patients is influenced by the amount of infusion adapted to the hourly production of urine. Apart from possible effects on the function of the kidney itself, this could be the reason for the lack of a relationship between muscle PO_2 and urine production. The muscle PO_2 prior to a period with subsequent shock was significantly lower than the value obtained before a period without subsequent shock. When muscle PO_2 was higher than 3 kPa, there was no drop in blood pressure during the subsequent period.

Muscle PO_2 may be normal under effective treatment of a severe burn. The development of burn shock is accompanied by hypoxia in the muscles, which generally manifests itself before the shock develops. Our investigation shows that measuring muscle PO_2 in severely burned patients makes it possible to detect such developing shock in an early stage and provides a parameter for the effectiveness of the therapy.

References

1. A.J.v.d.Kley, AJ, de Koning J, Beerthuizen GIJM, Goris RJA, Kreuzer FJA, Kimmich HP (1983) Early detection of hemorrhagic hypovolemia by muscle PO_2 assessment. Surgery 93: 518-524
2. de Koning J, v.d.Kley AJ (1980) An electrode for clinical PO_2 monitoring. Arzneimittelforsch Drug Res 30: 2215

Muscle Tissue PO_2 Before and After Pedal Ergometric Stress in Patients with Intermittent Claudication

A. M. Ehrly, H. Landgraf, and K. Saeger-Lorenz

Introduction

The oxygen pressure of the resting anterior tibial muscle tissue was directly measured in patients with chronic arterial occlusion at the stage of intermittent claudication. The results showed that the tissue PO_2 was only about half as high as the PO_2 values in healthy patients of the same age and that the histogram is shifted to the left [1]. This quantitative value allows an evaluation of the oxygen supply to the tissue to be made, provided both the oxygen consumption and the hemodynamic parameters are in a steady state [2].

Increases in the tissue oxygen pressure in a resting muscle, after therapeutic measures, for example, can be interpreted as a sign of improved oxygen supply to the tissue, regardless of which mechanism is responsible for the improved transport of oxygen and nutrients [3]. The measurements of tissue oxygen pressure under resting conditions make it possible to draw differential diagnostic conclusions and are taken into account when determining the level of amputation [4]. There is a correlation of $r = 0.72$ between the value of the tissue oxygen pressure of the resting muscle and the pain-free walking distance on the treadmill ergometer [5; see [6] for a summarized presentation of the tissue oxygen pressure measurements in patients with peripheral circulatory disturbances.]

Our own studies on tissue oxygen pressure in ischemic disease, as well as those of other authors, were carried out under resting conditions, i.e., without muscle stress in the extremity to be examined. The ischemic symptoms of intermittent claudication do not appear, however, until the muscle is under stress. It was therefore desirable to carry out measurements of tissue oxygen pressure under or after defined exercise conditions. We hoped to obtain information not only on the time dependent behavior of the tissue oxygen pressure after exercise, in comparison with the initial values, but also on the presence of maldistribution of blood [7, 8] in the tissue examined.

Since for technical reasons it is not possible to achieve ergometric stress of the lower leg muscles when the electrode is in place, the measurements of the tissue oxygen pressure in the anterior tibial muscle should be carried out in such a way that the tissue PO_2 is determined after the resting values are recorded, immediately after the end of the exercise period, and in the subsequent periods. It is also advisable to perform the measurements at very short intervals after the exercise period and to carry out the measurements - as with the microplatinum needle electrode technique - quasi atraumatically.

Patients and Methods

The measurements of tissue oxygen pressure in the anterior tibial muscle of the lower leg were carried out with coated microplatinum needle electrodes [9]. These electrodes, which have a tip diameter of 1 to 3 μm, are advanced continuously through the tissue by means of a hydraulic micromanipulator at a speed of 33 mm/s, during which recordings are done continuously. In this way, the distribution of oxygen pressure in the tissue can be established and graphically represented in a histogram [10].

Measurements were taken in 34 patients with intermittent claudication and a mean age of 66.4 years at walking distances of between 50 and 200 m (measured on the treadmill ergometer under standardized conditions) and in seven healthy, physically untrained volunteers of an average age of 22 years. First, two baseline values were measured at 10-min intervals, after which the lower leg muscles were exercised using a treadmill ergometer, which makes use of a mechanical brake. By pulling the brake, the amount of exercise can be roughly adjusted. Absolute data on the level of work and performance, however, could not be obtained. The level of exercise was empirically set in such a way that it was an effort for the patients to hold out for the required 4 min. Some of the patients had to stop halfway because of pain in the lower leg muscles. For the healthy volunteers, the brake was pulled more tightly, so that they were exhausted after the 4 min. The tissue oxygen pressure measurements were carried out immediately (approximately 2.5 min) and 10, 20, and 30 min after the ergometric work was over. In one series of patients, plethysmographic measurements were taken along with the tissue PO_2 measurements.

Results

Among the healthy volunteers, who were not athletically trained, the exercise resulted in a more or less marked increase in the tissue oxygen pressure. Subsequently, the values dropped, but they remained above the baseline even after 30 min.

Studies in 34 patients with chronic arterial occlusion at the stage of intermittent claudication of different degrees of severity and with different initial values of tissue oxygen pressure under resting conditions showed a completely different development of the PO_2 values. Three types of tissue PO_2 reactions after maximum stress could be determined:

1. In one group, there is a slower and less marked increase than among the healthy volunteers.

2. In a second group, the PO_2 values exceeded the baseline value only slightly after the end of the exercise period and remained at that level until the end of the investigation.
3. In a third group of patients, the PO_2 dropped to below the baseline immediately after the exercise period and only slowly returned to normal afterwards.

The plethysmographic tests carried out in a few patients showed delayed reactive hyperemia, as has also been described by Sheperd [11]. Only in very rare exceptions did the flow volume in the lower extremity decrease after exercise. Since these are only preliminary results, and as the amount of exercise could not be exactly determined, we are not presenting the data in absolute Torr values.

Discussion

Although the time and duration of ergometric exercise was not the same among all of the test persons and patients, a result of methodical difficulties, it can be said that in comparison with the young, untrained, healthy volunteers, the tissue oxygen pressure in patients with occlusions and stenoses in the wake of reactive hyperemia after exercise increases only very slowly. In some cases, it drops to below the initial resting value and slowly begins to rise again during the course of the investigation. There are certain correlations with the behavior of the total blood flow in patients with intermittent claudication. It appears, however, that the tissue PO_2 in a number of patients decreases immediately after exercise, even though there is an increase in resting circulation. This can be explained not only by the significantly raised oxygen consumption, but also by a metabolically induced maldistribution in the muscle tissue.

Current experiments on exercise with quantified performance will provide exact answers to these questions. It will also then be possible to arrive at a correlation between the reaction types of the tissue PO_2 after exercise, on the one hand, and the pain-free walking distance and the degree of severity of the intermittent claudication, on the other.

In the meantime, we have a device which makes it possible to carry out foot ergometric exercise in such a way that work and performance can be quantified. This new method, which corresponds to the conditions of the pathophysiology of intermittent claudication, is very valuable in the evaluation of pathophysiologic, diagnostic, and therapeutic problems.

References

1. Ehrly, A.M., H.-J. Köhler, W. Schroeder, R. Müller: Sauerstoffdruckwerte im ischämischen Muskelgewebe von Patienten mit chronischen peripheren arteriellen Verschlußkrankheiten. Klin. Wschr. 53 (1975) 687
2. Lübbers, D.W.: Grundlagen und Bedeutung der lokalen Sauerstoffdruckmessung und des PO_2-Histogramms für die Beurteilung der Sauerstoffversorgung der Organe und des Organis-

mus. In: A. M. Ehrly (Hrsg.): Messung des Gewebesauerstoffdrucks bei Patienten. Witzstrock-Verlag, Baden-Baden (1981) S. 11

3. Ehrly, A. M., W. Schroeder: Sauerstoffdruckmessungen im ischämischen Muskelgewebe als Kontrolle therapeutischer Effekte bei chronischen peripheren Durchblutungsstörungen. In: R. Hild, G. Spaan (Hrsg.): Therapiekontrolle in der Angiologie. Witzstrock-Verlag, Baden-Baden (1978) S. 83
4. Ehrly, A. M., W. Schroeder: Oxygen Pressure in the Ischemic Muscle Tissue: A new Diagnostic Method for Evaluating the Patients with periphereal arterial occlusion. Vasc. Surg. 12 (1978) 215
5. Ehrly, A. M., W. Schroeder, H.-J. Koehler: Zeitliche Beziehungen zwischen den Zunahmen des muskulären Gewebesauerstoffdrucks und der schmerzfreien Gehstrecke bei der Therapie der schweren Claudicatio intermittens mit Ancrod. VASA 8 (1979) 28
6. Ehrly, A. M. (Hrsg.): Messung des Gewebesauerstoffdrucks bei Patienten. Witzstrock-Verlag, Baden-Baden (1981)
7. Ehrly, A. M., W. Schroeder: Zur Pathophysiologie der chronischen arteriellen Verschlußerkrankung. I. Mikrozirkulatorische Blutverteilungsstörung in der Skelettmuskulatur. Herz/Kreislauf 11 (1979) 275
8. Ehrly, A. M.: New pathophysiological concept of ischemic diseases: Microcirculatory Blood Maldistribution (MBM). Bibl. Anat. 20 (1980)
9. Kunze, K., D. W. Lübbers: Die Messung des lokalen Sauerstoffverbrauches von Organen in situ mit der Platinelektrode. Pflüg. Arch. 74 (1961) 247
10. Kunze, K.: Das Sauerstoffdruckfeld im normalen und pathologisch veränderten Muskel. Neurology Monography Series Nr. 3, Springer-Verlag, Berlin/Heidelberg/New York (1969)
11. Sheperd, J. T.: The blood flow through the calf after exercise in subjects with arteriosclerosis and claudication. Clin. Science 9 (1950) 49-58

Tissue Oxygen Pressure in Ulcers of the Lower Leg

A. Creutzig, T. Wuppermann, U. Hanauske, and K. Alexander

Introduction

The development of leg ulcers is attributed to disturbed skin microcirculation, which is said to be the result of an increase in capillary pressure with the leaking of blood components through the capillary wall and the formation of a specific microedema [5]. Transcutaneous oxygen pressure measurements in the vicinity of ulcers yielded low values in both venous and arterial ulcers [1]. This is explained by a rarefication of capillaries [8]. In contrast, patients with liposclerosis or venous ulcerations show a marked proliferation of capillaries in the subcutaneous tissue [3]. Around the capillaries, fibrin walls are found, which act as diffusion barriers for oxygen [2]. Local hyperemia is observed after positron emission tomography and measurements of the blood flow in the skin using the laser-Doppler method [7, 10]. We have determined the tissue oxygen pressure in leg ulcers.

Patients and Methods

Eleven patients, aged between 22 and 85 years with chronic ulcers of the lower leg were examined. Eight of them suffered from chronic venous insufficiency. All of the patients had had a phlebothrombosis in the leg. Three patients had ulcers as a result of chronic venous insufficiency that had been complicated by an arterial occlusive disease (combined ulcers). They had the typical symptoms of intermittent claudication of the calf; the pulses of the popliteal arteries and the foot pulses were not palpable. The systolic ankle pressures had dropped to 60–80 mm Hg. The ulcers were at least 3×3 cm in size and showed no inflammatory changes.

The tissue oxygen pressure was determined using a multiwire surface electrode [9]. The technique is described in detail elsewhere [4, 6]. Prior to the measurements, the floor of the ulcer was cleaned with 0.9% saline solution. The PO_2 was measured both in the middle and at the edges of the ulcer. In five patients with chronic venous insufficiency and three patients with combined ulcers, the leg was first elevated at a 20° angle and the first histogram was recorded. Then the patient was instructed to sit up and let the leg hang down. A second histogram was recorded in this position.

In five patients with chronic venous insufficiency, the leg was wrapped with elastic bandages to find out the effect of bandaging on the tissue PO_2 in the ulcer. First, a histogram was recorded on the unbandaged leg hanging down. Then, a compression bandage was applied. The ulcer was covered with a piece of foam rubber with a hole cut in the middle. The compression bandage was put on in such a way that the hole was left open for the electrode to be inserted through it to the floor of the ulcer. To compare the histograms, the Kolmogoroff-Smirnoff test was used as a procedure independent of mean values.

Results

All of the histograms were shifted to the left, and low oxygen values were predominant. Almost a third of all the values registered with the leg hanging down were be-

Table 1. Values of oxygen histograms in ulcers (PO_2 values in mm Hg)

Pat. no.	Diagnosis			*n*	$\bar{x}$	SD	M	KS
1	CVI	↑		98	11.9	12.9	4.9	p<0.01
		↓		89	8.7	7.2	6.0	
2	CVI	↑		102	6.4	5.0	4.5	n. s.
		↓		106	7.7	6.5	4.7	
3	CVI	↑		99	10.8	8.9	5.2	p<0.01
		↓		98	7.9	6.5	5.4	
4	CVI	↑		103	15.7	7.6	14.5	n. s.
		↓		95	15.8	9.7	13.5	p<0.001
		↓	+C	88	26.6	5.3	26.6	
5	CVI	↑		110	13.2	10.6	10.9	p<0.05
		↓		99	12.3	13.4	14.9	p<0.001
		↓	+C	98	23.5	7.1	22.9	
6	CVI	↑		95	7.5	5.9	5.1	p<0.001
		↓	+C	104	12.5	10.6	12.2	
7	CVI	↓		89	12.3	8.5	11.0	n. s.
		↓	+C	101	13.8	10.8	12.0	
8	CVI	↓		93	10.9	6.8	10.4	n. s.
		↓	+C	95	12.9	9.4	10.5	
9	CVI+AOD	↑		103	16.4	14.3	18.7	p<0.001
		↓		97	36.8	15.4	42.8	
10	CVI+AOD	↑		106	13.6	10.7	13.0	p<0.001
		↓		102	27.9	10.3	26.8	
11	CVI+AOD	↑		98	17.8	10.9	16.1	p<0.001
		↓		100	25.6	10.0	14.4	

Abbreviations: n, number of oxygen pressure measurements; $\bar{x}$, mean; SD, standard deviation; M, median; KS, Kolmogoroff-Smirnoff test; CVI, chronic venous insufficiency; AOD, arterial occlusive disease; ↑, leg raised; ↓, leg lowered; +c, leg bandaged with elastic bandages; n. s., not significant

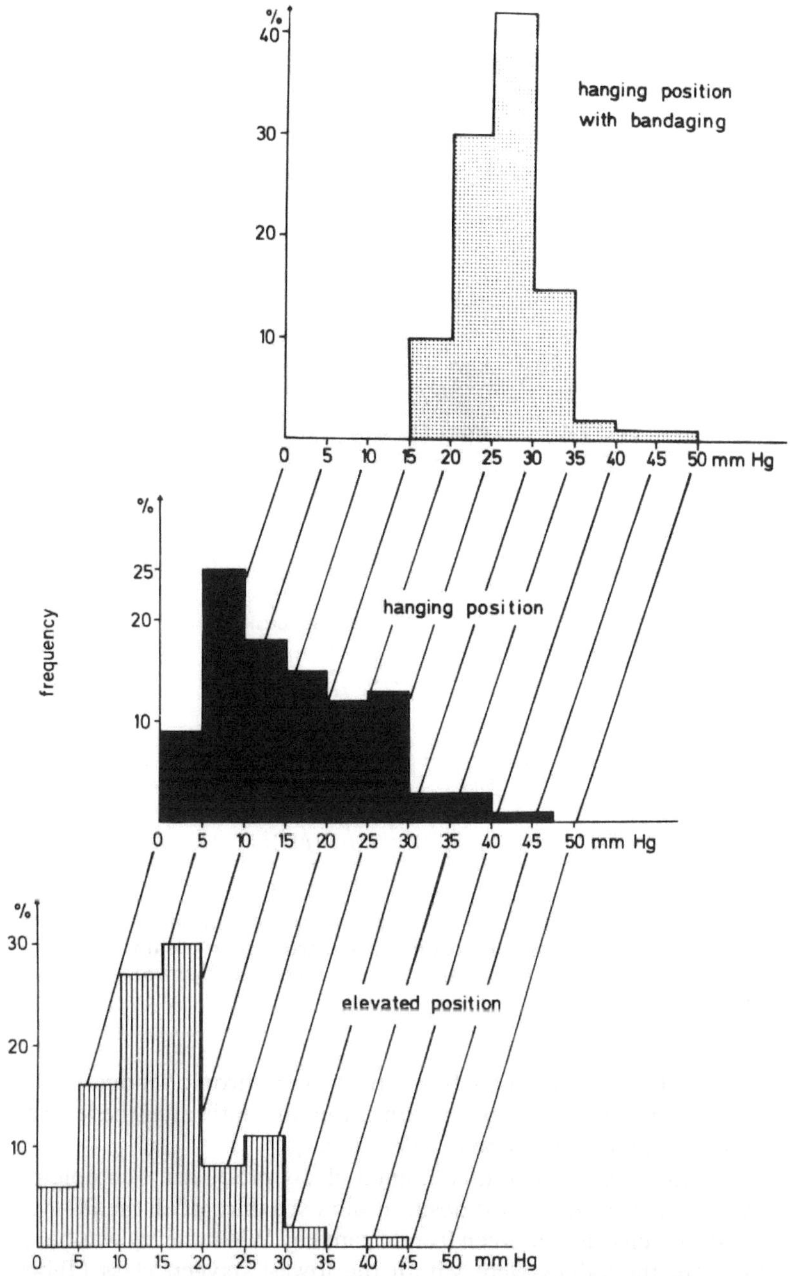

Fig. 1. Oxygen histogram in a venous ulcer according to the position of the leg (patient no. 4). Shift to the left in the histogram to lower values after the leg is lowered to a hanging position. After the leg is bandaged, marked shift to the right with a doubling of the median PO_2 value

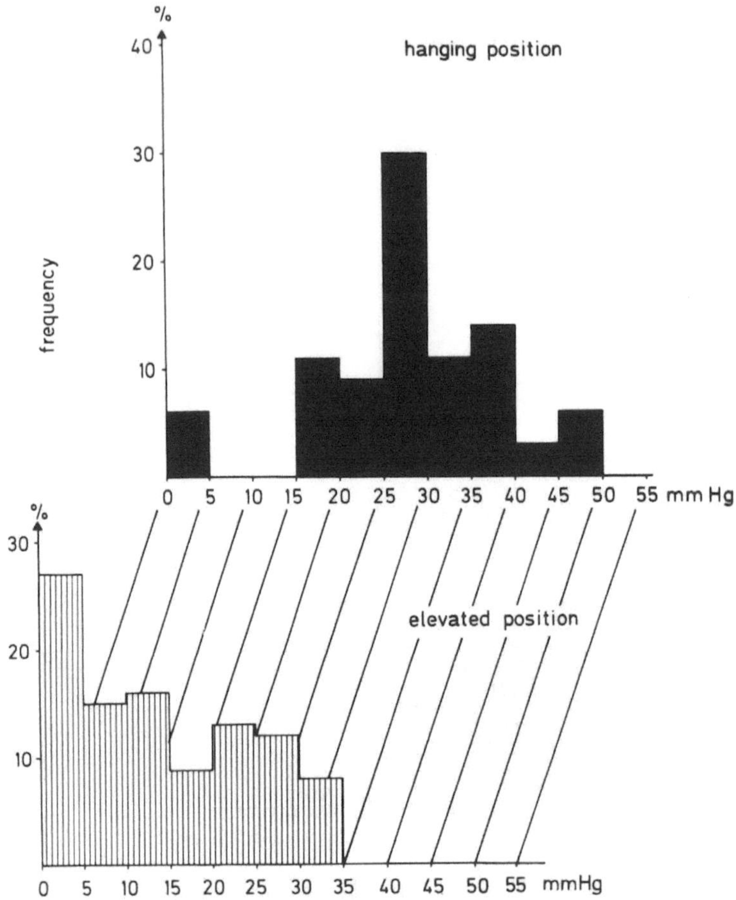

Fig. 2. Oxygen histogram of an lower leg ulcer in patient with chronic venous insufficiency and arterial occlusive disease (patient no. 10). Values according to position of the leg: doubling of the mean PO_2 value and significant change in the histogram configuration when the leg is lowered to a hanging position

low 5 mm Hg. In three of the five patients with chronic venous insufficiency, the distribution type of the histogram changed significantly after the leg was moved from a raised to a hanging position (Table 1, Fig. 1).

The three patients with combined ulcers produced a completely different picture: with the leg in a raised position, almost a third of all the PO_2 values were in the lowest oxygen class, between 0 and 5 mm Hg. When the leg was hanging down, only 7% of all the values were within the lowest oxygen class (Table 1, Fig. 2). The changes in the histogram were highly significant in all cases.

Bandaging the leg in a hanging position resulted in a significant shift to the right, with higher PO_2 values in three of the five cases. The median PO_2 in these patients had nearly doubled (Table 1, Fig. 1). In the other patients, the median PO_2 did not change; however, a second group was formed with higher oxygen values between 15 and 25 mm Hg.

Discussion

The oxygen histograms of all of the ulcers showed an inhomogeneous distribution with predominantly low PO_2 values as a sign of disturbed microcirculation on the surface of the ulcer. The reason for this could be the rarefication of the capillaries or pericapillary fibrin deposits [2, 8].

The results prove the relevance of the position of the extremity for the oxygen supply to the ulcer. The improved venous reflux when the leg is raised results in a shift to the right with higher PO_2 values in the histogram, which indicates that certain regions in the ulcer are better oxygenated.

The findings among patients with combined ulcers are quite the opposite: the PO_2 dropped considerably when the leg was raised. This can be explained by the reduced perfusion pressure owing to the impaired arterial flow. Bandaging the hanging leg with a compression bandage resulted in a marked increase in PO_2 in the ulcer tissue in three of five patients, and in the remaining two patients led to the formation of a second group with higher PO_2 values. The increase in PO_2 during compression and the changes taking place when the leg is raised are presumably caused by the improved nutritive blood flow in the ulcer tissue.

References

1. Borzykowski, M., B. Krähenbühl: Mesure non invasive de l'oxygénation cutanée en cas d'ulcères chroniques des membres inférieurs. Schweiz. med. Wschr. 111 (1981) 1972–1974
2. Browse, N. L., K. G. Burnand: The cause of venous ulceration. Lancet 2 (1982) 243–245
3. Burnand, K. G., I. Whimster, G. Clemenson, L. M. Thomas, N. L. Browse: The relationship between the number of capillaries in the skin of the venous ulcer-bearing area of the lower leg and the fall in foot vein pressure during exercise. Br. J. Surg. 68 (1981) 297–300
4. Creutzig, A., W. Wrabetz, M. Lux, K. Alexander: Muscle tissue oxygen pressure in patients with arterial occlusive disease. Microvasc. Res. 29 (1985) 350–359
5. Fagrell, B.: Die gestörte Hautmikrozirkulation als Ursache venöser Ulcera cruris. In: Trübestein, G. (Hrsg.): Arterielle Verschlußkrankheit und tiefe Venenthrombose. Thieme-Verlag, Stuttgart/New York (1984) S. 212–215
6. Hauss, J., K. Schönleben, H. U. Spiegel: Therapiekontrolle durch Überwachung des Gewebe-PO_2. Huber-Verlag, Bern (1982)
7. Hopkins, N. F., T. J. Spinks, C. G. Rhodes, A. C. Ranicar, C. W. Jamieson: Positron emission tomography in venous ulceration and liposclerosis: study of regional tissue function. Brit. med. J. 286 (1983) 333–336
8. Huch, A., U. K. Franzeck, R. Huch, A. Bollinger: A. transparent transcutaneous oxygen electrode for simultaneous studies of skin capillary morphology, flow dynamics and oxygenation. Int. J. Microcirc.: Clin. Exp. 2 (1983), 103–108
9. Lübbers, D. W.: The meaning of the tissue oxygen distribution curve and its measurement by means of Pt-electrodes. Progr. Resp. Res. 3 (1969) 112–123
10. Partsch, H.: Hyperaemic hypoxia in venous ulceration. Br. J. Derm. 110 (1984) 249–251

PO_2 Distribution in Resting Muscle and Pulmonary Gas Exchange in Patients with Cirrhosis

R. Heinrich, H. Schomerus, W. Fleckenstein, W. Grauer, A. Huber, and C. Weiss

Introduction

As is well known, many cirrhosis patients have elevated cardiac output and lowered peripheral vascular resistance [1, 2]. The demonstrable disturbances of pulmonary gas exchange [3] in these patients could be partially attributed to changes in pulmonary microflow [4]. Moreover, the capability of cirrhotics to withstand major surgery depends more on the extent of these circulatory changes than on liver function parameters [5, 6]. The cardiac output-O_2 consumption ratio, i.e., O_2 extraction rate, an index of microflow efficiency, has proved to be a prognostically important parameter for establishing this capability [6]. The findings tend to indicate that changes in peripheral blood flow distribution in these patients impair tissue oxygen supply. The following studies were carried out to test whether disturbances of gas exchange, such as those demonstrated in the lungs, are also detectable in peripheral tissue and whether they correlate significantly with impaired pulmonary function or severity of liver disease.

Method

12 patients with histologically confirmed liver cirrhosis between the ages of 40 and 74 and 6 hospitalized controls of comparable age were examined. 6 of the patients had Child classification B cirrhosis, 6 classification C. PO_2 distribution in the vastus lateralis muscle was measured with a fast responding polarographic needle electrode which was advanced stepwise into the tissue. The measured local pO_2 values were recorded as histogram using a microcomputer system (for details, see [1-10]).

The pO_2 histograms were recorded under ambient air and after inhalation of 40% O_2 for 15 minutes. In 7 patients with cirrhosis and 5 control patients, pulmonary diffusion capacity for CO (DL_{co}) and its components, alveolar membrane diffusion capacity (D_M) and pulmonary capillary volume (V_c), were also determined according to the method described by Roughton and Forster [11].

Research supported by a grant from the Deutsche Forschungsgemeinschaft (He 1239-1)

Results

The pooled histogram for the 6 control patients under ambient air and with 40% O_2 is presented in Fig. 1 a. Both the mean pO_2 value and the configuration of the histogram are normal for bedfast patients of this age. Arterial pO_2 rose from 74 ± 11 mm Hg to 134 ± 29 mm Hg after inhalation of the oxygen-rich mixture for 15 minutes, and, as a result, the pooled histogram shifted to the right (10 mm Hg) but remained bell shaped.

By contrast, the pooled histogram for the patients with Child B cirrhosis under ambient air (Fig. 1 b) showed signs of maldistribution, i.e., a broadening of the histogram and a higher mean. Inspite of an adequate rise of arterial pO_2 from

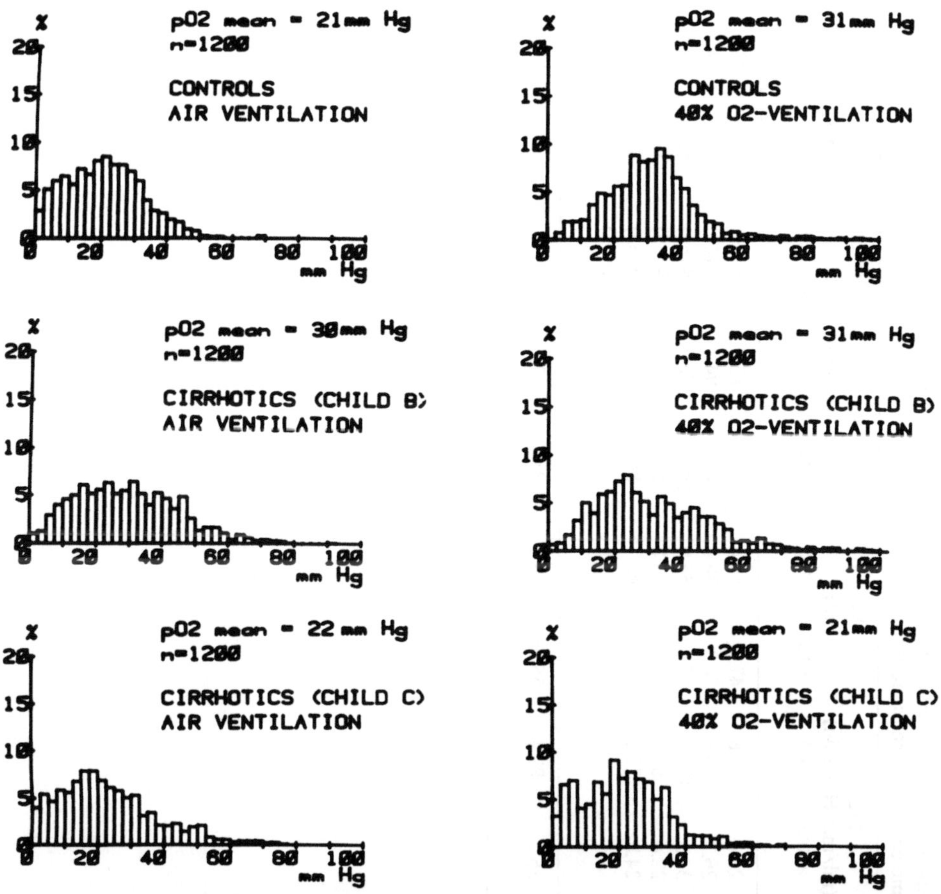

Fig. 1. pO_2 distribution in vastus lateralis muscle under ambient air (left) and after inhalation of 40% O_2 for 15 minutes (right); pooled histograms (a) 6 controls patients, (b) 6 patients with Child B cirrhosis, and (c) 6 patients with Child C cirrhosis. pO_2-mean = arithmetic mean, n = number of measurements (200 measurements/patient)

Table 1. Pulmonary diffusion capacity for CO (D_LCO) and its components V_c (pulmonary capillary volume) and D_M (alveolar membrane diffusion capacity) in controls and cirrhotics (males, no. 9 female)

Controls No. (Years)	DL_{co} (ml/mm HG/min) actual	% rated	V_c (ml) actual	% rated	D_M (ml/mm HG/min) actual	% rated	Cirrhotic Patients No. (Years)	Child Class.	DL_{co} (ml/mm HG/min) actual	% rated	V_c (ml) actual	% rated	D_M (ml/mm HG/min) actual	% rated
1 (60)	25.1	92.6	47.8	56.4	89.8	185.7	6 (50)	C	23.7	86.5	52.8	61.4	61.5	126.5
2 (56)	26.2	86.8	76.3	78.6	49.5	94.1	7 (40)	C*	14.2	44.8	56.6	61.8	21.8	41.1
3 (49)	32.5	106.3	57.4	56.9	144.5	276.9	8 (62)	C*	14.3	71.7	81.0	156.1	18.9	44.8
4 (61)	22.4	84.1	49.9	60.8	57.9	121.2	9 (65)	C	18.8	79.6	57.1	81.3	34.4	78.1
5 (35)	24.8	77.2	48.3	42.7	84.0	159.4	10 (63)	B*	12.5	51.1	68.3	92.9	16.6	37.2
							11 (71)	B	20.0	86.5	60.5	89.7	36.7	83.0
							12 (43)	C	19.7	66.3	50.6	50.6	42.4	84.5
X	26.2	89.4	55.96	59.1	85.2	167.5			17.61	69.51	50.99	83.4	33.19	70.4
δ	3.78	10.95	12.01	12.0	37.3	70.5			4.02	16.55	10.52	30.46	15.86	32.09

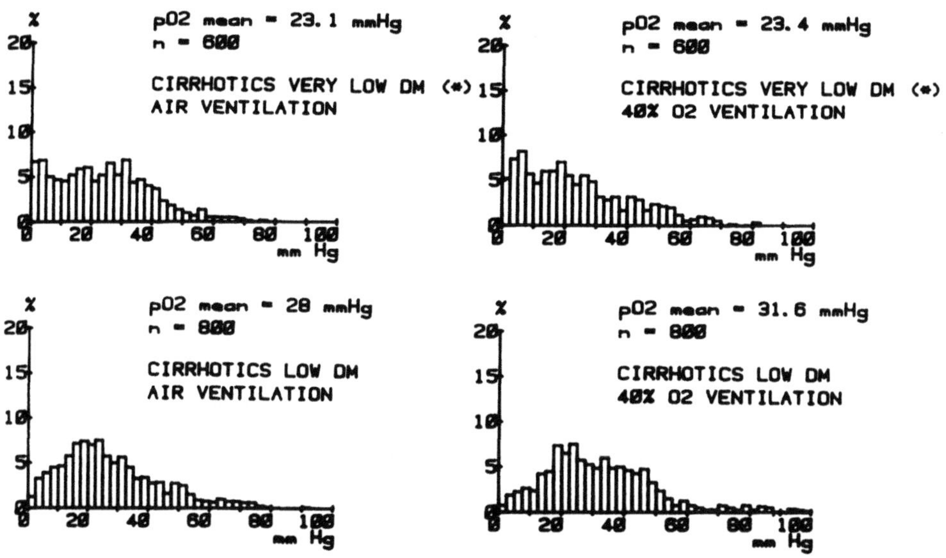

Fig. 2. pO_2 distribution in vastus lateralis muscle under ambient air (left) and after inhalation of 40% O_2 (right); pooled histograms of (a) 3 patients with cirrhosis and markedly lowered D_M, and, in comparison (b) of 4 patients with cirrhosis and almost normal D_M. Abbreviations as in Fig. 1

73 ± 6 mm Hg to 109 ± 14 mm Hg after inhalation of the hyperoxic mixture, neither mean pO_2 nor pO_2 distribution in muscle was altered.

The histogram for patients with Child C cirrhosis under ambient air (Fig. 1c) was maldistributed with a mean of 22 mm Hg. Arterial pO_2 rose from 73 ± 10 mm Hg to 116 ± 2.5 mm Hg after inhalation of the oxygen-rich mixture for 15 minutes, but the mean pO_2 value or pO_2 distribution in muscle remained unchanged.

The values for pulmonary diffusion capacity for CO (DL_{co}) and its components V_c and D_M in the cirrhotics and the controls are presented in Table 1. DL_{co} values were somewhat lower in the cirrhotics than in the controls, and D_M values were markedly lower (expressed as percentage of normal values for comparable age and weight); differences in pulmonary capillary volume, however, were negligible. The diminution of D_M did not correspond with the Child classification nor affect arterial pO_2 values under normoxia or hyperoxia.

Within the group of cirrhotics, however, comparison of pO_2 histograms for the 3 patients with a D_M less than 50% of normal (Fig. 2a) with those of the other 4 patients (Fig. 2b) revealed a more pronounced shift to the left of the maldistributed histogram for the group with a lower D_M than for the patients with an almost normal D_M.

Discussion

The findings indicate that, in contrast to the control patients, an increase of arterial pO_2 in our cirrhosis patients did not result in a rise of tissue pO_2 in resting muscle.

Impaired perfusion appears to be highly unlikely as the cause of disturbed oxygen transport into the tissue, since peripheral blood flow in these patients tends rather to be intensified [1, 12]. Higher metabolic activity, which could also lower tissue pO_2 given constant organ perfusion transport, appears to be equally improbable. One possible explanation is a disturbance of oxygen supply at the level of the microcirculation. The configuration of the pO_2 histogram corresponds best to that of a microcirculation. The configuration of the pO_2 histogram corresponds best to that of a microcirculatory maldistribution [13], which apparently increases in relation to the severity of the liver disease.

The disturbances of pulmonary gas exchange in our patients, particularly the marked diminution of alveolar membrane diffusion capacity, have no effect on arterial pO_2, possibly because of the hyperventilation characteristic of cirrhotics. Disturbances of pulmonary gas exchange and changes of pO_2 distribution in tissue were parallel in our small patient group. The results tend to indicate that disturbances of gas exchange in cirrhotics occur not only in the lung, but also in peripheral tissue, probably as a result of impaired microflow.

References

1. Kontos, H. A., W. Shapiro, H. P. Mauck, J. L. Patterson: General and regional circulatory alterations in cirrhosis of the liver. Am. J. Med. 37 (1964) 526–534
2. Bayley, T. J., N. Segel, J. M. Bishop: The circulatory changes in patients with cirrhosis of the liver at rest and during exercise. Clin. Sci. 26 (1964) 227–235
3. Schomerus, H., I. Buchta, H. Arndt: Pulmonary function studies and oxygen transfer in patients with liver cirrhosis and different degrees of portasystemic encephalopathy. Respiration 32 (1975) 1–20
4. Arndt, H., I. Buchta, H. Schomerus: Analysis of factors determining the resistance to diffusion in patients with liver cirrhosis. Respiration 32 (1975) pp 21–31
5. Gordon, M. J., L. R. M. del Guercio: Late effects of portal systemic shunting procedures on cardiorespiratory dynamics in man. Ann. Surg. 5 (1972) 672–679
6. Siegel, J. H., R. M. Goldwyn, H. J. Farrell, P. Gallin, H. Friedman: Hyperdynamic states and the physiologic determinants of survival. Arch. Surg. 108 (1974) 282–292
7. Fleckenstein, W., C. Weiss: A comparison of pO_2-histograms from rabbit hind limb muscle obtained by simultaneous measurements with hypodermic needle electrodes and with surface electrodes. Adv. Exp. Med. Biol. 169 (1984), 447
8. Fleckenstein, W., C. Weiss: Evaluation of pO_2-histograms obtained by hypodermic needle electrodes. Proc. World Congress on Med. Physics and Biomed. Eng. (1982)
9. Fleckenstein, W., R. Heinrich, W. Grauer, H. Schomerus, W. Dölle, C. Weiss: Fast local regulations of muscle pO_2-fields in patients suffering from cirrhosis of the liver. Adv. Exp. Med. Biol. (in press)
10. Fleckenstein, W., R. Heinrich, T. Kersting, H. Schomerus, C. Weiss: A new method for the bedside recordings of tissue pO_2-histograms. Verh. Dt. Ges. Inn. Med. 90 (1984), 439
11. Roughton, F. J. W., R. E. Forster: Relative importance of diffusion and chemical reaction rates determining rate of exchange of gases in the human lung, with special reference to true diffusing capacity of pulmonary membrane and volume of blood in the lung capillaries. J. Appl. Physiol. 11 (1957) 290–302
12. Silverstein, E.: Peripheral venous oxygen saturation in patients with and without liver disease. J. Lab. and Clin. Med. 47 (1956) 513–518
13. Ehrly, A. M., W. Schroeder: Zur Pathophysiologie der chronisch arteriellen Verschlußerkrankung. I. Mikroszirkulatorische Blutverteilungsstörung. Herz/Kreislauf 11 (1979) 275–281

Clinical Results of Local Tissue PO_2 Measurements on the Surface of the Brain During Neurosurgical Operations

R. Schultheiß, R. Wüllenweber, and E. Leniger-Follert

Introduction

The determination of local tissue PO_2 on the surface of the brain is now known and established as a method of dealing with physiologic problems in animal experiments [4, 1]. During neurosurgical operations, the local tissue PO_2 in the region being operated on was measured for the purpose of establishing a basis for comparable investigations.

Materials and Methods

For the measurements, a polarographic multiwire surface electrode (Kessler and Lübbers [3]) was used in conjunction with the MIT PO_2 system. With this combination, sufficiently short measuring times can be achieved by using electronic data processing to keep the operation from becoming unreasonably long. The system is sterilized using the sterilization technique described by Hauss et al. [2]. A special holder prevents interference from the pulsations of the brain and makes it possible to guide the probe for the histogram recording with a micromanipulator.

The measuring conditions were those of general anesthesia under slight hyperventilation, which resulted in arterial PCO_2 values of around 30 mm Hg and not under 25 mm Hg. Of the arterial PO_2 values, a value of 150–170 mm Hg could be most easily reproduced, so that this value was considered the standard value for the purpose of interindividual comparative studies. For an interpretable measurement, circulation must be stable and the blood composition constant. These parameters were carefully monitored and, if necessary, corrected.

Results

Even a normal cortex can show clear differences in the PO_2 histogram. Changes in the cortical structure, such as those that occur when a glioma breaks through the cortex, have an effect on the configuration of the PO_2 histogram.

The formation of a cerebral edema changes the cortical histogram considerably, although the type of change cannot yet be classified. A broader distribution of the values is the rule; a shift to the left to hypoxic values can occur.

The histogram of an astrocytoma grade II (WHO), which shows up as an avascular space-occupying lesion on both CT and angiography, is comparable to a normal cortical histogram (Fig. 1). The malignant version of this tumor, an astrocytoma grade IV (WHO), produces a narrower grouping of the values, as well as a higher mean value with comparable arterial blood gases (Fig. 2). On CT, this tumor showed the typical contrast staining in its marginal zone, where the measurement was carried out. Accordingly, in angiography a pathologic vascularization could be proved. In the case of meningiomas, all of the histograms were characteristic. With a mean value of approximately 16 mm Hg, the curve is shifted to the left with a narrow peak (Fig. 3). The homogeneity of the values for this tumor, which is well vascularized and has a uniform composition, even macroscopically, is not surprising. However, the low mean value is unexpected. Angiographically, these tumors are marked by their pronounced capillary staining and their arterial supply, which is usually via branches of the carotid artery. Intraoperatively, there is tendency for hemorrhaging to occur without any previous embolization of the feeding arteries.

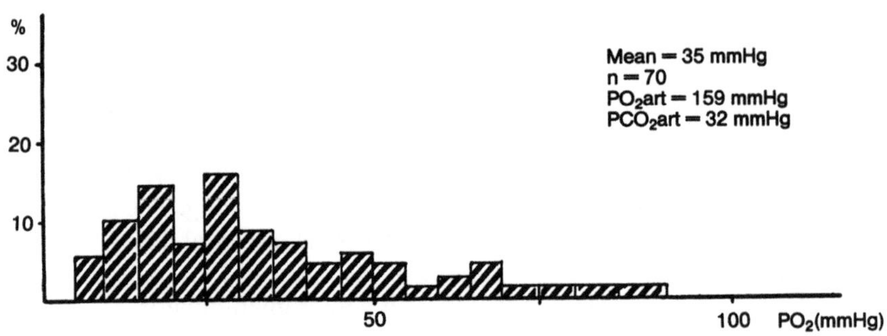

Fig. 1. Astrocytoma grade II (WHO). Histogram appears normal in comparison with normal cortex

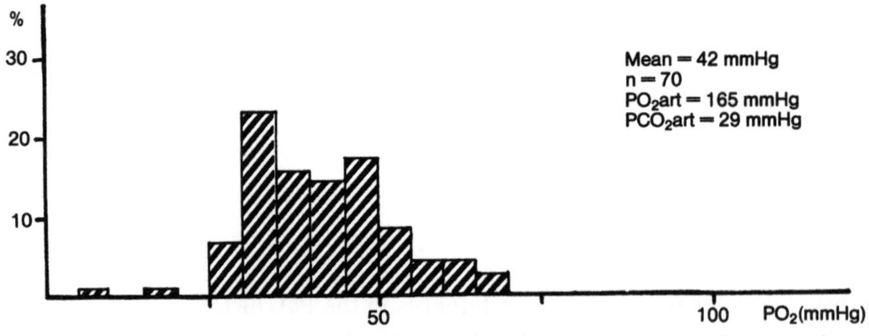

Fig. 2. Astrocytoma grade IV (VHO). Slight shift to the right and narrow, i.e., more homogeneous, grouping of the values in comparison with normal cortex

An explanation for this could be the altered diffusion conditions when there are changes in the capillary wall, as they can be observed histologically.

The measurement of a preoperatively embolized tumor shows the effect of this measure by an extreme shift to the left of the histogram and a mean value of 7 mmHg with narrow distribution. Necroses in the material of this tumor could be detected histologically. The histogram of a metastasis of an adenocarcinoma is also interesting (Fig. 4): the distribution is homogeneous and, compared with the normal cortex, a shift to the left is noticeable, as is a resemblance to a glioblastoma.

One physiologic question which could be dealt with in connection with the removal of cortical arteriovenous malformations (AVM) was how removing the AVM changes the PO_2 of the cortex in the immediate vicinity of the malformation. In a to-

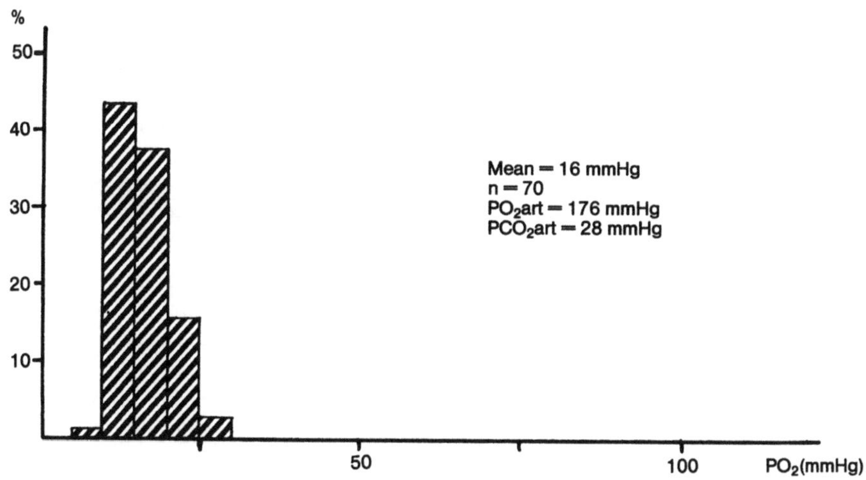

Fig. 3. Meningioma. Narrow distribution of PO_2 values with low mean value and marked shift to the left

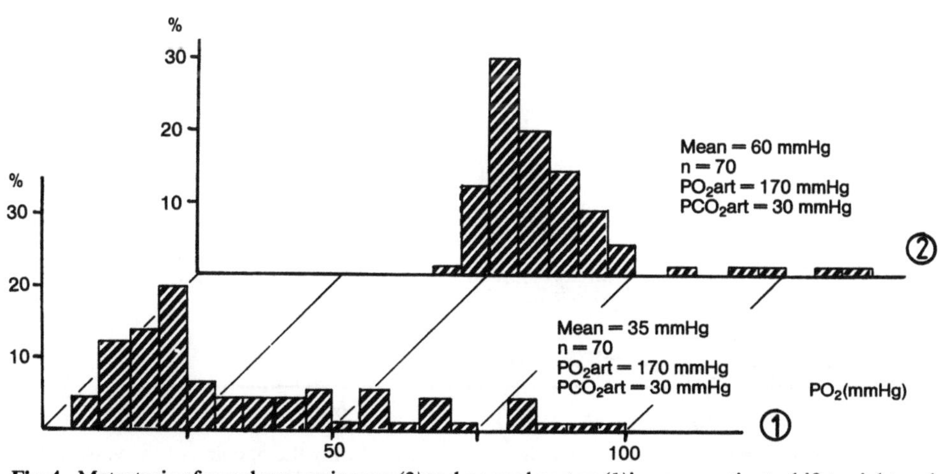

Fig. 4. Metastasis of an adenocarcinoma *(2)* and normal cortex *(1)* in same patient: shift to right and homogeneous distribution in histogram of the metastasis

tal of nine patients, a histogram was recorded in a cortical area that was near the angioma but not incorporated in the lesion. The recording was done before and after extirpation. There was a marked increase in the initially hypoxic values, which was indicated by a shift to the left. The mean difference in the mean values was 42 mm Hg, which is more than 100% of the initial value. A circulatory steal phenomenon caused by these malformations was also proved by flow measurements: parallel to this, a PO_2 steal phenomenon can be identified (Fig. 5).

The same question was studied in several patients in connection with a tumor removed while there was an increase in intracranial pressure. Here, too, the PO_2 values rose after the extirpation, and in association with this the cerebral pressure returned to normal.

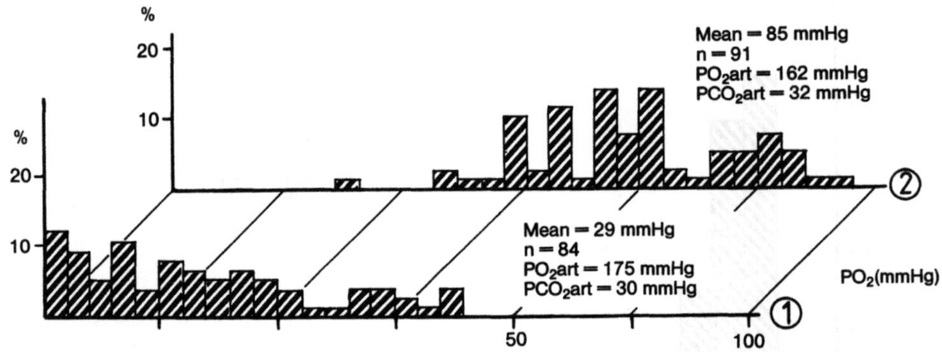

Fig. 5. Histogram of normal cortex in the vicinity of an arteriovenous malformation (AVM) before *(1)* and after *(2)* extirpation of the AVM. After removal of the malformation, the previously hypoxic values and shift to the left of the tissue PO_2 values returned to normal

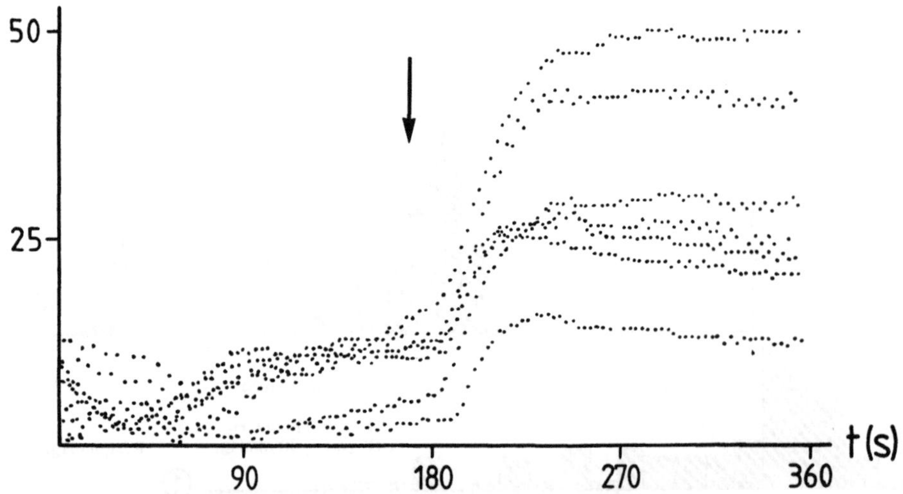

Fig. 6. Continuous measurement on the neighboring cortex during the opening of the vessels after an extracranial-intracranial anastomosis was performed for carotid artery occlusion. Marked increase of tissue PO_2 after release of the clip

Another possibility offered by this method of measuring can be seen in the following example: while the vessels were being opened after an extracranial-intracranial anastomosis was performed from the superficial temporal artery to the middle cerebral artery for carotid artery occlusion, measurements were performed continuously (Fig. 6). The measurement taken in a cortical part of the circulation area of the anastomosis shows an immediate increase after the vessels are opened. An interpretation of the results for a validation of the operation would be premature. It can be maintained, however, that rapid changes in the microcirculation can be detected if the conditions under which the measurements take place are defined.

Discussion

In summary, it can be said that the technique for measuring surface PO_2 is a practicable method for clinical research on the brain. Problems arise out of the necessity to surgically expose the area to be measured, which means that measuring can only be done during surgery. Measurements on brain tumors are only possible when the tumor is not subcortically located.

According to our initial results, there would seem to be a relationship between the nature of the glioma (benign or malignant) and the configuration of the PO_2 histogram, although we do not have statistical proof of this. The results from the studies on meningiomas are striking and require further clarification. Evidence of improvement in local tissue PO_2 in the cortex near AVM as a result of their extirpation justifies both the accuracy of one aspect of the indication for this operation and the possibility of solving physiologic questions with this method.

References

1. Grote, J., M. Krüger-Biegner, R. Schubert, K. Zimmer: Significance of brain surface PO_2 measurements in determining oxygen supply conditions in the brain cortex. In: H. P. Kimmich (ed): Monitoring of vital parameters during extracorporeal circulation. Proc. Int. Conf., Nijmegen. Karger-Verlag, Basel (1980) pp 101-105
2. Hauss J., K. Schönleben, H. U. Spiegel: Therapiekontrolle durch Überwachung des Gewebe-PO_2. Eine tierexperimentelle und klinische Studie. (Aktuelle Probleme in der Angiologie Bd. 41). Huber-Verlag, Bern/Stuttgart/Wien (1982)
3. Kessler, M., D. W. Lübbers: Aufbau und Anwendungsmöglichkeiten verschiedener PO_2-Elektroden. Pflüg. Arch. 291 (1966) S. 82
4. Leniger-Follert, E., D. W. Lübbers, W. Wrabetz: Regulation of local tissue PO_2 of the brain cortex at different arterial O_2 pressures. Pflüg. Arch. 359 (1975) pp 81-95

The Supply of Oxygen to the Human Small Intestine in Mechanical Ileus

M. Thermann, L. Jostarndt, F. Eberhard, H. Richter, and W. Saß

Introduction

Mechanical ileus results in an intraluminar increase in pressure proximal to the obstruction, with stretching of the intestinal wall. Aside from the role played by the intestinal contents and swallowed air, which are carried further from the proximal sections of the intestine, reasons for the intraluminar increase could be continuing secretion and the formation of gas from the fermentation of the congested intestinal contents [3]. The subsequent damage to the intestinal wall, which can lead to necrosis, and edema formation have been interpreted as the result of disturbed microcirculation and the hypoxia or anoxia caused by this disturbance. With the aid of the PO_2 surface electrode developed by Kessler and Lübbers [6], it is now possible to measure the local tissue PO_2 in the microvessels without causing any damage to the tissue. The surface of the organ must be smooth, the membrane must be $<20\ \mu m$ thick, and the capillary blood supply must reach below the membrane. These conditions exist on the serous membrane side of the wall of the small intestine, since the visceral serous membrane has a thickness of approximately 10 μm and the musculature beneath it is supplied with capillaries (1). An animal experiments, we were able to prove that an increase in intraluminar pressure results in a pressure dependent reduction in the tissue PO_2 of the intestinal wall [5]. The objektive of our clinical investigation was to measure, in a prospective study, the tissue oxygen pressure in the obstructed human intestine in order to obtain information on the oxygen supply.

Materials and and Methods

Investigations were carried out in 28 patients. Sixteen patients underwent emergency laparotomy for mechanical ileus. Table 1 lists the locations and causes of the intestinal obstruction. The mean age of the patients was 56 ± 16 years; eight patients were male and 20 were female. Prior to anesthesia, an electrolyte and fluid compensation was carried out by means of infusion and drug application to ensure that circulation was stable when anesthesia began. The patient was included in the study only if the serous membrane of the distended intestine appeared wet, and there was

Table 1. Location and cause of mechanical ileus

Diagnosis	n
Form of ileus:	
High obstruction of small intestine	5
Low obstruction of small intestine	8
Obstruction of large intestine	3
Causes of ileus:	
Adhesions	10
Tumors	5
Incarcerated Hernia	1

no indication of perforations or peritonitis. Patients who had generalized circulatory disturbances of the intestine were excluded, as were those whose poor general condition did not permit the operation to be prolonged for the measurements.

The PO_2 measurements were performed on a dilated intestinal loop, which without manual manipulation lay exposed after the area was held open with an abdominal retractor. Clarfication of the cause of the ileus was held off until after the PO_2 measurements had been taken so that there would not be any flow changes in the distended intestine as a result of the manipulation. Figure 1 shows an electrode in position.

A control group consisted of one male and 11 female patients (mean age 51 ± 12 years), who underwent elective cholecystectomy. After standardized preoperative evacuation with anthrachinon derivatives, laparotomy was carried out. Without intra-abdominal manipulation, a proximal intestinal loop was placed on the abdomen, which was moistened with a NaCl solution at body temperature. The PO_2 measurement was carried out on this intestinal loop.

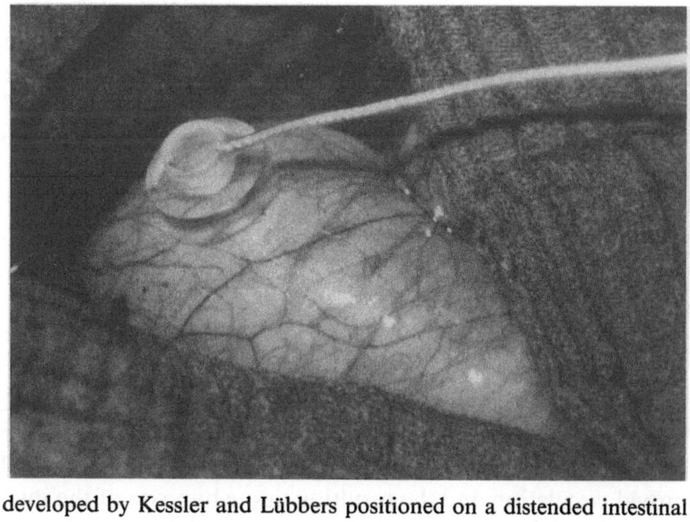

Fig. 1. Surface electrode developed by Kessler and Lübbers positioned on a distended intestinal loop (ileus patient)

All of the patients were premedicated with atropine and promethazine or diazepam (i. m., 30 min before beginning of anesthesia). Anesthesia was induced with thiopental sodium i.v. and maintained with 66% N_2O, 33% O_2, and halothane. As a relaxant, *d*-tubocurarine was given. At the beginning of the PO_2 measurement, blood gas was removed from the A. radialis for blood gas analysis. Blood pressure and pulse rate were measured at 15 min intervals.

Preoperative laboratory parameters were determined during the course of routine laboratory work up. Blood gas analyses were done with a Corning analyzer at actual patient body temperature. Arterial blood pressure was measured according to Riva-Rocci; the pulse rate was counted over a period of 15 s. A membrane-covered platinum surface electrode with three platinum wires (15 μm) developed by Kessler and Lübbers [6] was used for the polarographic PO_2 measurement of the intestinal tissue. The wires were recorded cumulatively. The electrodes and their parts were sterilized with 15% acetylene oxide, prepared for measurement under sterile conditions, and calibrated with ambient air and pure N_2 at actual patient temperature before and after the measurements were completed. After the electrode was placed in position, we waited for a steady state. The electrode was then placed at another site. A plastic holder (Fig. 1) made it possible to position the electrode under direct vision outside the small intestine. The measurements were performed at 20–25 different locations in the small intestine and took a total of 15–30 min. The Wilcoxon test was used for statistical evaluation.

Results

In the preoperatively obtained (in the ileus group following infusion and drug application) clinicochemical lab parameters (Table 2), a significantly raised serum creatinine level was found in the group of ileus patients; there were no differences among the other parameters. The intraoperatively defined circulatory parameters showed, with the mean arterial blood pressure the same, a significant tachycardia of the ileus patients of $\tilde{x} = 110$ (90–115) beats/min median, first to third quartile) as opposed to a pulse of $\tilde{x} = 90$ (80–110) in the control group ($P < 0.05$, Table 3).

The arterial blood gases showed no differences within either group. The local PO_2 measurements revealed, with a median of $\tilde{x} = 22.5$ (12.5–35) Torr vs $\tilde{x} = 61.2$ (53.0–71.0) Torr, a highly significant reduction in the local PO_2 of the intestinal wall distended by the mechanical ileus.

Discussion

The effect of intraluminar pressure increase on the circulation of the intestinal wall was already proved in animal experiments by Dragstedt et al. in 1929 [2]. The disturbed circulation and the resulting tissue hypoxia of the intestinal wall when there is an obstruction is assumed to be the essential cause of ileus [8, 10].

Table 2. Preoperative clinicochemical parameters, Significance test in Tables 2, 3: Wilcoxon test, median of first to third quartiles

Parameter	Ileus (n=16)	Controls (n=12)	*P*
Hct (%)	41.1 (38.0–44.3)	42.0 (38.9–45.0)	n.s.
Hb $\left(\frac{g}{100\,ml}\right)$	14.3 (11.8–15.4)	14.2 (12.2–15.7)	n.s.
Leukocytes $\left(\frac{1000}{mm^3}\right)$	8.65 (6.9–9.9)	8.95 (6.5–12.3)	n.s.
Creatinine $\left(\frac{mg}{100\,ml}\right)$	1.06 (0.79–1.45)	0.80 (0.70–0.88)	0.05
Na^+ (mval/l)	138 (135–141)	139 (136–141)	n.s.
K^+ (mval/l)	4.0 (3.6–4.2)	3.7 (3.6–4.2)	n.s.
Ca^{++} (mval/l)	4.2 (3.8–4.4)	4.5 (4.3–4.8)	n.s.
Bilirubin $\left(\frac{mg}{100\,ml}\right)$	0.8 (0.5–1.6)	0.6 (0.5–0.8)	n.s.

Table 3. Intraoperative flow dimensions and blood gas values at beginning of tissue oxygen pressure measurement

Characteristic	Ileus (n=16)	Controls (n=12)	*P*
Mean arterial pressure (Torr)	95 (76–107)	93 (77–98)	n.s.
Pulse (beats/min)	110 (90–115)	90 (80–110)	<0.05
Arterial pO_2 (Torr)	130.5 (106.0–196.3)	113.2 (92.2–150.6)	n.s.
Arterial pCO_2 (Torr)	35.2 (32.2–39.0)	31.6 (28.9–35.9)	n.s.

Our investigations show that the restricted O_2 supply in the human intestine in mechanical ileus can be proved. We reserve judgement on the intraoperative intraluminar pressure conditions: we did not measure them, since complications can arise when direct puncture is performed for the purpose of pressure measurement. We can, however, assume that, based on the clinical picture and the results of intraoperative palpation, there was pathologic intraluminar pressure in our group of patients.

The microcirculation rate of the intestinal tissue cannot be directly measured by the tissue PO_2 measurement. It can be expected, however, that with a normal arterial O_2 supply, a lower PO_2 tissue value is caused by a reduction in the perfusion rate. The PO_2 values of the distended ileus intestine are an indication of the reduced tissue perfusion rate as a result of distension, which can subsequently lead to local ischemia and tissue necrosis. A histogram with individual local PO_2 values might give more exact information on the hypoxic and anoxic tissue areas [7]; however, for methodical reasons this was not possible. We did not carry out tissue PO_2 measurements after decompression, since operative procedures such as the insertion of probes and the evacuation of intestinal contents would have resulted in local tissue reactions. For similar reasons, we dispensed with a comparative measurement of the distended and nondistended intestine. Here individual intestinal loops would have had to be pulled out, and these measurements would only have been possible in the case of relatively proximal intestinal occlusions (five of the 16 patients). Moreover, this procedure would have prolonged the measurements by more than an hour, which alone is an unreasonable strain on the patient.

The size of the catchment area of a platinum electrode in polarographic measurements is dependent on the diameter of the measuring wire and on the thickness of the membrane. This range is defined as that in which "a change in the PO_2 field induce(s) a distinct change of the measuring signal" [4]. In our case, the membrane thickness of the electrode is 15 μm; the serous and the subserous layers of the small intestine, which together have a maximum thickness of 20 μm when the intestine is distended, are regarded as additional "membrane" [1]. For our 15-μm electrode, the range is calculated as approximately 50 μm [4]. The values we determined thus provide information only on the oxygen pressure (and, indirectly, on the perfusion rate) of the outer layers of the musculature. The flow conditions of the mucosa are not registered in our measurement.

Using a comparable method in an earlier animal experiment, we proved the dependency of the local PO_2 of the intestinal wall on the amount of arterial flow [9]. Among our ileus patients, the stage of the ileus, the duration of the disease, and the local condition of the intestine varied. To rule out the possibility of these factors playing a role, we excluded all those patients from the investigation in whom results of disturbed circulation owing to a longer ileus condition could be macroscopically detected.

The proof that PO_2 values in the ileus intestine are lowered as the result of stretching emphasizes the importance of decompression for the purpose of interrupting the essential pathophysiologic mechanisms of intestinal wall damage.

Summary

In 16 patients with mechanical ileus and in 12 patients who underwent elective laparotomy, the local tissue oxygen pressure of the small intestine was measured with PO_2 surface electrodes. In the ileus patients, the pulse rate increased intraoperatively to 110 (90–115) vs 90 (80–110) beats/min, and the creatinine level was raised to 1.06 (0.79–1.45) vs 0.80 (0.70–0.88) mg% ($P<0.05$). The local PO_2 of the wall of the small intestine on the side of the serous membrane was $\tilde{x} = 22.5$ (12.5–35) Torr in the ileus group, and $\tilde{x} = 61.2$ (53.0–71.0) Torr in the control group. The reduction in flow and oxygen supply that is assumed to occur in the case of mechanical ileus was proved by the local PO_2 measurement, thus emphasizing the importance of early decompression of the ileus intestine.

References

1. Doerr, W., G. Seifert, E. Uehlinger: Spezielle patholog. Anatomie II/2. Springer-Verlag, Berlin/Heidelberg/New York (1976) S. 634
2. Dragstedt, C. A., V. F. Lang, R. F. Millet: The relative effects of distension on different portions of the intestine. Arch. Surg. 18 (1929) 2257–2263
3. Duthie, H. L.: Intestine. In: W. T. Irvine (ed.): The scientific basis of surgery. Churchill Livingstone, London (1972)
4. Grunewald, W.: How "local" is pO_2 measurement? In: M. Kessler et al. (eds.): Oxygen supply. Urban & Schwarzenberg, München/Berlin/Wien (1973)
5. Jostarndt, L., H. Richter, I. Tichai, M. Thermann: Auswirkungen der intraluminären Druckerhöhung auf die Sauerstoffversorgung des Kaninchenileus und ihre pharmakologische Beeinflußbarkeit. Langenb. Arch., Suppl. Chir. Forum (1976) 435–437
6. Kessler M., D. W. Lübbers: In: Lübbers, Luft (eds.): Oxygen transport in blood and tissue. Thieme-Verlag, Stuttgart (1967) pp 242–251
7. Kessler, M.: Oxygen supply to tissue in Normoxia and in Oxygen deficiency. Microvasc. Res. 8 (1974) 283–290
8. Noer, R. J., M. J. Robli, L. F. Jacobsen: Circulatory disturbances produced by acute intestinal distension in the living animal. Arch. Surg. 63 (1951) 520–528
9. Piasecki, C.: A new method for the assessment of gut viability. Brit. J. Surg. 68 (1981) 319–322
10. Wangensteen, O. H.: Alte und neue Ansichten über therapeutische Probleme beim Darmverschluß. Zbl. f. Chirurgie 100 (1975) 1089–1098

Effect of Infusion of 500 ml 5% Human Albumin on the Muscle Tissue Oxygen Pressure in Patients with Intermittent Claudication

A. M. Ehrly, H. Landgraf, and K. Saeger-Lorenz

Introduction

The infusion of plasma protein solutions or colloidal plasma substitutes is a well-known therapy not only for volume deficiency, but also for cerebral and peripheral ischemic diseases [1]. Besides replacement of the volume, the improvement in flow properties of the blood explains the improvement of clinical symptoms associated with hypovolemic conditions such as shock [2]. In the case of ischemic disease such as chronic arterial occlusion, however, the clinical effectiveness of an isovolemic infusion is attributed solely to an improvement in the flow properties of the blood [3, 4]; under hemodilution, the cardiac output is also raised [5].

There are diverging opinions on the clinical relevance of hemodilution for arterial diseases: the difficulty in assessing clinical effectiveness is finding suitable objective parameters. This becomes even more problematic when, as a result of hemodilution, the number of oxygen-carrying erythrocytes per unit of blood is reduced. This deficit has to be compensated for by an increase in flow volume in order to achieve an optimum hematocrit value and and optimum oxygen-transporting capacity. An objective parameter for quantifying the tissue oxygen supply, particularly with regard to questions on the effects of hemodilution in ischemic disease, is the tissue oxygen pressure. Such investigations were carried out by Meßmer et al. [6] and Hauss et al. [7] in animal experiments and on human patients.

Hauss et al. observed that when human albumin and various plasma substitutes were infused in patients undergoing intensive care, the tissue PO_2 reacted in different ways. Meßmer et al., however, found that after dextran solutions were administered in an animal model, the tissue PO_2 increased. The reason for the improved oxygen supply in the tissue was considered to be the improvement in flow properties of the blood.

To date, there have been no investigations with therapeutic hemodilution on patients with intermittent claudication, with the exception of experiments on an increase in tissue oxygen pressure following short-term infusion of 500 ml saline solution [8]. If an improvement in the oxygen supply of the tissue were the only result of improved, blood flow properties, the infusion of an iso-oncotic colloidal solution like human albumin, which remains in the vessels for a longer period of time, would have a particularly favorable effect on the supply of oxygen to the tissue of the lower leg musculature in the case of intermittent claudication. The objective of our investigation was to see whether the infusion of 500 ml 5% human albumin solution has a favorable effect on the tissue oxygen pressure and, consequently, on the oxygen supply. As a control solution we used a 10% hydroxyethyl starch solution with a molecular weight of 200 000.

Materials and Methods

In six patients with chronic arterial occlusive disease, 500 ml of a 5% human albumin solution was infused intravenously within 30 min. Another group of patients with the same stage of intermittent claudication was given an infusion of 500 ml of the 10% hydroxyethyl starch solution (HAES-steril, Fresenius), also within 30 min. Apart from measurement of the tissue oxygen pressure in the lower leg musculature with a microplatinum needle electrode [9], several hemorrheologic parameters were determined.

Results

Figure 1 shows a comparison of the pooled histograms (frequency distribution analysis) of six patients treated with human albumin recorded before and after infusion. There is a marked shift to the left of the histogram: during the course of the study the values approach the initial values and exceed them again 2 h after the infusion is ended. The medians of the tissue oxygen pressure behave accordingly.

After one patient had an allergic reaction to the infusion characterized by chills, backache, and urticaria, for ethical reasons we decided not to include any more patients in the study, especially since a decrease, not an increase, in the tissue oxygen pressure was registered after the administration of human albumin. The measurement of the blood flow properties, which cannot be described in detail here, revealed a statistically significant lowering of the hematocrit values, the whole blood viscosity, and the plasma viscosity, the minimum being reached immediately after infusion. The aggregation of erythrocytes also decreased.

In ten patients who also suffered from intermittent claudication, the infusion of 500 ml of a 10% HAES-200 solution within 30 min resulted in a moderate increase in the PO_2 values during the period following infusion. The properties of the blood flow showed a decrease in whole blood viscosity, a slight increase in the plasma viscosity, and a reduction in erythrocyte aggregation. There were no significant changes in blood pressure and heart rate during the investigation with either solution.

After the infusion of 500 ml 5% human albumin for 30 min, there was a temporary reduction in the oxygen pressure in the lower leg muscles tissue, even though all the factors of the blood flow properties had improved. In contrast, the infusion of 500 ml of the hydroxyethyl starch solution resulted in an increase in the tissue

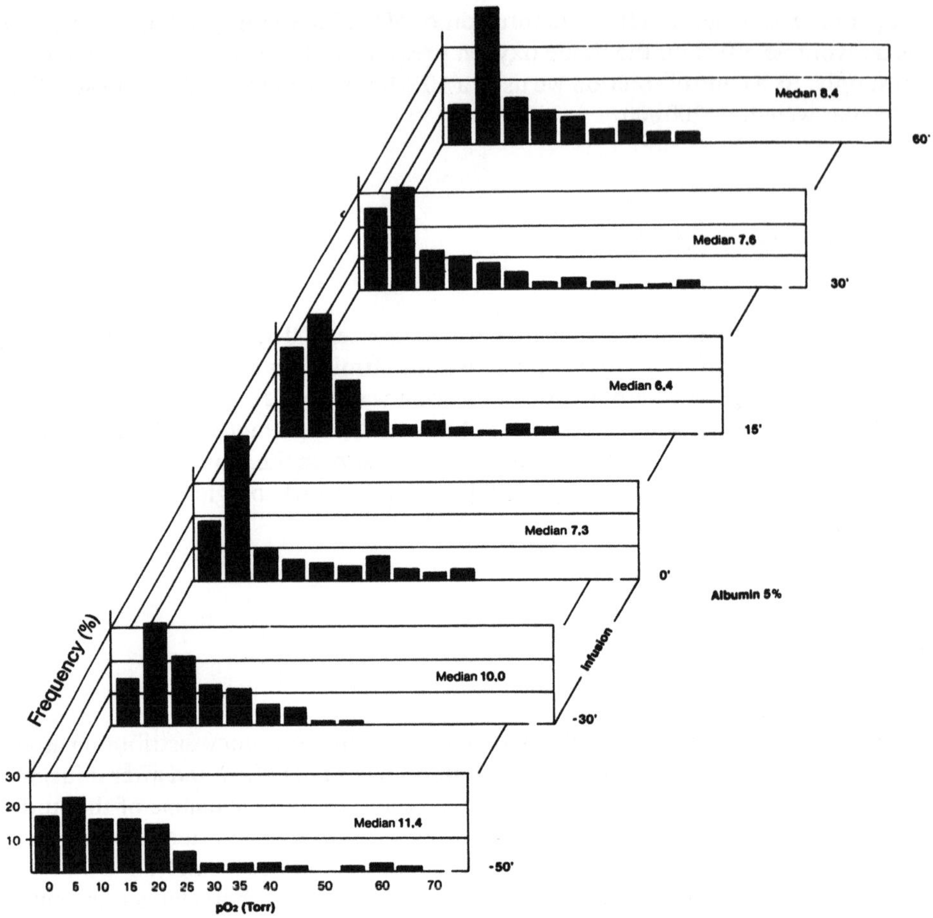

Fig. 1. Pooled histograms before and after infusion of human albumin

oxygen pressure, although the improvement in blood flow properties was not as pronounced under this drug as it was under human albumin. Because of the small number of cases, we have forgone a statistical confirmation of the albumin results. Nevertheless, it is interesting to note that in just about all of the patients who received human albumin, the tissue oxygen pressure dropped; only one patient showed a slight initial increase in tissue PO_2 after the albumin infusion.

Discussion

In discussing the reason for the varied reactions of the tissue oxygen pressure to the infusion of colloidal solutions, it first has to be emphasized that the hemodynamic parameters, systemic blood pressure and pulse rate, were not influenced. The improvement in blood flow properties known to be brought about by human albumin should, according to the Hagen-Poiseuille law, have led to increased perfusion in the capillaries with improvement in the oxygen supply to the tissue. We conjecture that vasoactive compounds such as kinins and serotonin in the microcirculation can lead to a microcirculatory steal phenomenon as a result to maldistribution [10, 11], so that even though the total perfusion is good, the nutritive circulation is reduced.

A moderate reduction in the hematocrit value caused by hypervolemic hemodilution with a plasma substitute (hydroxyethyl starch MG 200, 10%), on the other hand, did not result in a decrease in the tissue O_2 pressure, but rather led to a moderate increase in these values [12]. Apparently, the reduced number of oxygen-carrying erythrocytes after hemodilution is compensated for by the improved microcirculatory perfusion conditions resulting from the improvement in the blood flow properties.

The investigations show that the theoretical assumption of an improvement in microcirculation as a result of improved blood flow properties - as was found in extra vivum measurements performed under the application of human albumin - does not necessarily mean that the oxygen supply to the tissue is improved. Competing mechanisms in the microcirculation can apparently cause a reduction in the tissue oxygen supply. Other authors in some cases have been able to establish an improvement in tissue oxygen supply during infusion of human albumin [7]. In our opinion, however, this is because those intensive care patients had relative or absolute hypovolemia, so that besides the blood flow properties improving as a result of human albumin, the circulating blood volume returned to normal. Consequently, those findings cannot be compared with the present data on patients with chronic arterial occlusive disease.

Summary

The 30-min intravenous infusion of 500 ml 5% human albumin solution in patients with intermittent claudication causes a temporary reduction in the oxygen pressure values in the ischemic muscle tissue, while the blood pressure and pulse rate remain stable and the flow properties of the blood markedly improve. In contrast, the infusion of hydroxyethyl starch of medium molecular weight leads to a moderate increase in the tissue PO_2 values distal from the occlusion or stenosis, despite the reduced hematocrit values. The discrepancy between the findings obtained with the two drugs in our open investigation can be explained by the possibility that vasoactive kinins and serotonin in the human albumin cause a maldistribution in the microcirculation, which is not the case with the technically manufactured hydroxyethyl starch.

The investigation was carried out on normovolemic patients. The type of infusion was hypervolemic; in other words, there was no venesection prior to infusion. The data presented indicate that the transfer of hemorrheologic results obtained from extra vivum measurements to the in vivo situation with patients can be misleading and is therefore inadmissible as evidence. Here, the measurement of tissue oxygen pressure as an in vivo parameter on the tissue and cell level takes on particular significance.

References

1. Messmer, K., H. Schmid-Schönbein (eds.): Hemodilution. Theoretical basis and clinical application. Karger-Verlag, Basel (1972)
2. Gelin, L. E., B. Ingelman: Rheomacrodex - a new dextran solution for rheological treatment of impaired capillary flow. Acta Shir. Scand 122 (1961) 294-302
3. Zederfeldt, B.: Rheological disturbances and their treatment in clinical surgery. In: A. L. Copley (ed): Symposium and Biorheology. Pergamon Press (1963) pp 393-407
4. Rieger, H., W. Schoop, H. Schmid-Schönbein, R. Schneider, H. Malotta: Isovolemic hemodilution (IHD) in peripheral vascular occlusive disease (PVOD). Concepts, methods and preliminary results. Microvasc. Res. 13 (1977) 267
5. Landgraf, H., A. M. Ehrly: Hämorheologische Wirkungen hypervolämischer Hämodilution mit Dextran 40 und HAES 200 bei Patienten mit arterieller Verschlußkrankheit. In: Ehrly, A. M. (Hrsg.): Therapie mit hämorheologisch wirksamen Substanzen. Zuckschwerdt-Verlag, München (1984) S. 138-144
6. Messmer, K., L. Görnand, E. Sinagowitz, L. Sunder-Plassmann, F. Jesch, M. Kessler: Local oxygen tension in tissue of different organs during limited normovolemic hemodilution. Bibl. Anat. 12, Karger-Verlag, Basel (1973) p 327
7. Hauss, J., K. Schönleben, H.-U. Spiegel: Therapiekontrolle durch Überwachung des Gewebe-pO_2. Aktuelle Probleme in der Angiologie, Bd. 41, Huber-Verlag, Bern (1982)
8. Landgraf, H., A. M. Ehrly: Verhalten des Muskelgewebesauerstoffdrucks bei Patienten mit Claudicatio intermittens nach Placebo- respektive Kochsalz-Infusionen. In: Ehrly, A. M., J. Hauss, R. Huch (Hrsg.): Klinische Sauerstoffdruckmessung. MWP-Verlag, München (1985) S. 63-69
9. Ehrly, A. M., H.-J. Köhler, W. Schroeder, R. Müller: Sauerstoffdruckwerte im ischämischen Muskelgewebe von Patienten mit chronischen peripheren arteriellen Verschlußkrankheiten. Klin. Wschr. 53 (1975) 687
10. Ehrly, A. M., W. Schroeder: Zur Pathophysiologie der chronischen arteriellen Verschlußerkrankung. I. Mikrozirkulatorische Blutverteilungsstörung in der Skelettmuskulatur. Herz/Kreislauf 11 (1979) 275
11. Ehrly, A. M.: New pathophysiological concept of ischemic diseases: Microcirculatory Blood Maldistribution (MBM) Bibl. Anat. 20 (1980) 456
12. Ehrly, A. M., H. Landgraf: Effects of muscle tissue oxygen tension in claudicants: infusions of 5% albumin versus 10% hydroxyethylstarch solution (200/0,5). Third World Congress for Microcirculation, Oxford 1984 (abstract)

Tissue PO_2 Measurements in Critical Care: The Effects of Dopamine on Muscular Oxygen Pressure Fields

T. Kersting, K. Reinhart, W. Fleckenstein, R. Dennhardt, K. Eyrich, and C. Weiss

Introduction

In critical care medicine, dopamine is used for treatment and prevention of circulatory insufficiency and shock [5, 7, 13, 20]. At low doses (2 to 5 μg/kg per min) renal blood flow and sodium excretion are increased [1, 5, 6, 12, 13, 14, 18, 19, 22]; these effects are attributed to the action of specific dopamine receptors [5, 16, 23]. At higher doses (5 μg/kg per minute), sympathic β_1- and α-receptors are stimulated, so that cardiac output and (especially at doses over 10 μg/kg per min) vascular resistance are increased [5, 6].

The effects of dopamine have been studied in several organs by a number of authors [8, 9, 12, 17, 19]. Still its effects on the blood flow in skeletal muscle have not yet been clearly established. In hypoxic and normoxic dogs, an insignificant increase in muscle blood flow can be demonstrated with 10 μg/kg dopamine per minute [9], whereas in humans, there is evidence of a diminution of blood flow in the limbs following the administration of 2 to 6 μg/kg dopamine per minute [17, 19]. It is generally assumed, however, that dopamine increases the perfusion of the kidney and mesenteric organs – partly at the expense of the perfusion of peripheral organs [5, 6, 17, 19].

A considerable part of cardiac output is consumed by muscle blood flow. The effects of dopamine on human muscular blood flow and the related effects on muscular PO_2 are of interest with regard to its clinical use. In order to obtain further information on the peripheral action of systemically applied dopamine, the distribution of PO_2 in the m. vastus lateralis of m. quadriceps femoris of nine healthy adult volunteers and seven critically ill patients following dopamine administration has been studied.

Methods

The hypodermic needle PO_2 probes used in this study were inserted in a stepwise manner into the tissue; PO_2 was determined within the standstill periods (1.5 s) [3, 4]. The probes had a stainless steel shaft with a diameter of 350 μm and a response time of <500 ms (37 °C). The sharply ground tip of the probes contained a membranized, recessed microcathode which was 15 μm diameter. Details about this recently developed method of tissue PO_2 measurement were published elsewhere [3, 4] (see also a companion report in this book).

The recording of 200 individual local PO_2 values for each histogram took 6 min. During the procedure, the single PO_2 values were displayed on a monitor screen. Thus it was possible to adjust the step unit if the time course of the PO_2 values indicated penetration of the probe into an intramuscular space or into a blood vessel or if bleeding occurred into the probe channel.

PO_2 measurements were carried out in patients and in volunteers before, during, and after intravenous infusion of dopamine. As the healthy volunteers were not willing to accept the unpleasant sensations brought about by higher doses of the drug, only the effects on muscle PO_2 of "low-dose" dopamine (3 μg/kg per minute) were studied. In order to obtain information on the specificity of the observed dopamine effects on muscular PO_2, five volunteers received metoclopramide [an antagonist of peripheral dopamine effects (0.17 mg/kg)] at the end of the dopamine infusion.

The dopamine doses given to the patients (2 to 10 μg/kg per minute) depended on individual therapeutic requirements; thus no systematic study of the dose/efficiency relationship with regard to muscular PO_2 could be carried out.

In all patients, arterial and venous catheters were inserted for measurements of blood gases and blood pressure values. In four of the seven patients, cardiac output, pulmonal arterial wedge pressure, and mixed venous O_2 saturation were determined using a Swan-Ganz thermodilution catheter.

Results

Volunteers

In Fig. 1 histograms pooled from the PO_2 data of nine healthy volunteers are shown before (A) and during (B–D) dopamine infusion (3 μg/kg per minute). The mean tissue PO_2 increased from 27 mm Hg PO_2, the "reference value", (before dopamine) to a maximum of 36 mm Hg PO_2 (after dopamine) within 20 to 30 min. Subsequently, tissue PO_2 fell for about 30 min and reached a nearly constant, slightly elevated level from the 50th min to the end of the measuring period.

Figure 2 shows pooled PO_2 histograms from five volunteers before and during infusion of dopamine and after the injection of metoclopramide (at continuing dopamine infusion). After the metoclopramide injection, the mean tissue PO_2 decreased by 9.7 mm Hg (SD, 3.7). The mean blood pressure of the volunteers was not altered by the infusion of dopamine or metoclopramide.

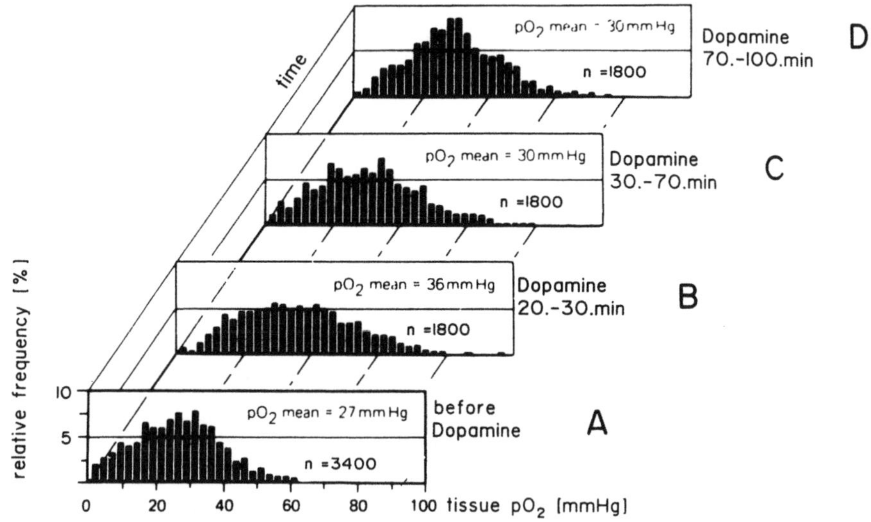

Fig. 1. PO_2 histograms from skeletal muscle before *(A)* and during *(B, C, D)* dopamine infusion of 3 µg/kg per minute; data are pooled from nine healthy volunteers

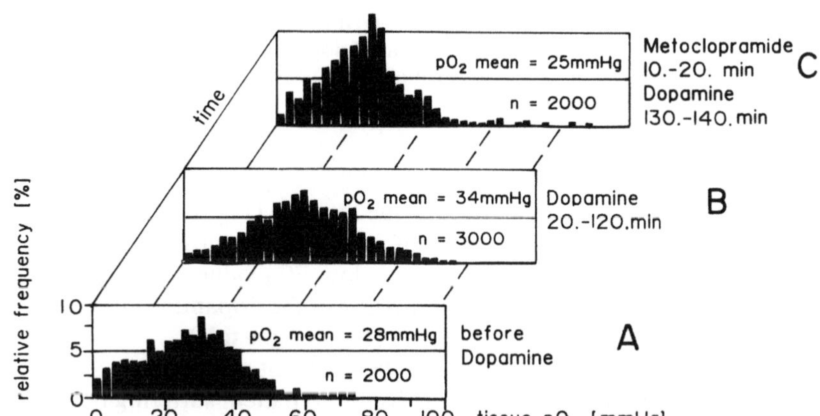

Fig. 2. PO_2 histograms from skeletal muscle before *(A)* and during *(B, C)* dopamine infusion of 3 µg/kg per minute before determination of histogram *C* metoclopramide (0.17 mg/kg) was injected i.m.; data are pooled from five healthy volunteers

Critically Ill Patients

Figure 3 shows the effects of dopamine on the mean tissue PO_2 in seven critically ill patients. From the start of the dopamine infusion, mean tissue PO_2 rose slowly but steadily. After 50 min, a plateau was reached which was maintained for several hours (not shown in Fig. 3). A mean PO_2 increase of 11.6 mm Hg (SD, 4.3) was observed within 100 min. After discontinuing the infusion, PO_2 values decreased slowly and returned to the reference values (Fig. 3, right side). Before the administration

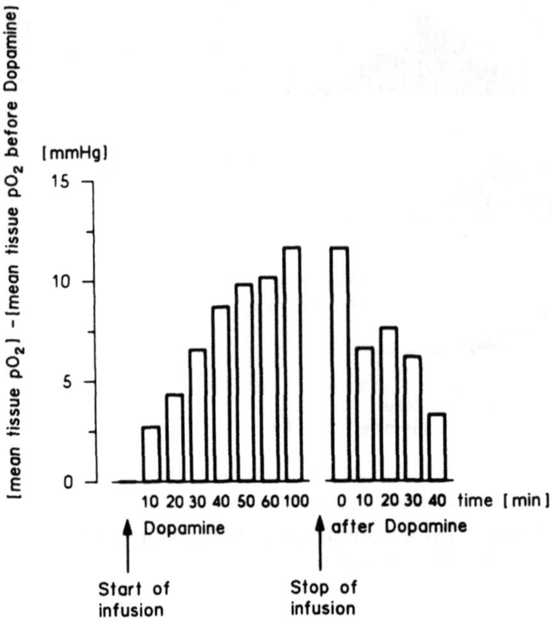

Fig. 3. *Ordinate* differences in mean PO_2 (mm Hg) during and after dopamine infusion measured in skeletal muscle. *Abscissae,* time (min) after start of dopamine infusion *(left)* and after stop of dopamine infusion *(right).* Data are pooled from seven intensive care patients; dopamine doses depended on therapeutic requirements

of dopamine, four patients showed tissue hypoxia (Fig. 4, left side) characterized by an unphysiologically large number of single PO_2 values below 5 mm Hg. During dopamine infusion, the number of such low PO_2 values decreased to physiologic levels (Fig. 4, middle). After discontinuation of the dopamine infusion, the reference values were regained relatively quickly (Fig. 4, right side).

As can be seen in Fig. 5, the tissue PO_2 did not decrease in all patients after the end of the dopamine infusion. The 15-year-old patient suffering from septicemia after pneumonectomy was artificially ventilated. Before dopamine infusion, systemic circulation parameters were not pathognomonically affected, but an elevation of serum lactate content was observed (2.3 mmol/liter). Although nitroglycerin was infused (2.5 μg/kg per minute), the muscular PO_2 histogram was shifted to the left (mean PO_2, 17.2 mm Hg), as shown in Fig. 5 A. In this situation, dopamine was given in order to increase urine production. Shortly after starting dopamine infusion, tissue PO_2 and cardiac output rose, while mean arterial and pulmonal arterial wedge pressure were reduced; hypoxic tissue zones disappeared (Fig. 5 B). Though the dopamine infusion was disconnected after 110 min (while nitroglycerin infusion continued) the PO_2 histogram (Fig. 5 C) and the cardiac output remained at an elevated level until the end of the observation period (90 min after stop of infusion). At this time the serum lactate content was reduced (1.7 mmol/liter).

During dopamine infusion, no significant correlation was observed between the changes in arterial PO_2, mixed venous PO_2, arterial HbO_2 saturation, mixed venous

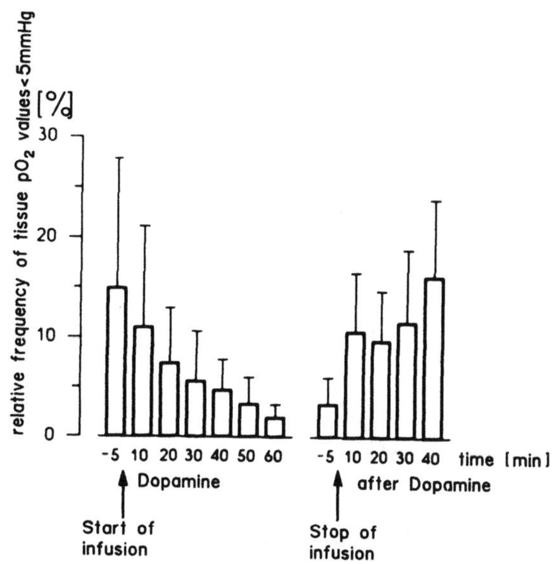

Fig. 4. *Ordinate,* mean relative frequency of muscular PO_2 values below 5 mm Hg measured from four patients with left-shifted histograms before dopamine infusion. Dopamine doses depended on therapeutic requirements. Abscissae, time (min) after start of dopamine infusion *(left)* and after stop of dopamine infusion *(right)*

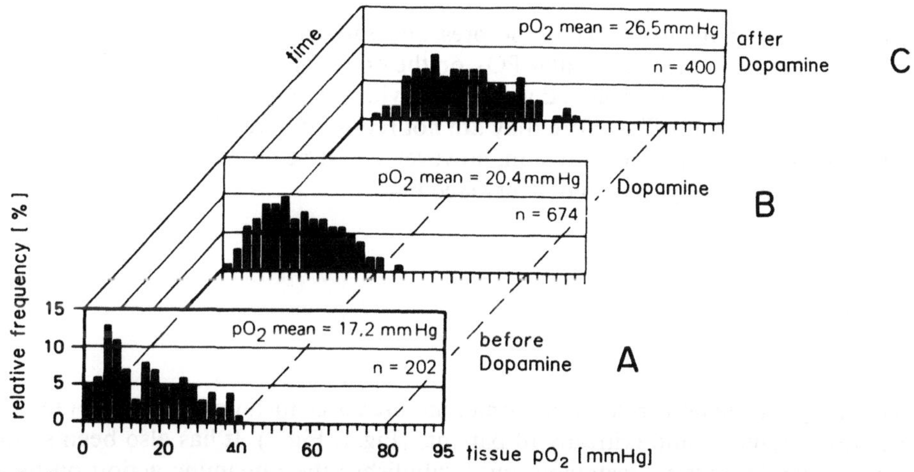

Fig. 5. Case report: muscular PO_2 histograms from a patient suffering from septicemia after pneumonectomy before *(A)*, during *(B)*, and after *(C)* dopamine infusion of 3.6 µg/kg per minute. All the time nitroglycerin (2.5 µg/kg minute) was infused. After stop of dopamine infusion the dopamine induced changes of hemodynamic parameters persisted

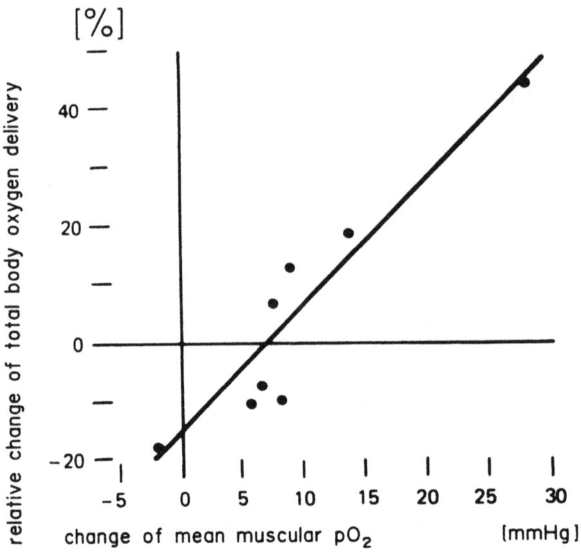

Fig. 6. *Ordinate,* difference between total body oxygen delivery (TBOD) during dopamine infusion and TBOD before dopamine infusion, given as percentage. *Abscissae,* difference between mean muscular PO_2 during dopamine infusion and mean muscular PO_2 before infusion (mm Hg). Values of tissue PO_2 and TBOD were recorded "simultaneously" (maximum time interval 6 min). Data were obtained from four patients

HbO_2 saturation, mean arterial blood pressure, and cardiac index, on the one hand and the changes in mean muscular PO_2, on the other. The data obtained by invasive hemodynamic monitoring agreed with the previously reported circulatory effects of dopamine [5, 6, 7, 13]. Therefore they are not reported here. The elevation of muscular PO_2 during dopamine infusion, however, could be observed at either reduced or elevated total body oxygen delivery (see Fig. 6).

Discussion

The study has demonstrated that dopamine increases mean muscular PO_2 in both healthy volunteers and critically ill patients (Fig. 1, Fig. 3). It has also been shown that in healthy people, metoclopramide abolishes the dopamine action on tissue PO_2 (Fig. 2). In critically ill patients we studied, dopamine administration produced a significant rise in muscular PO_2 (Fig. 3), especially in formerly low PO_2 regions (Fig. 4), at unchanged mean arterial blood pressure. Moreover, the data have shown that mean muscular PO_2 rises significantly (Fig. 6) not only when systemic oxygen delivery is increased or unchanged but also when it is reduced.

The observed changes in the shape of the PO_2 histograms show that during dopamine infusion, the initially maldistributes histograms become progressively more symmetric in most of our patients thus indicating a redistribution of microflow toward a physiologic pattern [10, 21]. A dopamine induced increase in muscular blood flow would be the simplest and most plausible explanation for our findings. However, the available evidence [15, 17, 19] indicates that in the healthy human, dopamine leads to a decrease in blood flow in the extremities - to the benefit of renal and splanchnic perfusion.

With the methods used in humans by previous authors, no conclusions could be drawn as to whether the observed increase in blood flow in the extremities was the result of an increase in blood flow in skin or muscle tissue.

Blood flow in peripheral muscle has been shown to increase initially upon infusion of dopamine in the dog [9]. In this study, during continuing dopamine infusion, muscular blood flow increased after 15 min and fell below initial values within 30 to 50 min [15].

At unchanged total organ flow, a redistribution of the muscular microflow - as indicated by the observed changes in the shapes of the PO_2 histograms - were able to explain the PO_2 increases in formerly low PO_2 tissue areas; it cannot explain our finding of a drastically increased mean muscular PO_2 in the critically ill patients. So far, we explain the marked elevation of mean muscular PO_2 in the intensive care patients by an increased muscular blood flow since there is no evidence available for a dopamine-induced depression of tissue O_2 consumption. It should be mentioned in this context that the effects of the different doses of dopamine used in our study may vary not only quantitatively but also qualitatively. It is known that the vasodilator effect of dopamine at low doses may be obscured by its α-mimetic activity at higher doses [5, 6, 8].

The blocking effect of the dopamine antagonist metoclopramide [2, 15] on the dopamine induced increase in muscular oxygen pressure suggests that the observed rise in muscle PO_2 is, at least partly, due to the action of dopamine on specific peripheral receptors [2, 11].

Obviously it is necessary to carry out systematic animal studies with monitoring of blood flow in a muscle artery, monitoring of muscular oxygen consumption, and measurements to determine dopamine dose/efficiency with regard to muscle PO_2.

References

1. Barnado, D.E., Baldus, W.P., Maher, F.T.: Effects of dopamine on renal function in patients with cirrhosis. Gastroenterology 58 (1970) 524-528
2. Cavero, I., Massingham, R., Lefevre-Borg, F.: Peripheral dopamine receptors, potential targets for a new class of antihypertensive agents. Part I: subclassification and functional description. Life Sciences 31 (1982) 939-948
3. Fleckenstein, W., Weiss, Ch.: A comparison of PO_2-histograms from rabbit hindlimb muscles obtained by simultaneous measurements with hypodermic needle electrodes and with surface electrodes. Adv. Exp. Med. Biol. 169 (1984) 447-455
4. Fleckenstein, W., Weiss, Ch.: Ein neues Gewebe-pO_2-Meßverfahren zum Nachweis von Mikrozirkulationsstörungen. Focus MHL 1, Heft 2 (1984) 74-84
5. Goldberg, L. I.: Cardiovascular and renal actions of dopamine: potential clinical applications. Pharmacol. Rev. 24 (1972) 1-29
6. Goldberg, L. I.: Dopamine - clinical uses of an endogenous cathecholamine. New Eng. J. Med. 291 (1974) 707-710
7. Gunar, R., Lieb, H., Pietras, R., Dritz, J., Tobin Jr., J.: Hemodynamic effects of dopamine compared to norepinephrine and isoproterenol in clinical shock. Circulation 37/38. Supp. VI (1968) 6

8. Higgins, C. B., Millard, R. W., Braunwald, E., Vatner, S. F.: Effects and mechanisms of action of dopamine on regional hemodynamics in the conscious dog. Am. J. Physiol. 225 (1973) 432-437
9. Jackson, L. K., Key, B. M., Cain, S. M.: Total and hindlimb O_2 uptake and blood flow in hypoxic dogs given dopamine. Crit. Care. Med. 10 (1982) 327-331
10. Kessler, M.: Lebenserhaltende Mechanismen bei Sauerstoffmangel und bei Störungen der Organdurchblutung. Mitt. Max-Planck-Ges. (1974) 444-463
11. Laubie, M., Schmitt, H., Falq, E.: Dopamine receptors in the femoral vascular bed of the dog as mediators of a vasodilator and sympathoinhibitory effect. Europ. J. Cardiol. 42 (1977) 307-310
12. Leier, C. V., Heban, P. T., Huss, P., Bush, C. A., Lewis, R. P.: Comparative systemic and regional hemodynamic effects of dopamine and dobutamine in patients with cardiomyopathic heart failure. Circulation 58 (1978) 466-475
13. Loeb, H. A., Einslow, E. B., Rahimtoolla, S. H., Rosen, K. M., Gunnar, R. M.: Acute hemodynamic effects of dopamine in patients with shock. Circulation 154 (1971) 163-173
14. Maskin, C. S., Kugker, J., Sonnenblick, E. H., Lejemtel, T. H.: Acute inotropic stimulation with dopamine in severe congestive heart failure: Beneficial effect at rest but not during maximal exercise. Am. J. Cardiol. 52 (1983) 1028-1032
15. McCannell, K. L., Giraud, G. F., Hamilton, P. L., Groves, G.: Hemodynamic responses to dopamine and dobutamine infusion as a function of duration of infusion. Pharmacology 26 (1983) 29-39
16. McNay, J. L., McDonald, R. H., Goldberg, L. I.: Direct renal vasodilation produced by dopamine in the dog. Circul. Res. 16 (1965) 510-517
17. Mostbeck, A., Pratsch, H., Peschl, L.: Der Einfluß von Dopamin auf die regionalen Blutvolumina in Leber, Abdomen, Thorax und Extremitäten. Wiener Klin. Wschr. 89 (1977) 505-512
18. Mostbeck, A., Peschl, L., Schüller, J., Neumayr, A.: Simultane Untersuchungen von Funktionsparametern der Leber und der Niere unter dem Einfluß von Dopamin. Wiener Klin. Wschr. 87 (1975) 639-642
19. Ramdohr, B., Biamino, G., Schröder, R.: Vergleichende Untersuchungen über die Wirkung von Dopamin und Orciprenalin am gesunden Menschen: Muskeldurchblutung, Nierendurchblutung, Nierenfunktion. Klin. Wschr. 50 (1972) 1499-157
20. Ramdohr, B., Biamino, G., Schüren, K. P., Schröder, R.: Dopamin bei der Behandlung der akuten schweren Herzinsuffizienz. Med. Welt 26 (1975) 540-546
21. Scheoeder, W.: Die Messung des Sauerstoffdruckes in der Skelettmuskulatur - eine quantitative Methode zur Kontrolle der Sauerstoffversorgung und der Funktion der terminalen Muskelstrombahn. Herz/Kreisl. 10 (1978) 146-153
22. Talley, R. C., Forland, M., Beller, B.: Reversal of acute renal failure with a combination of intravenous dopamine and diuretics. Clin. Res. 18 (1970) 518-522
23. Yeh, B., McMay, J. L., Goldberg, L. I.: Attenuation of dopamine on renal and mesenteric vasodilation by haloperidol: evidence for specific dopamine receptor. J. Pharmacol. Exper. Ther. 168 (1969) 303-309

Changes in Muscle Tissue PO_2 Histograms of Patients with Arterial Occlusive Disease Following the Intravenous Injection of Calcium Dobesilate

A. M. Ehrly and H. Landgraf

Introduction

Dexium or Doxium (generic name calcium dobesilate; chemically calcium 2,5-dihydroxy benzenesulfonate) is a vasoactive compound, which until now has been used particularly in the treatment of diabetic retinopathy and venous diseases [1, 2, 3]. The sealing effect the drug is assumed to have on the capillaries also results in a reduction in capillary permeability in diabetics [4]. After there were indications of effectiveness in chronic arterial occlusive disease, Mörl and coworkers found a statistically significant increase in the walking performance in patients with intermittent claudication after treatment with Dexium [5].

Within the scope of an open investigation with patients suffering from chronic arterial occlusion stage IIb according to Fontaine (without diabetes mellitus), the muscle tissue oxygen pressure was measured before and after an acute intravenous injection of calcium dobesilate. Under resting conditions and in a steady state, an increase in the tissue oxygen pressure in the musculature is considered to be a sign of improved oxygen supply to the tissue [6].

Materials and Methods

Tissue oxygen pressure was measured with microplatinum needle electrodes in ten patients with proximal arterial occlusions or stenoses and clinical symptoms of severe intermittent claudication [7]. The tissue PO_2 was measured immediately before and 15, 30, 60, 90, and 120 min after in i. v. acute infusion (20 ml) of 500 mg calcium dobesilate via a motor pump. Besides the tissue oxygen pressure, the hemorrheologic parameters (hematocrit, whole blood viscosity, plasma viscosity, erythrocyte aggregation, and erythrocyte filterability) and the hemodynamic parameters (systemic blood pressure and pulse rate) were determined.

Table 1. Muscle oxygen partial pressure (Torr) before and after i.v. administration of 500 mg calcium dobesilate (Dexium, Doxium; Delalande, Cologne)

	Before		After				
	I	II	15 min	30 min	60 min	90 min	120 min
Median	8,3	7,8	8,1	8,7	9,7	9,2	11,7
Mean	11,0	9,6	10,8	12,2	13,0	12,8	14,2
SD	6,59	5,69	6,72	7,9	8,45	9,15	9,29
Wilcoxon test 2 α ≦				0,01			

Results

In ten patients with chronic arterial occlusive disease, there was initially a slight increase in tissue PO_2 after 30 min, starting from mean values of around 11 Torr. Table 1 shows that a statistically significant change can be observed 1 h after the injection (according to the Wilcoxon test). The PO_2 histogram at this time shows a shift to the right in favor of the higher PO_2 values. Among the rheologic parameters, the hematocrit decreased slightly (not significantly) and the whole blood viscosity dropped slightly but significantly toward the end of the investigation. The other hemorrheologic parameters did not change. The systemic blood pressure dropped slightly; the pulse rate (see Fig. 1) showed a marked and significant decrease.

Discussion

Following an acute intravenous injection of 500 mg calcium dobesilate in patients with intermittent claudication, there is a statistically significant increase in the tissue oxygen pressure with a maximum reached toward the end of the investigation period. It can be concluded that the oxygen supply to the tissue in these patients is improved. To date, no concrete data on the mechanism of this increase in tissue PO_2 have been provided. The statistically significant reduction in whole blood viscosity alone is too small to be responsible for the tissue PO_2 increase. We were not able to establish an improvement in the erythrocyte flexibility. It is theoretically feasible that the drug has a favorable effect on the maldistribution (impaired microcirculatory blood distribution [8]) in claudicants, so that a more homogeneous capillary perfusion may explain the rise in tissue PO_2. Of further significance is the fact that during the course of the investigation, the systolic blood pressure dropped by about 10 mm Hg, but the diastolic blood pressure remained practically at the same level. Normally, when the systolic blood pressure drops, the poststenotically measured muscle tissue PO_2 also decreases. In the case presented, the PO_2 values rose instead of falling, despite the drop in blood pressure.

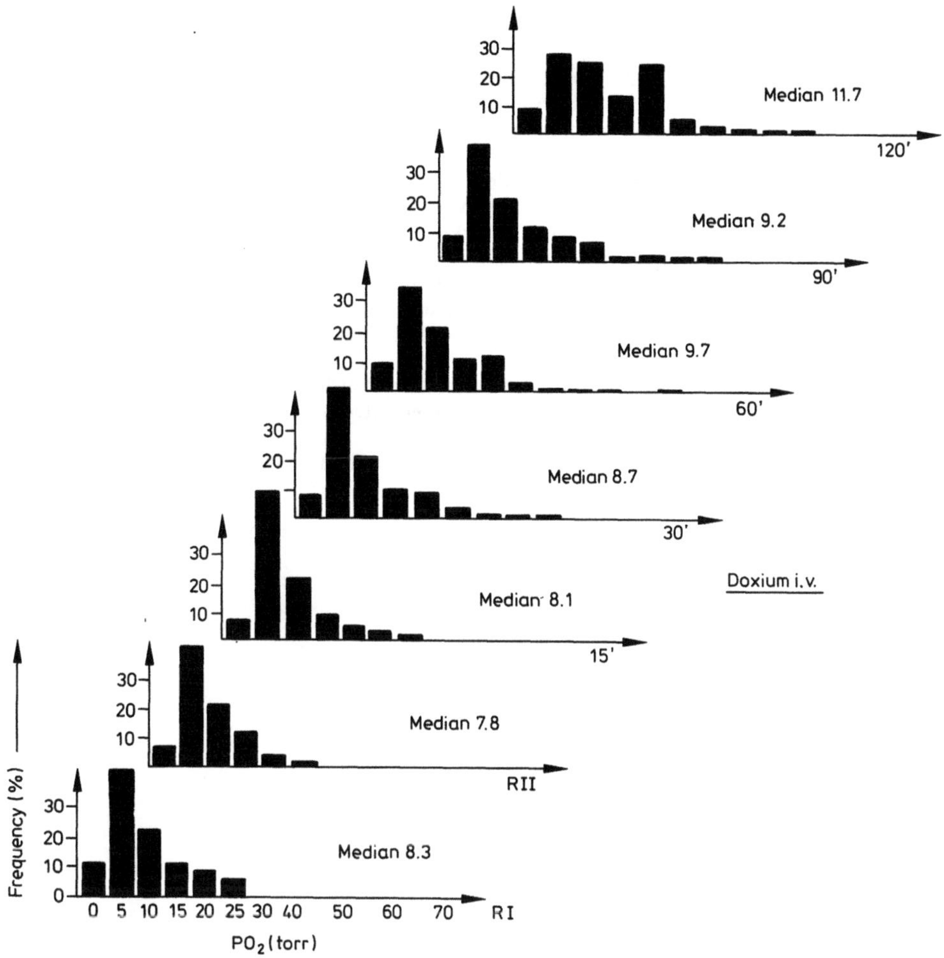

Fig. 1. Muscle tissue PO_2 histograms before *(R I/II)* and after i. v. injection of 500 mg calcium dobesilate

Summary

The intravenous injection of 500 mg calcium dobesilate in (non-diabetic) patients with arterial occlusive disease led to a statistically significant increase in the tissue oxygen pressure, with a maximum reached towards the end of the 2 h investigation. An increase in the tissue PO_2 under steady state conditions can be interpreted as a sign of improved oxygen supply to the tissue. This result is remarkable, since the tissue PO_2 rose despite a drop in the systolic blood pressure. There is no concrete evidence of the mechanism of this effect on the tissue PO_2. It is conceivable that the maldistribution in claudicants is favorably influenced.

References

1. Sevin, R., J. F. Cuendet: Calcium dobesilate in diabetic retinopathy. Ophthalmologica 159 (1969) pp 126-135, et Mémoires de la Société Française d'Ophthalmologie 82 (1969) pp 170-179
2. Berson, I.: A propos d'un nouveau veinotrope de synthèse. Praxis 37 (1970) pp 1305-1306
3. Sevin, R.: Treatments for diabetic retinopathy. Adv. Ophthalmol. 24 (1971) pp 315-374
4. Beyer, J., K. O. Meißner, J. Happ und U. Cordes: Wirkung von Calciumdobesilat auf die Permeation von Plasmaproteinen beim Diabetiker. Dt. Med. Wschr. 105 (1980) S. 1604-1608
5. Mörl, H., H. Hacker und H. C. Bierewitz: Behandlung arterieller Durchblutungsstörungen mit Dexium® 500. Therapiewoche 34 (1984) S. 2009-2014
6. Lübbers, D. W.: Grundlagen und Bedeutung der lokalen Sauerstoffdruckmessung und des pO_2-Histogramms für die Beurteilung der Sauerstoffversorgung der Organe. In: A. M. Ehrly (Hrsg.): Messung des Gewebesauerstoffdrucks bei Patienten. Witzstrock-Verlag, Baden-Baden (1981)
7. Ehrly, A. M., H.-J. Köhler, W. Schroeder, R. Müller: Sauerstoffdruckwerte im ischämischen Muskelgewebe von Patienten mit chronischen peripheren arteriellen Verschlußkrankheiten. Klin. Wschr. 53 (1975) S. 687
8. Ehrly, A. M.: New pathophysiological concept of ischemic disease: Microcirculatory Blood Maldistribution (MBM). Bibl. Anat. 20 (1980) p 456

Initial Results of Intracutaneous Measurements of PO_2 Profiles

H. Baumgärtl, A. M. Ehrly, K. Saeger-Lorenz, and D. W. Lübbers

Introduction

In order to understand transcutaneous PO_2 measurements, it is of importance to know the PO_2 profile in the epidermis and the cutis (stratum papillare corii) lying directly beneath it, since the O_2 exchange can be recognized directly from the PO_2 profile. Intracutaneous measurements, which up to now have been carried out only with relatively large polarographic electrodes, are not able to give information about PO_2 profiles [3, 4, 11]. Using a model, Grossmann and Lübbers tried to calculate the distribution of oxygen pressure within the skin, which is important for the gas exchange in transcutaneous PO_2 measurements, and, using simulation calculations, attempted to predict the behavior of $tcPO_2$ under different conditions [5, 6].

Figure 1 a gives a rough schematic presentation of the vessel distribution in the epidermis [7, 12]. The arteries can be seen ascending from the cutis and forming capillary loops in the stratum papillare corii. The upper part of the loop (capillary dome) supplies the epidermis with oxygen. The oxygen-poor blood (drawn in black) collects in the venous plexus and is transported through the veins. The model illustrated in Fig. 1 b of a capillary loop embedded in the surrounding tissue is used for simulation calculations. Figure 2 shows a PO_2 field calculated for this kind of capillary loop. The epidermis is in contact with ambient air (PO_2=21.3 kPa=160 Torr). The arterial PO_2 of the blood flowing into the capillary loop amounts to 13.3 kPa (100 Torr). The level above the area corresponds to the magnitude of the local PO_2. One can see how the oxygen pressure decreases along the capillaries.

The lowest point is reached just past the halfway mark of the arterial capillary limb at 7.3 kPa (54.8 Torr). Then the oxygen pressure increases again, as this part of the skin is supplied from the outside via the epidermis. In the capillary dome, the PO_2 amounts to 7.5 kPa (56.3 Torr). It then increases to 8.4 kPa (63 Torr). A marked increase takes place in the stratum corneum as a consequence of its high O_2 diffusion resistance. The resistance is particularly high in dry skin; it becomes lower when the skin is humidified, but nevertheless remains high.

Since oxygen flows from higher to lower oxygen pressures, the PO_2 profile can reveal how the individual layers of the skin are supplied with oxygen. Figure 2 shows that at moderate skin blood flow (0.01 ml/(g × min)), a major part of the outer layer of the skin is supplied from outside. When the skin is covered and made

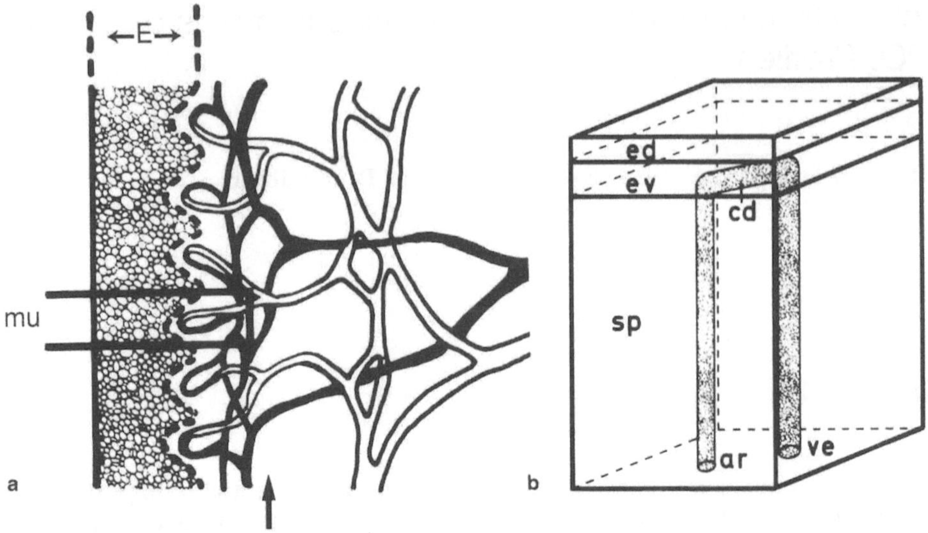

Fig. 1 a, b. Schematic drawing of the skin vasculature (capillary loop model according to Grossmann and Lübbers). **a** arteries ascending from the cutis, terminal arteries, and arterioles (vessels with margins) from capillary loops, out of which a venous network and draining veins (filled-in vessels) are derived. *E*, epidermis (only on the palms of the hand and soles of the feet it is as thick as in the drawing). The *arrow* represents a distance of 400 to 300 μm. **b** A piece of tissue with the dimensions 140 × 140 × 240 μm [6] as capillary loop model; *ar*, arterial limb of capillary (diameter 10 μm); *cd*, capillary dome; *ve*, venous limb of capillary (diameter 15 μm); *sp*, stratum papillare corii (thickness 200 μm); *ev*, viable part of epidermis (thickness 25 μm); *ed*, dead part of epidermis (thickness 15 μm) [5, 6]

gastight, this O_2 supply is cut off. The PO_2 drops continuously from the arterial vessels to the surface of the epidermis, where it reaches values close to zero.

Huch et al., to our knowledge, have been the only researchers to try to experimentally determine, although their method is an indirect one, the decrease in O_2 pressure between the surface of the epidermis and the capillary dome [8]. They placed a PO_2 electrode on the hyperemic surface of the skin (hyperemia was induced by the nicotinate compound Finalgon) of the forearm (Fig. 3.). While the volunteer breathed oxygen, the skin surface PO_2 values increased to more than 200 Torr. After circulation was stopped by an occlusive cuff, the PO_2 of the skin surface decreased linearly as a result of skin respiration. When the PO_2 reached values at which the chemically bound oxygen is released from the hemoglobin of the capillary blood, the linear curve bends. Since the PO_2 value at which this occurs can be taken from the oxygen dissociation curve of the blood, the PO_2 of the capillary blood can be determined by the course of the surface PO_2. This method, however, is not very reliable, because the point at which the curve no longer drops linearly can only be approximately determined. The difference between PO_2 in the capillary dome and on the surface of the skin is about 5 Torr [8]. The simulation calculations carried out on the covered epidermis yielded similar values.

In the following, attempts to measure the epidermal and intracutaneous PO_2 profile directly with PO_2 needle electrodes are discussed. It proved to be extremely

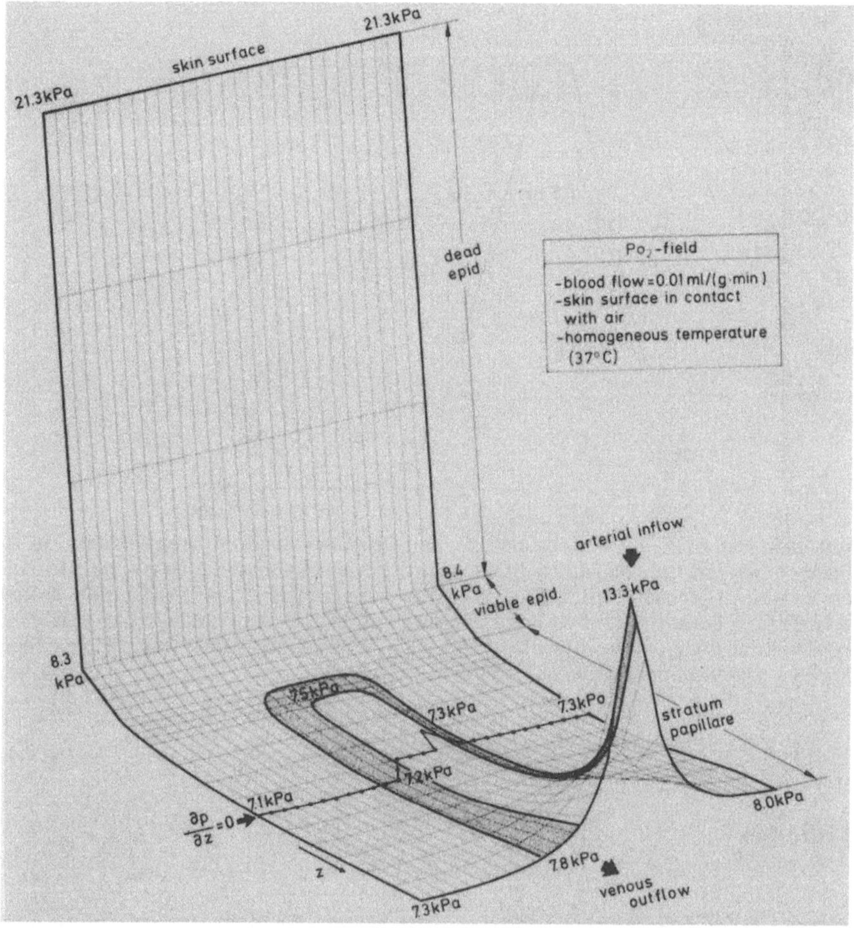

Fig. 2. PO_2 field in epidermis and stratum papillare corii when skin surface is in contact with air (capillary loop model). *z axis*, local tissue PO_2 (1 kPa = 7.5 Torr). On the skin surface, the tissue PO_2 is the same as the PO_2 of the air (21.3 kPa = 160 Torr). In the dead layer of the epidermis, the PO_2 decreases considerably as a result of the high diffusion resistance. Oxygen supply of the viable cell layer by diffusion through the epidermis layer with reduction in PO_2 from 8.4 kPa (63 Torr) to 7.5 kPa (56.3 Torr). In the area of the capillary dome, there is a further reduction in PO_2 as a sign of an oxygen supply from outside (minimum $\delta p/\delta z = 0$, marked by a *dotted line*). Blood inflow into the arterial capillaries has a PO_2 of 13.3 kPa (100 Torr). There is a distinct decrease in the PO_2 along the arterial capillary; continuous decrease in oxygen pressure lines between artery and vein is a sign of direct oxygen diffusion from the artery into the vein ("O_2-shunt diffusion"). PO_2 at the point of venous outflow is 7.3 kPa (54.8 Torr) [6]

difficult to penetrate the epidermis without either destroying the needle electrode or rendering it incapable of measuring. Therefore, we only had success in registering two PO_2 profiles. We believe, however, that in view of the basic importance of the problem, the temporary measuring data gathered and the experience gained in performing the measurements are worth discussion.

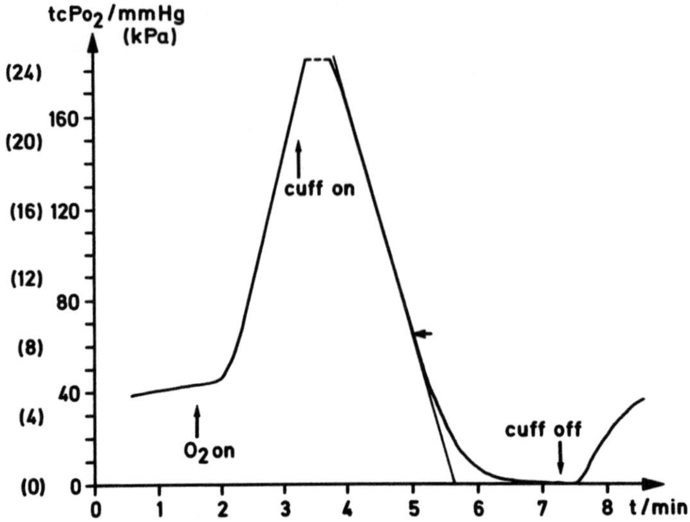

Fig. 3. Skin surface PO_2 on forearm and PO_2 in the capillary dome of the hyperemic skin. After inhalation of oxygen, the skin surface PO_2 increased to values above 200 Torr. After circulation was stopped with an occlusive cuff, the PO_2 first decreased linearly. *Arrows* mark visible deviations of the curve from linearity after 5 to 7 min as a result of the release of additional oxygen chemically bound to hemoglobin. In the presence of constant O_2 consumption, the PO_2 curve describes the O_2 dissociation curve of the capillary blood [8]

Methods

The PO_2 needle electrodes were used as described by Ehrly [2] and Baumgärtl and Lübbers [1]. The reference electrode was a broad-based Ag/AgCl electrode which was affixed to the surface of the skin. The needle electrode was inserted under microscopic supervision. Puncturing with the needle electrode caused deep dents in the skin. To reduce these dents, an approximately 10 × 10 mm large Lucite disk with a cone-shaped hole 1 mm in diameter in the center was fastened to the skin. Nevertheless, depressions continued to form, which caused the tip of the needle electrode to break. Needle electrodes with small diameters (0.2 to 1 μm) broke more easily than those with diameters up to 5 μm. Moreover, inserting the electrode often destroyed the electrode membrane, which made recalibration impossible. Puncturing was carried out manually with a nanostepper (Bachofer, Reutlingen). In most cases in this study, manual puncturing produced better results. It is most expedient to pierce the stratum corneum under microscopic control with a presharpened cannula tip. Unfortunately, however, the puncture is often so deep that bleeding results, which makes measurements impossible.

Because of the skin's elasticity, measuring the distance between the electrode tip and the skin surface is difficult and not very exact. Additional markings were put on the shaft of the electrode for orientation.

Of the many attempts, only two PO_2 profiles could be evaluated. They were obtained using the following technique:

1. Local affixation of the lower leg skin with a lucite plate (hole diameter 1 mm)
2. Air-saturated water drop placed on the skin
3. Prepuncturing with sharpened cannula
4. Deep insertion of needle electrode in prepierced opening
5. Slow, stepwise withdrawal with the nanostepper to skin surface

The depth of the puncture was calculated from the succession and the size of the individual puncturing steps.

Results

Figure 4 shows a copy of an original recording obtained by the stepwise withdrawal of the electrode. In the cutis, the PO_2 remained relatively constant between 40 and 50 Torr. At a depth of approximately 400 μm, the PO_2 toward the skin surface began to decrease and reached a value just below 20 Torr (the arrows mark points where the electrode was withdrawn). When the electrode came to about 200 μm from the skin surface, the size of the step was reduced to 12.5 μm. During these small steps the recording were less stable because the electrode moved slightly inside the skin between each step. This made it difficult to determine the size of the step and the distance from the skin surface, so that the figures given on the abscissa in this range can only be considered estimates. It can be clearly seen, however, that the PO_2, after having reached a low point at ca. 100 μm from the skin surface, begins to increase as the skin surface is approached, first gradually, then quickly. The reason why this value is lower than that of the air PO_2 is that the skin is covered by the water drop at the site of the puncture.

Figure 5 shows a second puncture through the same epidermis opening. This time, the PO_2 in the cutis is somewhat higher (between 50 and 60 Torr); the configuration of the PO_2 profile, however, is similar to that in Fig. 4. The PO_2 minimum occurs at a distance of 150 μm from the skin surface. The further PO_2 profile up to the surface is less stable, but principally it is similar to the PO_2 profile in Fig. 4.

Unfortunately, the measurements in this area are not precise, due to the unstable position of the electrode. One has to depend on the step sizes given by the nanostepper, as the position of the electrode tip in the skin cannot be unequivocally identified. The electrode can be bent in the area of the tip. Multiple punctures through the same opening in the epidermis show the same profile in the deeper region of the cutis. The measurements become less stable near the surface, probably because the opening in the epidermis is widened by puncturing with the electrode, thus causing the conditions of oxygen transport to change.

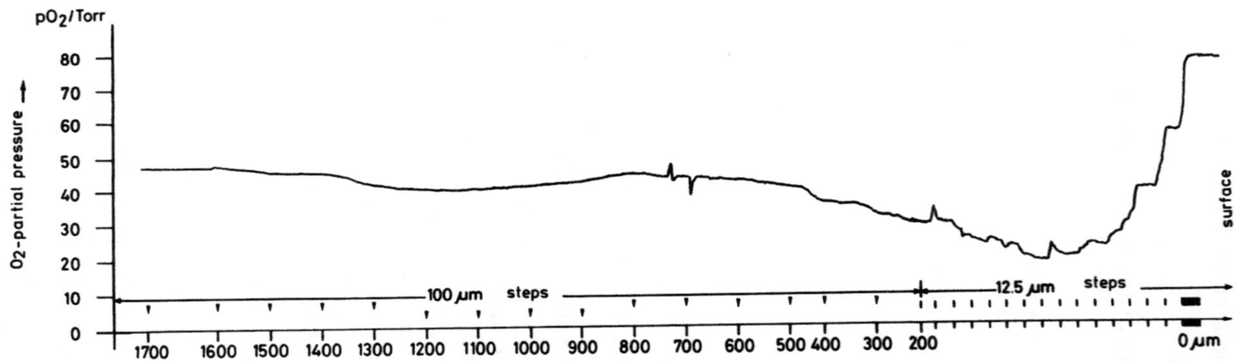

Fig. 4. Oxygen pressure profile in cutis and epidermis. The electrode was withdrawn between 1700 µm and 200 µm in steps of 100 µm. From 200 µm to the skin surface, steps were 12.5 µm (first puncture)

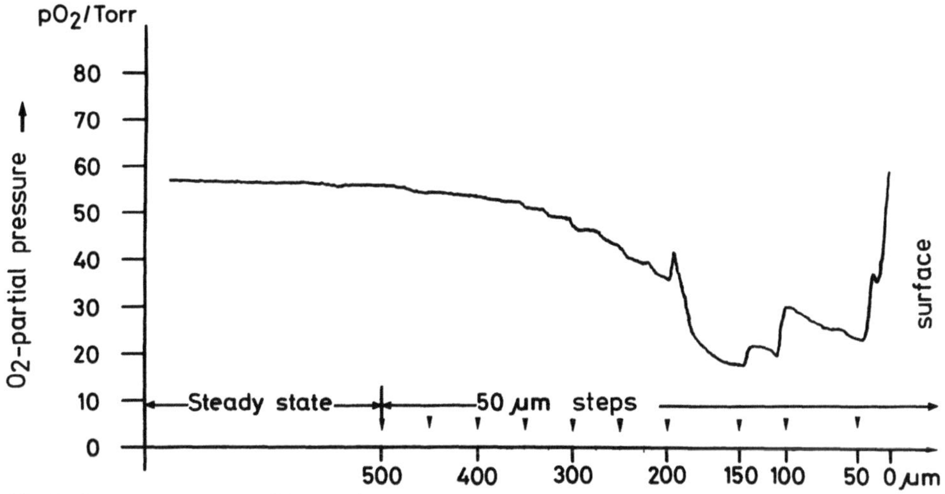

Fig. 5. Oxygen pressure profile in cutis and epidermis. The recording begins with a steady measurement at a depth of about 500 µm; from the *arrow* onward the electrode is pulled out in 50 µm steps. *Arrow tips* mark the 50-µm step (second puncture)

Discussion

Under our experimental conditions (normal skin circulation at ca. 22 °C) the PO_2 in the cutis lies well below that of the arterial blood: the PO_2 remains the same over a greater distance. The smaller cutaneous PO_2 could be due to the fact that in the cutaneous vessel networks, in which venous and arterial vessels lie close togehter [7], an O_2 shunt perfusion takes place. Another possibility is that the feeding vessels already contribute to the gas exchange in the subcutis. The constant PO_2 in the cutis is an indication that metabolism is low there and that the larger vessels participate in the gas exchange.

The decrease in PO_2 toward the skin surface begins at a depth of 400 to 300 µm. In the schematic drawing, this corresponds to a zone which lies just below the collecting venulae (Fig. 1, arrow). In this area of the stratum papillare corii, circulation at room temperature is set in such a way that the PO_2 along the capillary decreases rather sharply. A part of the stratum papillare corii and the epidermis are supplied with oxygen through the skin surface.

It is not likely that exact PO_2 profiles will be obtained with the technique described here. To be able to determine the details, a fold of skin will probably have to be pulled up and the electrode inserted in such a way that the PO_2 profile in the cutis and epidermis can be measured from inside to outside. In summary, it can be concluded from these initial measurements that the measured PO_2 profiles confirm our hypothesis about local circulation and O_2 transport conditions in the skin (epidermis and cutis).

References

1. Baumgärtl, H., D.W. Lübbers: Microcoaxial needle sensor for polarographic measurement of local O_2 pressure in the cellular range of living tissue. Its construction and properties. In: E. Gnaiger, A. Forstner (eds): Polarographic Oxygen Sensors. Springer-Verlag, Berlin/Heidelberg (1983) pp 37-65
2. Ehrly, A.M., H.-J. Köhler, W. Schroeder, R. Müller: Sauerstoffdruckwerte im ischämischen Muskelgewebe von Patienten mit chronischen peripheren arteriellen Verschlußkrankheiten. Klin. Wschr. 53 (1975) 687
3. Evans, N.T., P.F. Naylor: Steady state of oxygen tension in human dermis. Respir. Physiol. 2 (1966/67) 46-60
4. Gottrup, F., R. Firmin, N. Chang, W.H. Goodson III, T.K. Hunt: Continuous direct tissue oxygen tension measurement by a new method using an implantable silastic tonometer and oxygen polarography. Am. J. Surg. 146 (1983) 399-403
5. Grossmann, U.: Simulation of combined transfer of oxygen and heat through the skin using a capillary loop model. Math. Biosci. 61 (1982) 205-236
6. Grossmann, U., D.W. Lübbers: Analysis of epidermal oxygen supply by simulation of oxygen partial pressure fields under varying conditions. Crit. Care Med. 9 (1981) 734-735
7. Horstmann, E.: Blutgefäße der Haut. In: V. Möllendorf (Hrsg.): Handbuch der mikroskopischen Anatomie des Menschen (Bd. 3, 1). Springer-Verlag, Berlin (1957) S. 198-207
8. Huch, R., D.W. Lübbers, A. Huch: Quantitative continuous measurement of partial oxygen pressure on the skin of adults and newborn babies. Pflüg. Arch 337 (1972) 185-198
9. Lübbers, D.W.: Theoretical basis of transcutaneous blood gas measurements. Crit. Care Med. 9 (1981) 721-733
10. Lübbers, D.W., U. Grossmann: Gas exchange through the human epidermis as a basis of $tcpO_2$ and $tcpcO_2$ measurements. In: R. Huch, A. Huch (eds.): Continuous Transcutaneous Blood Gas Monitoring. M. Dekker, New York (1983) S. 1-34
11. Montgomery, H.: Oxygen tension of skin and muscle. Fed. Proc. 16 (1957) 697-699
12. Spalteholz, W.: Die Verteilung der Blutgefäße in der Haut. Arch. Anat. Physiol. (1983) 1-54

Transcutaneous Oxygen Pressure Measurements

Transcutaneous Oxygen Pressure Measurements in Arterial Occlusive Disease

D. Abendroth and L. Sunder-Plassmann

Introduction

The transcutaneous measurement of oxygen in patients with vascular disease seems to pose some problems, as the value measured gives completely different information than that obtained in measurements on healthy patients. This is especially true with regard to the hyperemia generated in the skin capillaries. In persons without vascular disease, this hyperemia causes so much oxygen to be delivered that the O_2 consumption of the skin layers between the capillary and the electrode can be ignored without the results being altered [1]. This results in a transcutaneous oxygen pressure which is comparable to the capillary PO_2.

In patients with vascular disease, however, proximal vascular occlusions restrict hyperemia and O_2 delivery so severely that owing to the O_2 consumption in the skin layers between these occlusions, the O_2 diffusion to the electrode is significantly reduced, and the electrode then measures a value lower than that which would correspond to the capillary. The absolute transcutaneous oxygen pressure therefore depends on the capillary PO_2, on the intensity of the hyperemia, and on the O_2 consumption in the skin layers [2].

If O_2 consumption and capillary PO_2 remain constant during the measuring procedure - as we assume they do in patients with vascular disease - the changes in the local hyperemia flow can be detected by measurement of the transcutaneous oxygen pressure [3]. Since O_2 consumption and diffusion properties of the skin are subject to considerable local fluctuation, very different values are obtained when the position of the electrode is shifted even just slightly. Transcutaneous oxygen measurements made in patients with stage IV occlusive disease with acral necrosis or gangrene are particularly revealing. Of particular interest are the measurements taken in the direct vicinity of the necrosis at the transition from the perfusion zone to the necrotic zone.

Method

We use a combination of transcutaneous oxygen pressure measurement and contact-free telethermography. The infrared thermography system provides digital re-

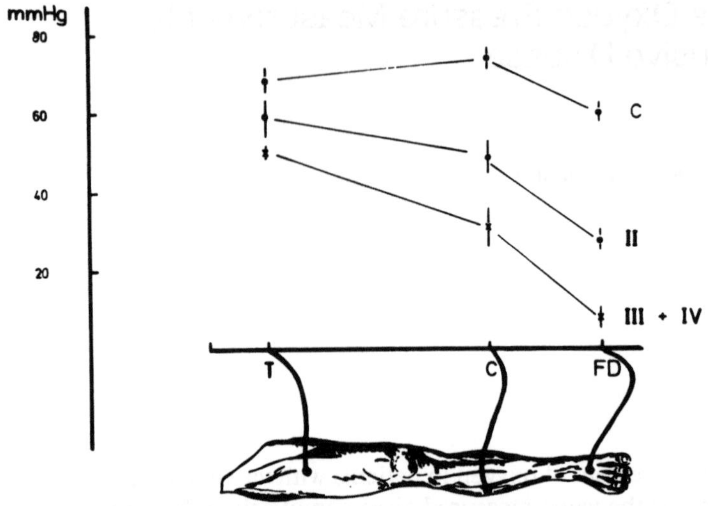

Fig. 1. Transcutaneous oxygen pressure along the lower extremity *T* (thigh; *C*, calf; *FD*, foot dorsum) in control persons *(C)* and patients with occlusive disease stages II, III, and IV [5]

cordings of temperature distributions and their time sequence with high resolution. By quantitatively evaluating heat recordings and sequences of recordings, as they are made after mandatory cooling, for example, one can draw conclusions about the irregularities in the tissue blood flow near the surface [4]. Thermography first determines the distal extent (usually to the forefoot) of the perfusion zone. The transcutaneous electrode is then placed in this distal perfusion zone and registers the drug-induced changes in oxygen pressure.

In contrast to investigation techniques such as Doppler pressure measurement or plethysmography, the $tcPO_2$ measurement, alone or preferably in combination with telethermography, detects even slight local changes in the microcirculation [5, 6, 7], as they can be effected by vasoactive drugs, for example (Fig. 1).

If the $tcPO_2$ is measured stepwise, beginning from the thigh over the calf to the back of the foot, the peripheral reduced flow can be clearly determined by the $tcPO_2$ profile. Under corresponding location of the occlusion, both correlate with the clinical symptoms (Fig. 2).

If the arterial occlusion is in both the angle of vision of the feeding transport artery, and is visible in the microcirculation of the muscles and the skin, additional therapeutic evidence can be obtained concerning the influence of the microvascular system. In this connection, indirect vascularization procedures are of interest, which only become effective secondarily as a result of induced collateral growth. The combination of profoundaplasty and lumbar sympathectomy is an example of this. The indication for lumbar sympathectomy combined with a measure to improve inflow arises from the desire for rapid growth of the collaterals distal from the level operated on [5]. The increase in skin blood flow achieved can be determined by the $tcPO_2$ measurement (Fig. 3).

The transcutaneous oxygen pressure measurement clearly shows the distal effect of the sympathectomy, which still cannot be measured by the Doppler method

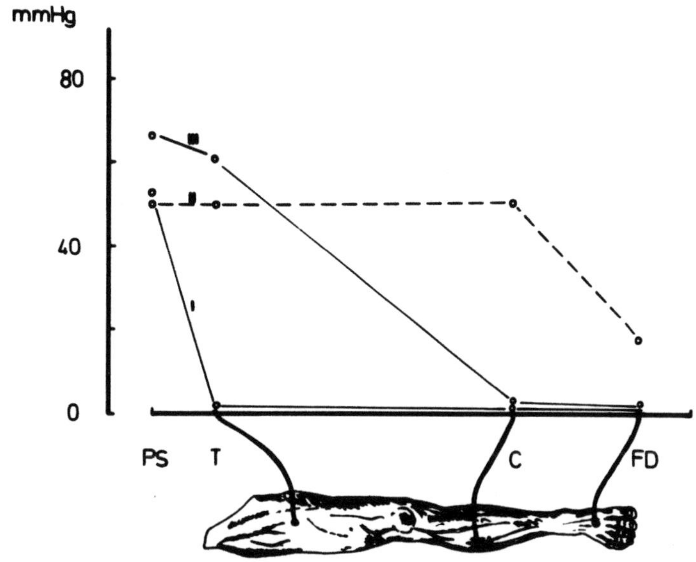

Fig. 2. Behavior of $tcPO_2$ using example of three different occlusion locations. *Curve I*, Leriche's syndrome with initial steep drop of transcutaneous pressure in thigh region; *curve II*, vascular occlusion in adductor canal; *curve III*, lower leg arterial occlusion [5]

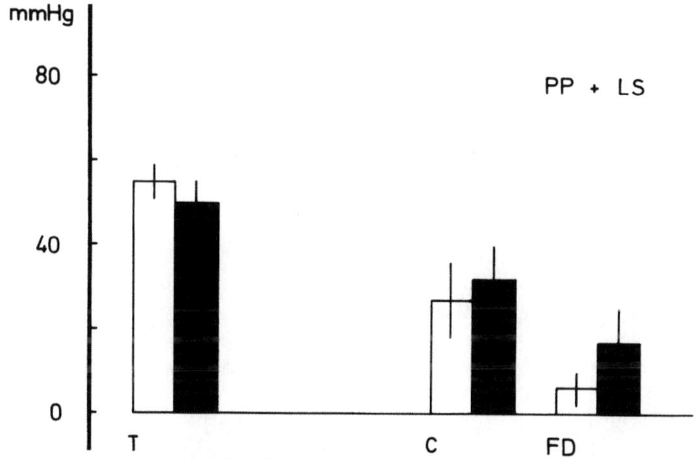

Fig. 3. Effect of lumbar sympathectomy prior to surgery *(light)* and after surgery *(dark)* on the $tcPO_2$ measured at levels *T*, thigh; *C*, calf; and *FD*, dorsum of foot ($n = 10$)

or plethysmography, for instance, at one week after surgery [6]. To be able to estimate the effect of the lumbar sympathectomy before the operation, we temporarily block the sympathetic trunk: using a peridural catheter, the sympathetic trunk is blocked with a drug (0.5% carbostesin 10 ml) at the L2-3 level.

Results

Figures 4–7 illustrate the recordings of transcutaneous oxygen pressure and the corresponding thermogram of an 80-year-old patient with stage III occlusive disease on the right; on the left temporary blockage of the sympathetic trunk prior to surgery is shown. Figure 7 shows the drug-induced improvement in microcircula-

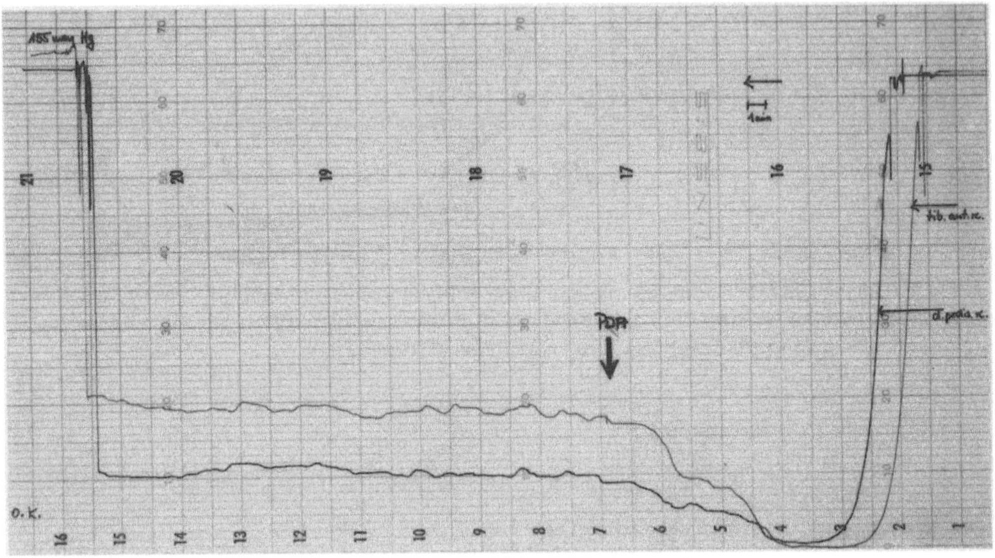

Fig. 4. Patient A: preoperative $tcPO_2$ curve during temporary sympathectomy

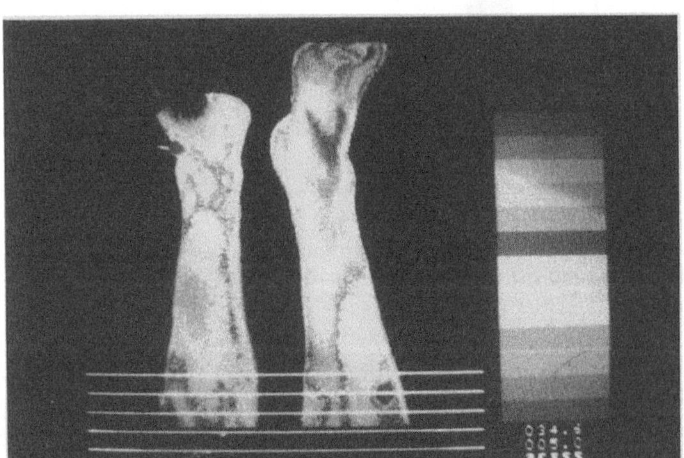

Fig. 5. Patient A: thigh/foot thermogram prior to sympathectomy; (*left* = right foot; all color thermograms shown here in black and white)

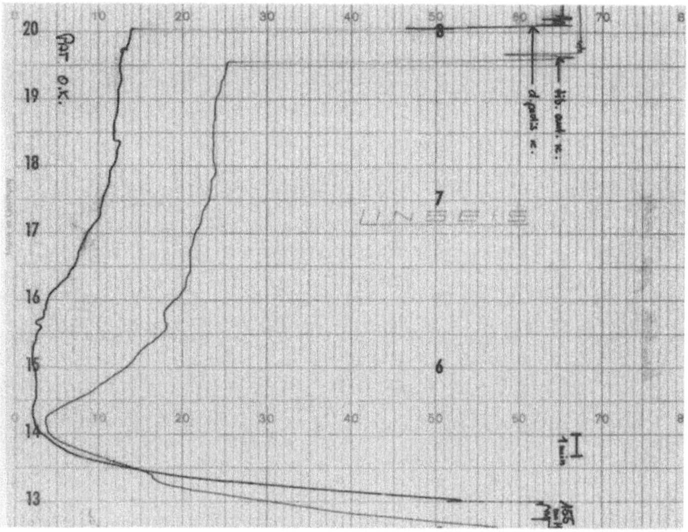

Fig. 6. Patient A: postoperative $tcPO_2$ curve after right-sided sympathectomy

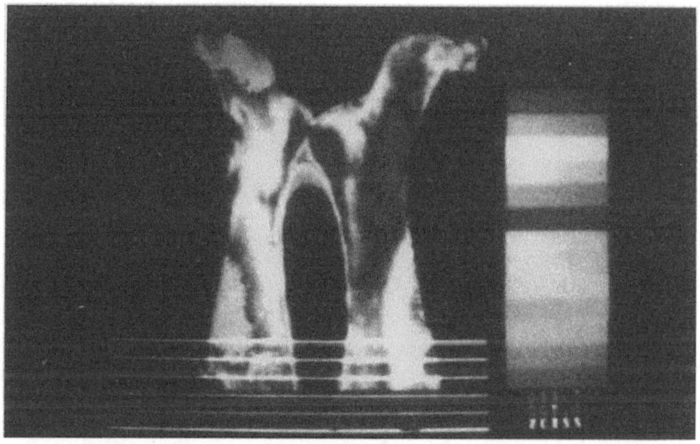

Fig. 7. Thermogram after sympathectomy

tion. Figures 8 to 11 demonstrate the anticipated effect, i. e., improvement in oxygen saturation and supply to the periphery. The patient was an 80-year-old woman with bilateral arteriosclerotic disease stage III–IV prior to an infusion of Buflomedil (Deutsche Abbott) 300 mg for 40 min on 5 consecutive days.

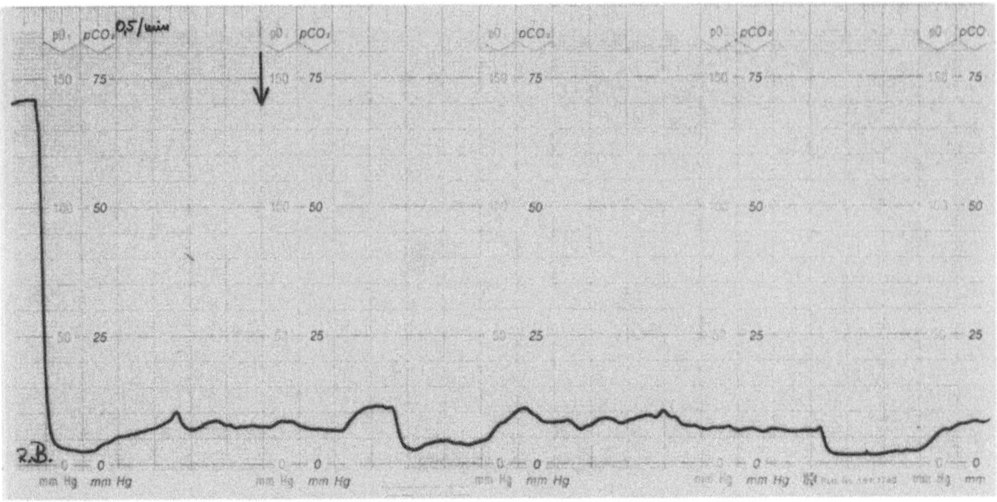

Fig. 8. Patient B: $tcPO_2$ findings before and during Buflomedil infusion

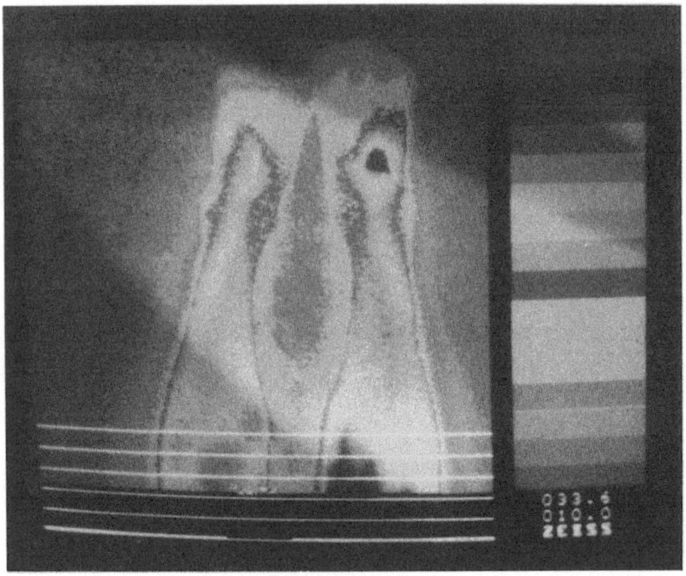

Fig. 9. Patient B: thermogram before Buflomedil infusion

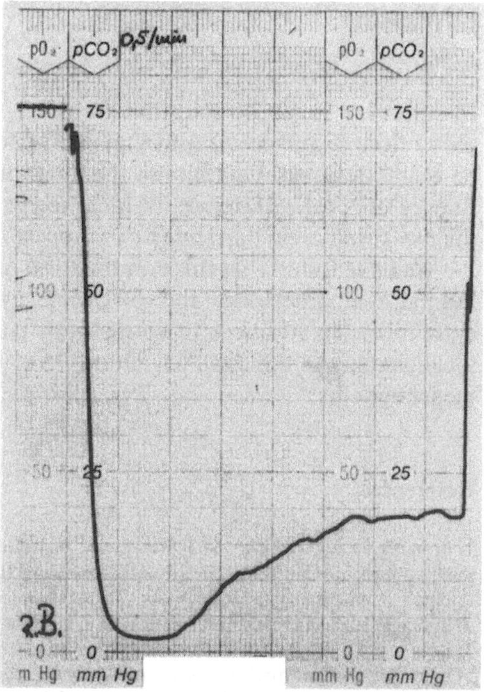

Fig. 10. Patient B: $tcPO_2$ curve after end of infusion therapy

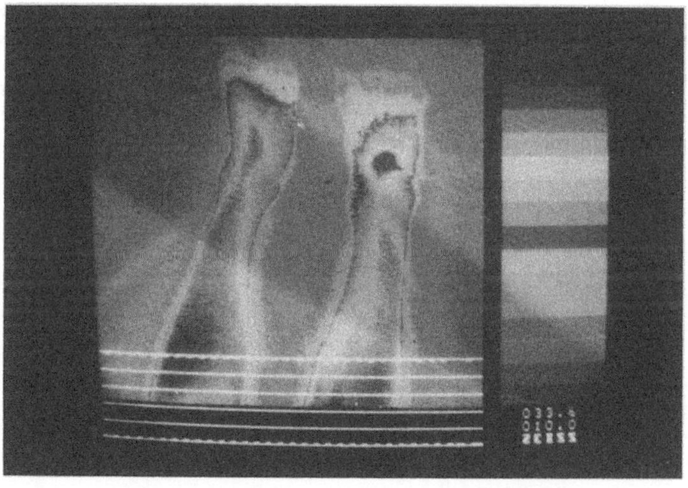

Fig. 11. Patient B: thermogram after end of infusion therapy

Discussion

In summary, it can be said that the $tcPO_2$ value corresponds to the severity of the blood flow disturbances and that subtle, local changes in the microcirculation can be easily detected. Combining this method with telethermography reduces the influence of local differences in local oxygen transport and oxygen consumption in the measured area, particularly in arterial occlusive disease.

We have found that this combination of methods has made it possible to predict lumbar sympathectomy and, of course, sympathetic trunk resection on the upper extremities. In addition, these methods enable the detection of the effect of subtle local changes in the capillary blood flow, which escape conventional pressure flow measurements.

References

1. Franzek U. K., P. Talke, E. F. Bernstein, F. L. Golbranson, A. Fronek: Transcutaneus pO_2 measurements in health and peripheral arterial occlusive disease surgery 91 (1982) 156
2. Huch P., A. Huch, D. W. Lübbers: Transcutaneous measurements of blood pO_2 ($tcpO_2$)-method and application in perinatal medicine J. perinat. Med. 1 (1973) 183
3. Lübbers, D. W., R. Huch, A. Huch: Unblutige kontinuierliche Erfassung des Säure-Basen-Status beim Menschen durch transcutane Messung des arteriellen pO_2 und $tcpO_2$ sowie der Hämoglobinkonzentration bei Körpertemperatur. Klin. Wschr. 51 (1973) 411
4. Hartmann, M., J.-M. Engel (Hrsg.): Thermologie-Fachberichte in nota-med, Baden-Baden (1980)
5. Sunder-Plassmann, L., K. Meßmer, H. M. Becker: Tissue pO_2 and Transcutaneous pO_2 as Guidelines in Experimental and Clinical Drug Evaluation. Angiology 32 (1981) 686
6. Meßmer, K., B. Fagrell (Hrsg): Kongreß über Mikrozirkulation und arterielle Verschlußkrankheit. Fortschr. Diagn. Ther., Karger-Verlag, Basel (1981)
7. Sunder-Plassmann, L.: Chirurgisch-therapeutisches Vorgehen bei Verschlußkrankheit der unteren Extremität. In: G. Trübestein (Hrsg): Arterielle Verschlußkrankheit und tiefe Venenthrombose. Thieme-Verlag, Stuttgart/New York (1984)

Transcutaneous Oxygen Pressure Measurements with an Electrode Core Temperature of 37 °C

A. Creutzig, D. Dau, and K. Alexander

Introduction

The oxygen pressure of the healthy skin fluctuates between 0 and 4 mm Hg [5]. The application of hyperemizating ointment or heat causes skin oxygen pressure to rise; maximum hyperemia results at temperatures above 40 °C, while the blood flow remains constant [8]. A further increase in blood flow, such as one induced by drugs, cannot be expected, just as a reactive hyperemia cannot be measured. In earlier investigations, a uniform decline in $tcPO_2$ was observed when an electrode was used at 44 °C in patients with arterial occlusive disease under an intra-arterial infusion of various drugs. This we interpreted as a steal phenomenon [2]. If, however, electrode core temperatures between 37 °C and 39 °C are selected, changes in the cutaneous blood flow can be detected [6]. In the following, the results of $tcPO_2$ investigations using an electrode core temperature of 37 °C are reported.

Patients and Methods

The skin oxygen pressure in 18 healthy volunteers and in 25 patients with arterial occlusions of the lower extremities was determined with a commercially available measuring system (TCM 2, Radiometer). The occlusive disease had been documented arteriographically in 18 patients. Three of the patients had been undergoing hospital treatment for severely shortened pain-free walking distances (Fontaine stage II b), one patient for rest pain (stage III), and 21 for trophic lesions (stage IV). Eight of the patients were treated for diabetes mellitus with insulin or tablets. Three patients had occlusions at three levels; a combination of thigh and lower leg occlusions was established in nine patients. Eleven patients had only main occlusions in the lower leg, and four patients suffered a predominantly acral type of occlusion.

The systolic brachiopedal pressure difference was between 0 and 115 mm Hg. Measurements were carried out on the medial thigh, on the medial calf, on the forefoot, and on the medial thigh of the opposite leg. Further details of the method are given elsewhere [3].

The healthy volunteers were given infusions of prostaglandin E_1. Patients with arterial occlusive disease received the following intra-arterial infusions: 50 ml dextran (50 min), xantinol nicotinate (600 mg ad 50 ml 0.9% NaCl/50 min), 10 ml of a mixture of adenosine triphosphate (ATP) and adenosine diphosphate (ADP), as well as other nucleotides and nucleosides (ad 50 ml 0.9% NaCl/50 min), and prostaglandin E_1 in doses between 0.1, and 8.0 ng/kg body weight per minute (ad 50 ml 0.9% NaCl).

Results

By nature, the absolute PO_2 values are much lower at a hyperemic temperature of 37 °C than at an electrode core temperature of 44 °C and are usually below 10 mm Hg. They are especially dependent on the properties of the skin, such as the thickness of the epidermis. If the skin oxygen pressure of a healthy person is measured 100 times at various points on the thigh in a 10 × 10 cm area under constant measuring conditions, values between 0 and 11 mm Hg are registered (Fig. 1).

After the venous outflow is occluded in healthy persons, the skin oxygen pressure drops significantly (Fig. 2). After arterial blockage, the $tcPO_2$ (37 °C) decreases significantly, and reactive hyperemia results when the blood flow is restored. The values measured in the thigh and the calf of the test persons prior to the infusion (5.0 mm Hg) do not significantly differ from the values derived in prestenotic regions in the patients (6.6 mm Hg), however, they do differ from those measured in poststenotic areas (2.9 mm Hg, $P<0.01$). A comparison of the prestenotic and

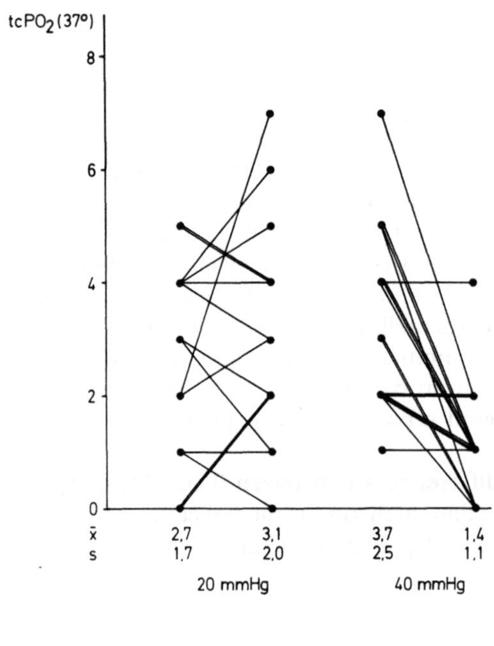

Fig. 1. Transcutaneous oxygen pressure histogram of the thigh of a healthy test person. (All the figures are based on a temperature of 37 °C)

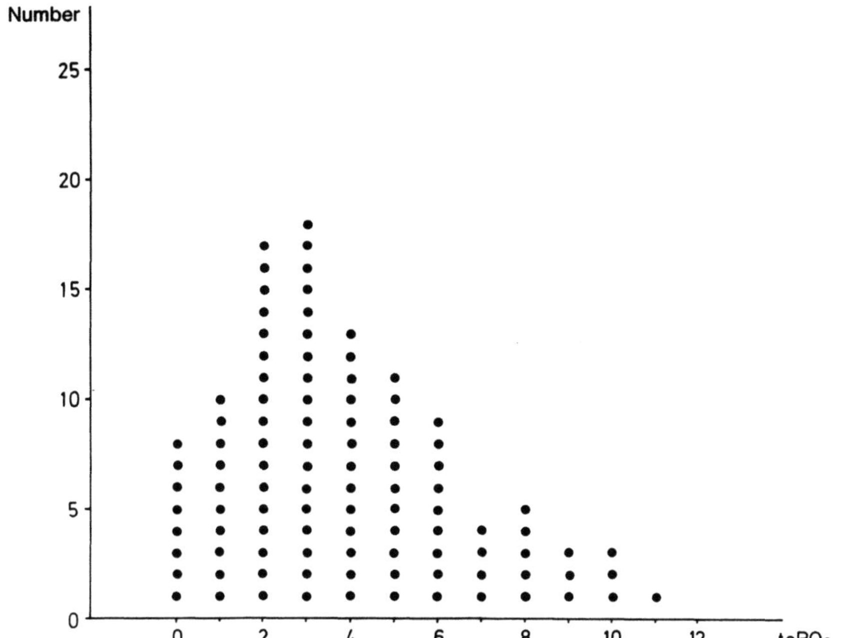

Fig. 2. Influence of compression (20 and 40 mm Hg) of the calf of a volunteer with a cuff on the $tcPO_2$ of the forefoot. Compression with 40 mm Hg results in significant reduction in $tcPO_2$ ($P<$ 0.001)

poststenotic values between diabetics and nondiabetics shows that the prestenotic regions do not significantly differ. In the poststenotic region, however, the diabetics have a significantly higher $tcPO_2$ at 37 °C: 4.0 vs 2.1 mm Hg ($P<0.005$).

During the intra-arterial infusion of xantinol nicotinate, the $tcPO_2$ rose significantly at the prestenotic measuring points from 6.6 to 10.2 mm Hg and then dropped back to 7.5 mm Hg after the infusion had been stopped. In the poststenotic regions, on the other hand, there was a significant decrease from 4.3 to 3.5 mm Hg during the infusion; when the infusion was discontinued, the $tcPO_2$ dropped further to 2.6 mm Hg. During the infusion of the ATP/ADP mixture, there was no change in the prestenotic region: however, in the poststenotic area, the $tcPO_2$ pressure decreased from 2.8 to 1.2 mm Hg. When the infusion was stopped, the oxygen pressure rose to 4.8 mm Hg and was thus significantly higher than the initial PO_2 ($P<$ 0.05).

The intra-arterial infusion of dextran 40 did not cause any measurable changes in the $tcPO_2$ in either the prestenotic or the poststenotic areas.

The intra-arterial infusion of prostaglandin E_1 in normal test persons resulted in a two- to threefold increase in the $tcPO_2$ in both the thigh and the calf (thigh: before infusion, 5.5 mm Hg, during infusion, 11.1, after infusion, 4.6 mm Hg; calf: before infusion, 5.1, during infusion, 16.6, after infusion, 7.0 mm Hg). The $tcPO_2$ increase was linearly dependent on the applied dose. In the patients with arterial occlusive disease, the infusion of prostaglandin E_1 also led to a dose-dependent increase in

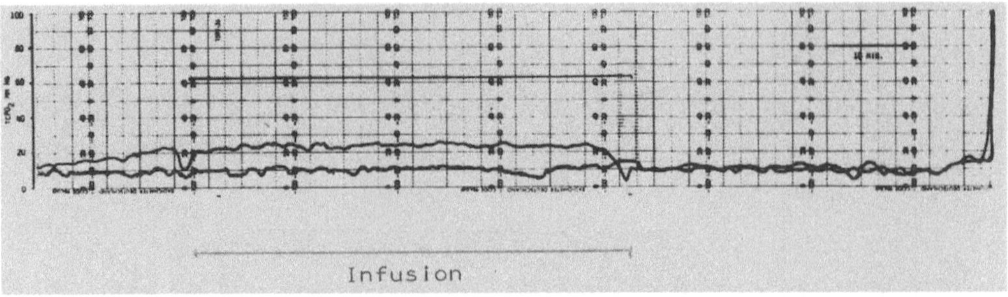

Fig. 3. Transcutaneous oxygen pressure measurement on the thigh and calf of a patient with peripheral arterial occlusive disease (to be read from right to left). Immediately after intra-arterial infusion of prostaglandin E_1 begins, $tcPO_2$ in the thigh increases and the calf shows no change. After the infusion is stopped, the $tcPO_2$ in the thigh decreases slowly

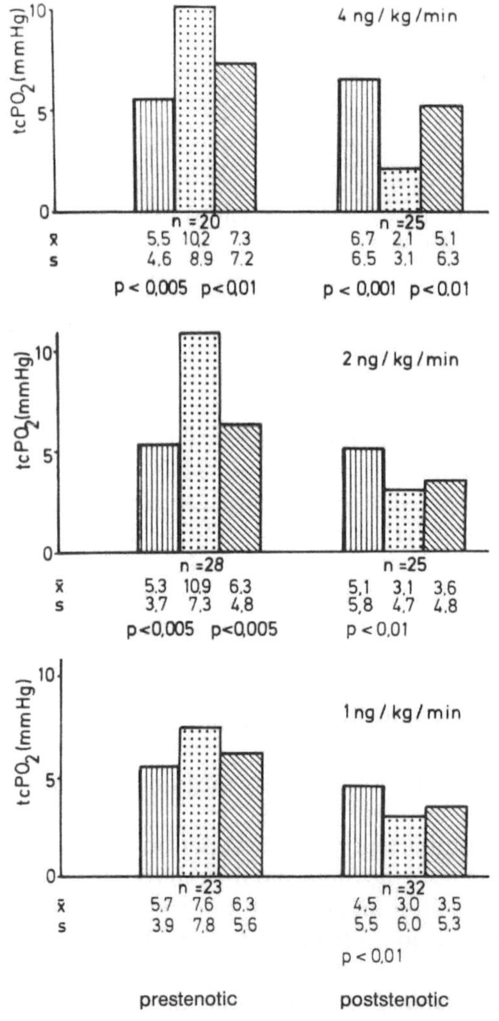

Fig. 4. $tcPO_2$ before, during, and after intra-arterial prostaglandin E_1 infusion at various doses. At every dose, there is an increase prestenotically and decrease poststenotically

$tcPO_2$ (37 °C), which, however, could only be confirmed for the prestenotic areas. Poststenotically, however, there was a significant decrease (Figs. 3, 4).

Discussion

The oxygen pressure of the skin is subject to considerable intraindividual fluctuation, which is largely caused by the varied local blood supply and the diffusion conditions. To measure pharmacologic effects on the skin blood flow, it therefore seems best to carry out continuous recording at one measuring point. For this reason, it is difficult to compare individual measurements of skin oxygen pressure for the purpose, for example, of judging the course of a therapy at various points in time. Changes in $tcPO_2$ at 37 °C reflect changes in the cutaneous blood flow if one assumes that the other factors influencing this value, such as arterial PO_2, blood pressure, local oxygen consumption, and diffusion conditions are not altered by the intra-arterial infusion. The arterial blood pressure dropped under xantinol nicotinate and initially under prostaglandin E_1, so that the $tcPO_2$ under infusion should drop slightly or at least not increase as much. The favorable effect of both drugs on the skin blood flow is well known [7, 9] and correlates with the increase in $tcPO_2$ in the prestenotic areas. Poststenotically, however, $tcPO_2$ dropped significantly under both drugs. Here, a reverse distribution from poststenotic into prestenotic regions or from the skin into the musculature apparently plays a role. Other authors also found no increase or decrease in skin circulation under the application of xantinol nicotinate in patients with advanced stages of arterial occlusive disease [1]. It should be noted, however, that the two drugs induce completely different changes in the oxygen pressure of the muscle: under xantinol nicotinate, the histogram of the m. gastrocnemius does not change, while under prostaglandin E_1 there is a marked increase in muscular PO_2 [4]. The intra-arterial application of dextran 40 apparently has no measurable effect on the circulation of the skin.

We have found that the continuous measurement of the $tcPO_2$ at a temperature of 37 °C is suitable for the evaluation of pharmacologic effects on the skin blood flow. Although the skin blood flow cannot be quantified, relative data can be obtained.

References

1. Bollinger, A., E. Lüthy: Kompensationsgrad arterieller Verschlüsse und Wirkung intravenös verabreichter vasoaktiver Medikamente. Schweiz. med. Wschr. 97 (1967) 1220-1225
2. Creutzig, A., K. Alexander: Die Bestimmung des transkutanen Sauerstoffdruckes der durchblutungsgestörten Extremität bei konservativer Therapie der AVK. Dt. Ges. Inn. Med. 89 (1983) 536-540
3. Creutzig, A., D. Dau, K. Alexander: Transkutaner Sauerstoffdruck während intraarterieller Infusionen. VASA 13 (1984) 207-212
4. Creutzig, A., M. Lux, K. Alexander: Muscle tissue oxygen pressure fields and transcutaneous oxygen pressure in healthy men during intra-arterial Prostaglandin E_1 infusion. Intern. Angiol. 3 (1984) 105-110

5. Evans, N. T. S., P. F. D. Naylor: The systemic oxygen supply to the surface of human skin. Resp. Physiol. 3 (1967) 21–37
6. Ewald, U., G. Rooth, T. Tuvemo: Postischemic hyperaemia studied with a transcutaneous oxygen electrode used at 33–37 °C. Scand. J. Clin. Lab. Invest. 41 (1981) 641–645
7. Gottstein, U., R. Felix, H. D. Flad, I. Sedlmayr: Untersuchungen zur Wirkung von Nikotinsäure und Adenosinmonophosphat auf Haut- und Muskeldurchblutung von Gefäßgesunden und Kranken mit peripheren Durchblutungsstörungen. Z. Kreisl. Forsch. 55 (1966) 970–987
8. Huch, R., A. Huch, D. W. Lübbers: Transcutaneous pO_2. Thieme-Verlag, Stuttgart (1981)
9. Nielsen, P. E., S. L. Nielsen, P. Holstein, H. L. Poulsen, E. H. Hansen, N. A. Lassen: Intraarterial infusion of Prostaglandin E_1 in normal subjects and patients with peripheral arterial disease. Scand. J. Clin. Lab. Invest. 36 (1976) 633–640

Control of the Conservative Therapy of Arterial Occlusive Disease by Means of Transcutaneous PO_2 Measurement[1]

J. M. Steinacker, F. Brock, R. E. Wodick, and F. Nobbe

Introduction

The transcutaneous method of measuring PO_2 according to Huch et al. [5] is based on the fact that the consumption of oxygen of the skin and the electrode is considerably less than the amount of oxygen supplied by the blood in a heat-induced hyperemic area of the skin. If an equilibration between O_2 consumption and O_2 transport is established after the heating-up period, the transcutaneous PO_2 ($tcPO_2$) follows the fluctuations of the arterial PO_2. In this state, the measuring system electrode/skin/capillary is not dependent on small fluctuations of the arterial and capillary blood flow, since the total flow far exceeds the necessary volume (excessive hyperemia according to Lübbers and Großmann [7]; Fig. 1).

As long as the circulation is not disturbed, this measuring system can be used to continuously check the course of the arterial PO_2. This has been reported from

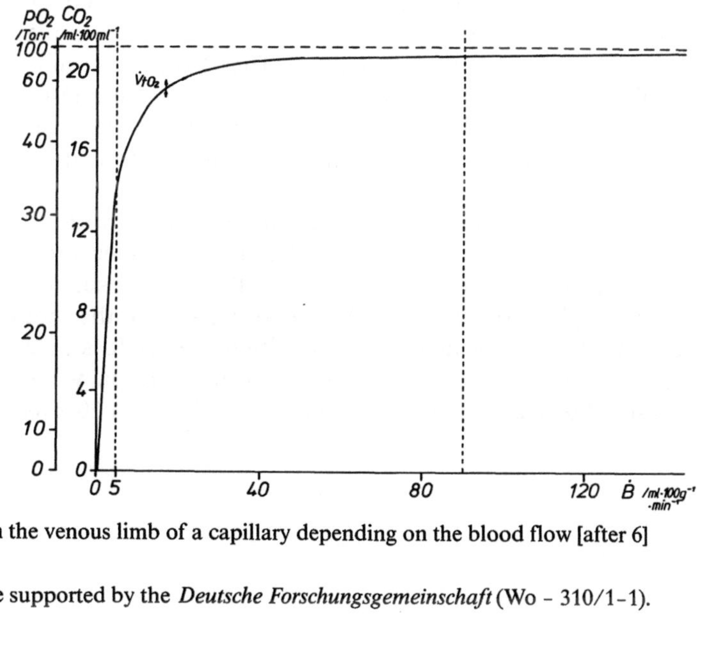

Fig. 1. PaO_2 and C_VO_2 in the venous limb of a capillary depending on the blood flow [after 6]

[1] Parts of this study were supported by the *Deutsche Forschungsgemeinschaft* (Wo - 310/1-1).

studies in perinatal medicine [5, 6], but measurements can also be carried out successfully in adults. There is, however, a difference from the arterial PO_2 values, depending on the skin site and the electrode [6, 11, 12].

Transcutaneous PO_2 can also be an aid in diagnosing a circulatory disturbance in an area of skin (Fig. 1). The transcutaneously measured values give an indication of the severity of the vascular disease [2, 3, 10]. The measurements enable a prognosis to be made concerning the affected extremity's capability for survival, and correlate well with the clinical results of amputations [1] and angioplastic reconstruction [10].

Considerable progress has been made in the conservative therapy of arterial occlusive disease over the past few years [8, 9]. Apart from rheologic treatment, thrombolytic therapy of vascular occlusions [2, 9] and dilatation of vascular stenoses [4, 9] have been carried out successfully. Doppler sonographic measurement of foot artery blood pressure and of the pain-free walking distance, and venous occlusion plethysmography are used to control the success of therapy, although specific sources of error are associated with each method [8, 9].

The transcutaneous measurement of PO_2 has not been widely used as a therapy control. The present study deals with questions concerning this technique.

Methods

The $tcPO_2$ is determined on skin in which hyperemia has been induced with a measuring electrode at a temperature of 45 °C and is registered with a pen (Electrode: Biochem) [5]. The electrodes are positioned laterally and medially on the dorsum of the foot and on the calf. The skin is stripped of cutaneous scales with adhesive tape, and oil is removed with benzine. The electrode is placed on the skin with a drop of water as a contact medium. The PO_2 is determined after 20 min or after a stable value has been reached, whichever is the latest. At the same time, the PaO_2 in a sample of capillary blood taken from the hyperemic ear lobe is measured in a blood gas analyzer and the transcutaneous index $tcPO_2/PaO_2$ is calculated.

Examinations were carried out on six patients with arterial occlusive disease stage IIb-IV of the lower extremity. Ten legs were measured before and 3 days after percutaneous transluminal angioplasty (PTA) of the superficial femoral artery with a Grüntzig balloon catheter. The blood pressure in the foot arteries foot was measured by Doppler sonography: the difference $\bar{\Delta}RR$ is determined from the mean difference from the arterial blood pressure. The pain-free walking distance was measured on a treadmill ergometer at 3 km/h with a 10% ascent.

Results

Figure 2 shows the mean values of the $tcPO_2$ values obtained from 17 measuring sites on the forefoot and on the lower leg before ($\bar{x}$ = 44.8 mm Hg) and after PTA (56.9 mm Hg). Despite successful treatment, only slight changes in the values can be observed in individual measurements. The course of the mean blood pressure difference $\bar{\Delta}$RR is given in Fig. 3 (before PTA, 0.59 mm Hg; after PTA, 0.79 mm Hg). The $\bar{\Delta}$RR decreases from 87 mm Hg before treatment to 37 mm Hg after treatment. The mean walking distance increases from 145 m to 250 m.

Discussion

If the arterial inflow into a vascular region is restricted by a stenosis of a previously occluded artery, the flow of an excessive hyperemia will hardly be reached by the measuring system, since most of the dependent vascular regions below the stenosis are dilated as a result of hypoxia. This is the case in the flow diagram in Fig. 1 in the area of the ascending limb of the curve, where the venous PO_2 is dependent on changes in arterial flow. Therefore, the more severe the restriction of the arterial inflow, the lower the $tcPO_2$ is (Fig. 1). For this reason, transcutaneous measurements can be used to differentiate between areas in danger of becoming ischemic and regions with sufficient oxygen supply. The PaO_2 represented in Fig. 1 does not correspond exactly to the $tcPO_2$: the electrode causes the gas pressure to rise as a result of

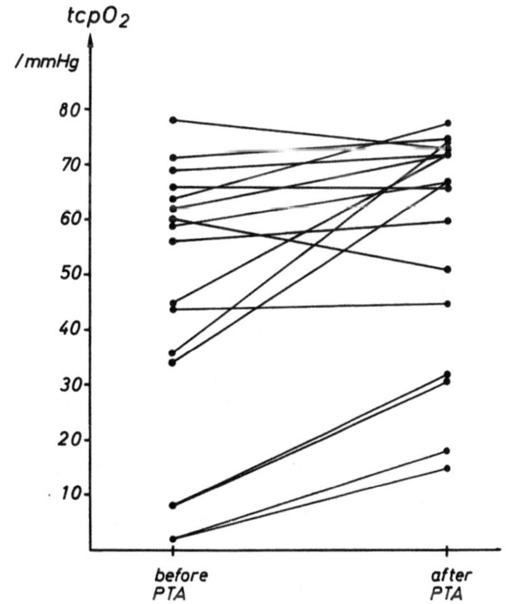

Fig. 2. Transcutaneous PO_2 before and after PTA in ten treated legs

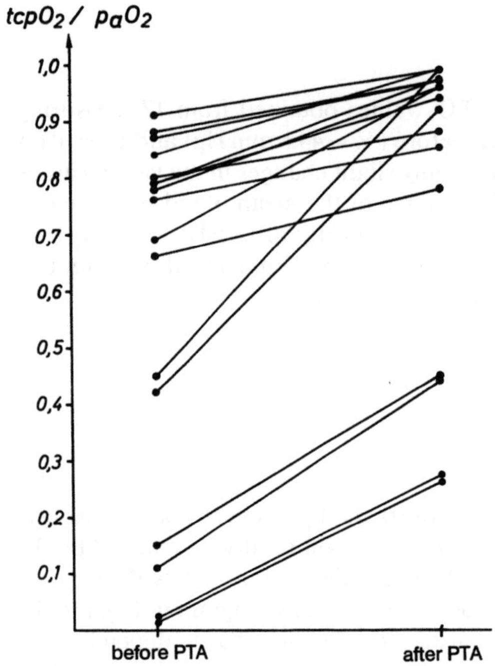

Fig. 3. Transcutaneous index $tcPO_2/PaO_2$ before and after PTA on ten treated legs

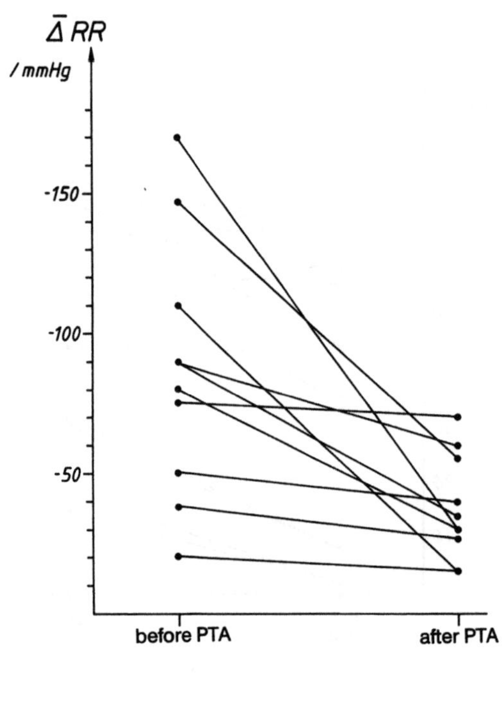

Fig. 4. Doppler sonographically measured mean difference $\bar{\Delta}$RR of systemic and peripheral blood pressure in ten legs before and after PTA

the heating and, because of its oxygen consumption, effects a diffusion gradient in the skin [7]. Thus a $tcPO_2$ value of zero does not necessarily mean anoxia of the skin. The $tcPO_2$ is therefore a special value, which, when determined on the "test organ - skin" (Lübbers), provides information on the circulation of an extremity [13].

While marked improvement in the very low $tcPO_2$ can be observed after successful PTA in patients with clinical stages III-IV, only slight changes are found in patients with lower clinical stages and higher $tcPO_2$ values. In one case, the $tcPO_2$ even declines somewhat (Fig. 2), which was probably effected by altered arterial PO_2 (Fig. 3).

The success of the PTA is documented angiographically and by the blood pressure difference $\bar{\Delta}$RR. The clinical improvement is demonstrated by the improved walking distance. If the arterial PO_2 is determined along with the $tcPO_2$, introducing the transcutaneous index $tcPO_2/PaO_2$ makes the measurement more reliable. This index increases following PTA, even when the initial values were good, and is an indication of improved circulation (Fig. 3). When the blood flow is increased, the $tcPO_2$ is essentially determined by the arterial PO_2; this is not as obvious when circulation is low, because here even small flow alterations bring about changes in the level of $tcPO_2$ [6, 7, 12, 13]. In our study, we examined patients whose arteries were compressible for Doppler sonographic flow measurements. In others, Doppler flow measurements are inadequate [8, 9].

The transcutaneous measurement of PO_2 can be an important supplement to the established angiologic examination procedures. It offers a noninvasive control of the course of therapy for peripheral arterial occlusive disease. The significance of the results presented above requires further verification.

References

1. Brock, F. E., F. Nobbe, P. Hausding: Ergebnisse der lokalen Katheterlyse mit Urokinase bei chronisch arteriellen Verschlüssen älterer Patienten. Verh. 3. Tagung Angiol. Ges. Deutschland/Österreich/Schweiz, Wien 1984
2. Burgess, E. M., F. A. Matsen, C. R. Wyss, C. W. Simons: Segmental Transcutaneous measurements of PO_2 in Patients Requiring Belowthe-Knee Amputation for Peripheral Vasular Insufficiency. J. Bone Joint Surg. 64 (1982) 378-382
3. Franzeck, U. K., P. Talke, E. F. Bernstein, F. L. Golbranson, A. Fronek: Transcutaneous PO_2-Measurements in Health and Arterial Occlusive Disease. Surgery 91 (1982) 156-163
4. Grüntzig, A., H. Hopf: Perkutane Rekanalisation chronischer arterieller Verschlüsse mit einem neuen Dilatationskatheter: Modifikation der Dotter-Technik. Dtsch. Med. Wschr. 99 (1974) 2502-2505
5. Huch, R., A. Huch, D. W. Lübbers: Transcutaneous Measurement of Blood PO_2 ($tcPO_2$) - Method and Application in Perinatal Medicine. J. Perninat. Med. 1 (1973) 183-191
6. Huch, R., A. Huch, D. W. Lübbers: Transcutaneous PO_2. Thieme-Verlag Stuttgart/New York (1983)
7. Lübbers, D. W., U. Großmann: Gas Exchange through Human Epidermis as a Basis of $tcPO_2$ and $tcPCO_2$ Measurements. In: R. Huch, A. Huch: Continuous Transcutaneous Blood Gas Monitoring. M. Deccer, Basel/New York (1983)
8. Mannick, J. A.: Evaluation of Lower-Extremity Ischemia. New Engl. J. Med. 309 (1983) pp 841-843
9. Nobbe, F., F. E. Brock, G. Rudowsky: Neue Aspekte in der konservativen angiologischen Therapie. Verh. Jubil. Sympos. Ulm 1984, 97-120

10. Oghi, S., K. Ito, T. Mori: Quantitative Evaluation of the Skin Circulation in Ischemic Legs by Transcutaneous Measurement of Oxygen Tension. Angiology 32 (1981) pp 833-839
11. Steinacker, j. M., R. E. Wodick: Die transkutane Messung des arteriellen Blutes beim Erwachsenen: Forderungen an eine geeignete Elektrode. Biomed. Techn. 26, Suppl. (1981) S. 6
12. Steinacker, J. M., R. E. Wodick: Transcutaneous PO_2 During Exercise. In: Oxygen Transport to Tissue (VI). Plenum Press, New York (1984) pp 763-774
13. Steinacker, J. M., W. Spittelmeister, R. E. Wodick, F. Fallenstein, D. W. Lübbers: Examinations of the dependence of transcutaneous PO_2 on cutaneous blood flow. Pflügers Arch 405 (1985) Suppl. 2, p. R63

Combined Capillary Microscopy and Transcutaneous PO_2 Measurements in Patients with Chronic Venous Insufficiency

U. K. Franzeck, A. Bollinger, R. Huch, and A. Huch

Introduction

Intravital microscopic examinations have been able to demonstrate that in advanced and severe stages of chronic venous insufficiency (CVI), capillary changes are brought about by venous hypertension [3]. In the centers of white atrophy spots, which are frequently found in CVI, this can result in the formation of avascular zones [1].

The present study was performed to investigate to what extent microangiopathy in CVI patients leads to an insufficient oxygen supply to the skin.

So far it was impossible to examine capillary morphology, capillary density, or flow dynamics during transcutaneous PO_2 ($tcPO_2$) measurements. The newly developed transparent PO_2 electrode [6] enables $tcPO_2$ measurements and studies of the nutritive skin capillaries to be carried out at the same time and at the same measuring site with the aid of videomicroscopy. This combined examination procedure was applied for the first time in patients with chronic venous insufficiency.

Patients and Methods

The CVI group comprised 17 patients (six women, 11 men) with a mean age of 56 (22–76), in whom 44 measurements were carried out (Table 1). Five patients had primary varicosities (insufficiency of perforating veins) and 12 had secondary varicosities (partial obstruction and/or insufficiency of perforating veins due to deep venous thrombosis). Nine patients had trophic lesions in the medial region of the lower leg, but without ulcers, and eight patients had crural ulcers or ulcers that had healed. In none of the patients evidence of peripheral arterial occlusive disease could be established. The pulse and the light plethysmographic (big toe) findings were normal; no vessel murmurs could be auscultated. The control group consists of 17 men and seven women with a mean age of 48 (19–75) years on whom 35 measurements were made.

The principle and the technique of the $tcPO_2$ measurement are discussed in detail elsewhere [7]. The transparent oxygen electrode used has also already been de-

Fig. 1. Prototype of transparent transcutaneous oxygen electrode (view from the skin side). In the middle the glass part with the platinum electrode

scribed [6]. Figure 1 shows a prototype: a platinum cathode with a tip diameter of 15 μm is melted in a glass cylinder. The electrode is covered with a 25-μm Teflon membrane, which produces only a negligible optic barrier. The video microscopy system [4] consists of an incident light microscope, a video camera, a television monitor, and a tape recorder (Fig. 2).

Simultaneous $tcPO_2$ measurements and videomicroscopic recordings were carried out at various locations on the lower leg. The morphology revealed was characteristic of CVI: skin sites above the insufficient perforating veins of the medial ankle region with no marked trophic disturbances of the skin, hyperpigmentation areas, border zones of venous ulcers and white atrophy spots, scar tissue of healed ulcers, and central regions of white atrophy spots.

Results

The mean $tcPO_2$ values of the healthy volunteers (35 measurements) on the lower leg was 56.8 ± 9.9 (37–79) mm Hg. The superficial (papillary) capillaries of the skin are distributed evenly and have a mean density of 45 per mm^2. In contrast to the nailfold, where the capillary loops run parallel to the skin surface, only the dome of the capillary can be seen on the lower leg.

The mean $tcPO_2$ in patients with CVI was 47.7 ± 14.5 (21–83) mm Hg when measurements were made above the insufficient perforating veins and the skin trophism

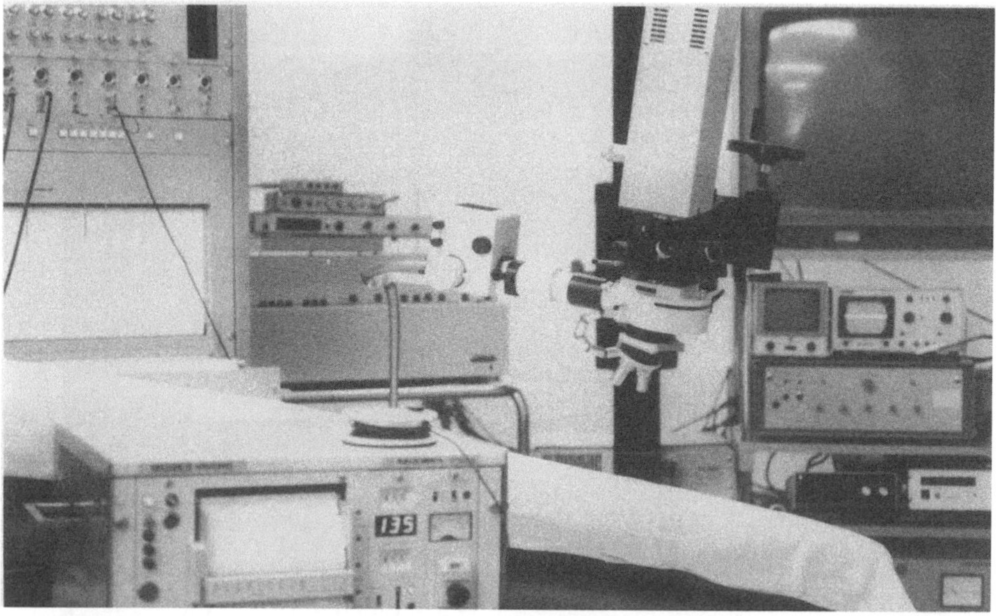

Fig. 2. Equipment for $tcPO_2$ measurements with simultaneous videomicroscopy. In front, the oxymonitor; in the middle, the incident light microscope with television camera; in background, the television monitor and the tape recorder. For details see [3]

was not substantially disturbed. A significant difference in comparison with the healthy volunteers could not be confirmed ($P > 0.05$). Likewise, the mean capillary density of 40 per mm^2 showed no difference to the normal findings ($P > 0.05$). Frequently, however, the capillaries were dilated and surrounded by a capillary halo (Fig. 3). The $tcPO_2$ measured above skin areas with hyperpigmentation, hyperkeratosis, and induration was reduced ($P < 0.001$), and similarly, the capillary density was less than 10 per mm^2. The morphological changes in the capillaries increased proportionally to the decrease in the capillary density. The remaining capillaries were extremely dilated and had pathologically shaped arterial and venous capillary limbs. The mean $tcPO_2$ was 22.5 ± 7.0 (13–36) mm Hg at these measuring sites.

There are hardly any more capillaries to be seen in the center of white atrophy spots, and the $tcPO_2$ is at zero and no longer measurable. The papillary capillaries are greatly proliferated at the borders of the venous ulcers and of the white atrophy spots, and some of them have a glomerulus appearance. The mean $tcPO_2$ is reduced (24.0 mm Hg), as is the capillary density. The range of distribution (12–44 mm Hg) is a result of the various distances of the platinum cathode (O_2 sensor) from the adjacent capillaries, which is due to the inhomogenous distribution of the capillaries and their reduced number. The measuring distance is of significant importance for the $tcPO_2$ measurement, as a long distance causes low PO_2 pressure and a short distance higher PO_2 values in CVI-patients.

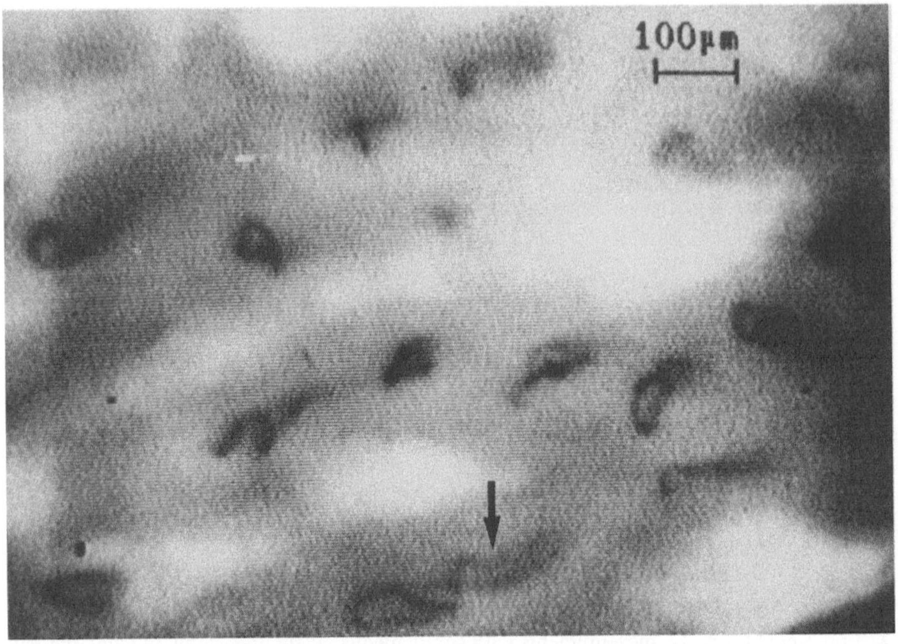

Fig. 3. Capillary microscopy in a CVI patient above a skin area with no marked trophic lesions (single frame photo from television monitor); view through the transparent oxygen electrode. The *arrow* marks the tip of the O_2 sensor; $tcPO_2$ = 73 mm Hg at normal capillary density and only slightly increased capillary coiling

Discussion

The results of the investigation show that microangiopathy in patients with advanced and severe stages of chronic venous insufficiency [1, 3] results in reduced $tcPO_2$ values in the medial region of the lower leg. This is corroborated by findings of other investigations [2, 8]; however, no data have been provided on the role of the microvessels. The use of a transparent $tcPO_2$ electrode allows a correlation between the capillary morphology and capillary density and the transcutaneous PO_2 values of the measuring site. The results of this study demonstrate the close dependency of $tcPO_2$ on the capillary density, which in the case of CVI is very variable: normal to subnormal values are measured above the skin areas where there is no capillary rarefication, reduced values above dilated and increased coiled capillary loops, and $tcPO_2$ of about 0 mm Hg over avascular areas. The local oxygen supply to the skin can thus be transcutaneously quantified. However, the diffusion distance above skin areas with induration and hyperkeratosis, might be enlarged, which represents an additional factor responsible for the reduced oxygen pressure. An attempt to diminish this effect was made by stripping the skin several times.

The extremely low PO_2 values measured above avascular areas show that CVI results in local hypoxia. The pronounced proliferation of the capillaries at the transition zone may represent an attempt to compensate for the reduced capillary density. Nevertheless, $tcPO_2$ values are measured which before were only obtained in stages III and IV of peripheral arterial occlusive disease [5].

Hypoxia occurring with CVI, however, is a locally very confined phenomenon which is caused by microangiopathy resulting from venous hypertension. In arterial occlusive diseases, larger skin areas such as the foot are affected, and the low oxygen pressure and the restricted oxygen supply is a result of reduced perfusion.

References

1. Bollinger, A., K. Jäger, A. Geser, F. Sgier, Seglias: Transcapillary and interstitial diffusion of Na-Fluorescein in chronic venous insufficiency with white atrophy. Int. J. Microcirc., Clin. Exp. 1 (1982) 5
2. Borzykowski, M., B. Krähenbühl: Mesure noninvasive de l'oxygenation cutanée en cas d'ulceres chroniques des membres inférieurs. Schw. med. Wschr. 111 (1981) 1972
3. Fagrell, B.: Local microcirculation in chronic venous incompetence and leg ulcers. Vasc. Surg. 13 (1979) 217
4. Franzeck, U. K., G. Isenring, J. Frey, K. Jäger, F. Mahler, A. Bollinger: Eine Apparatur zur dynamischen intravitalen Videomikroskopie. VASA 12 (1983) 233
5. Franzeck, U. K., P. Talke, E. F. Bernstein, F. L. Golbranson, A. Fronek: Transcutaneous PO_2-measurements in health and peripheral arterial occlusive disease. Surgery 91 (1982) 156
6. Huch, A., U. K. Franzeck, R. Huch, A. Bollinger: A transparent transcutaneous oxygen electrode for simultaneous studies of the skin capillary morphology, flow dynamics and oxygenation. Int. J. Microcirc. Clin. Exp. 2 (1983) 103
7. Huch, R.: Transkutaner PO_2: Prinzip, Handhabung, klinische Erfahrung und Grenzen der Methode. In A. M. Ehrly, J. Hauss, R. Huch: Klinische Sauerstoffdruckmessung. MWP-Verlag, München (1985) S. 53-59
8. Neumann, H. A. M., P. J. M. Berrety, M. v. Leuwen, M. J. v. d. Broek: Transcutaneous PO_2 measurement in chronic venous insufficiency syndrome. Microvasc. Res. 26 (1983) 262

Transcutaneous PO_2 Measurements in the Surrounding of Venous and Neurotrophic Ulcers

H. Partsch

Introduction

Transcutaneously measured partial oxygen pressure ($tcPO_2$) is dependent on the systemic oxygen pressure in the arterial blood, the local perfusion pressure, the capillary density, and the local metabolic consumption. When the uptake of oxygen is not disturbed and its transport through the blood stream is not hindered, the $tcPO_2$ apparently then has something to do with the nutrition of the tissue. It was therefore decided to perform $tcPO_2$ measurements in patients with venous and neurotrophic ulcers, i.e., substance defects of the skin which cannot be expained by local ischemia alone.

Patients

Forty-nine patients had a venous ulcer of the lower leg (post-thrombotic syndrome or varicosis with incompetence of the saphenous and perforating veins). Fifteen patients (16 legs) had neurotrophic ulcers on the parts of the sole of the foot bearing the most weight. The causes of the peripheral sensory neuropathy present in every case were the following: diabetes (five cases), alcohol (five cases), diabetes plus alcohol (two cases), hereditary sensory neuropathy (two cases), post-traumatic nerve damage (one case) [4].

Doppler sonography and electronic oscillography ruled out any relevant blood stream hindrances in the transport arteries of the legs.

Methods

The $tcPO_2$ measurement in patients with venous ulcers was carried out at the border of the ulcer. Patients with neurotrophic ulcers (as well as with arterial gangrene) were measured on the dorsum of the foot with a TCM 1 oxygen monitor (Radiometer, Copenhagen) while the electrode was heated to 44 °C. After a plateau had developed, the patient inhaled pure oxygen for 3 min. Because of the ulcer, determination of the reoxygenation time after 3 min of arterial blockage was often omitted.

To establish a standard value for the local blood flow in the upper skin layers, a laser-Doppler examination was carried out in most patients before and after direct heating of the measuring site to 40 °C (Periflux) [3].

Results

In Table 1, the results of the $tcPO_2$ measurements in patients with venous and neurotrophic ulcers are compared with those obtained in healthy subjects (measured on lower leg) and in patients with arterial occlusive disease stage IV (measured on foot). In the last group, a distinction is made between patients with and without diabetes.

In comparison with the healthy control group, statistically significantly reduced values were obtained in patients with venous ulcers and arterial gangrene without diabetes. Figure 1 gives a frequency analysis of the values measured in each case shown in the histograms. In the case of venous ulcers, 63% of the values obtained are below the critical range of 20 mm Hg, 93% for nondiabetic arterial gangrene. Among the neurotrophic ulcers, however, there was no value below 20 mm Hg; 94% of all the values were above 40 mm Hg. Table 2 shows the $tcPO_2$ values after 3 min pure oxygen inhalation.

Figure 1 shows that in comparison with the healthy volunteers, the $tcPO_2$ values after the oxygen inhalation remain significantly reduced only in patients with arterial gangrene. In the groups with venous and neurotrophic ulcers, the values rise markedly.

The laser-Doppler examinations (Fig. 2) yielded significantly higher initial values of the resting blood flow in patients with venous and neurotrophic ulcers than in the healthy group. Direct heating does not effect any further blood flow increase in those patients with venous ulcers. In the control group and in the patients with neurotrophic ulcers, the laser-Doppler initial values rose.

Table 1. The $tcPO_2$ values under resting conditions (supine) in various groups of patients

	Healthy controls	Venous ulcers	Neurotrophic ulcers	AOD IV without diabetes	AOD IV with diabetes
	Calf		Dorsum of foot		
n	10	49	16	28	13
Mean ± SD (mm Hg)	59.6 ± 11.2	19.7 ± 18.4	54.7 ± 12.6	7.5 ± 6.9	23.5 ± 20.8
*P**		< 0.05	n. s.	< 0.05	n. s.

* Kruskal-Wallis test (see also Table 2)

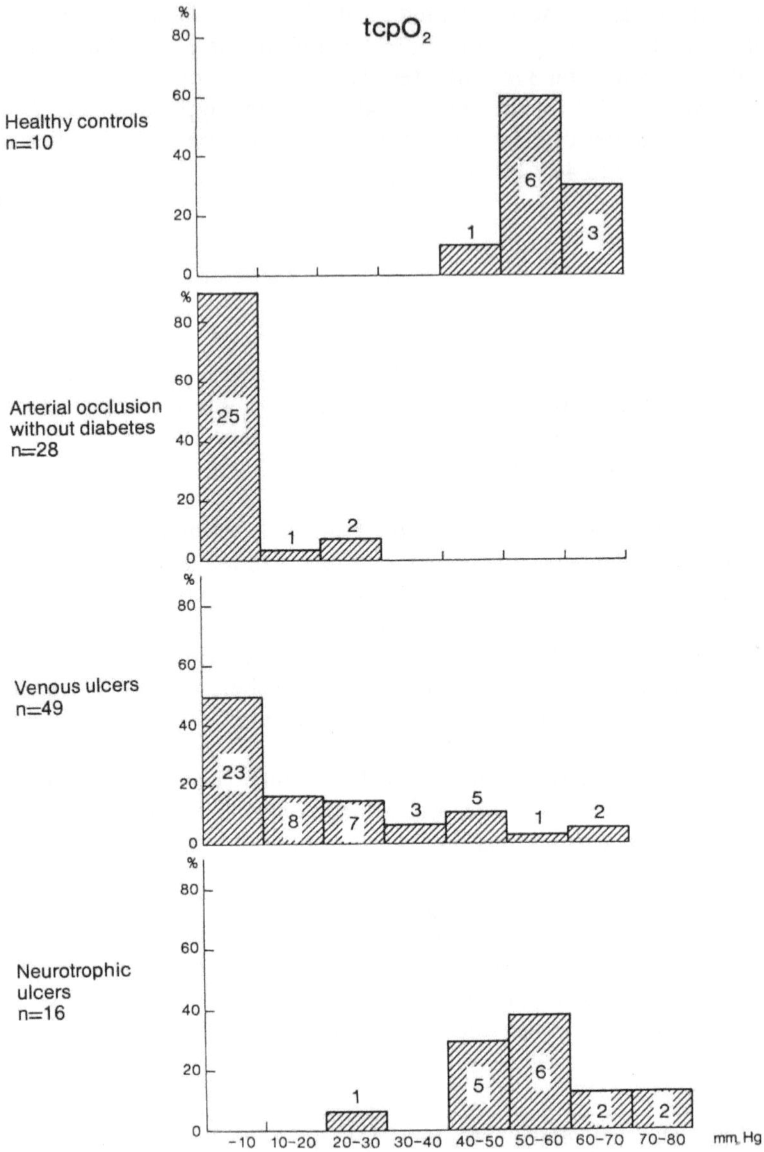

Fig. 1. Frequency analysis of the $tcPO_2$ values in the vicinity of the ulcers. Measurement on lower leg in healthy volunteers and patients with venous ulceration, on the foot of patients with arterial occlusion stage IV and with neurotrophic ulcers

Table 2. The $tcPO_2$ values after 3 min oxygen inhalation

	Healthy controls	Venous ulcers	Neurotrophic ulcers	AOD IV without diabetes	AOD IV with diabetes
n	10	49	16	27	13
Mean ± SD (mm Hg)	197.5±61.1	123.8±97.8	135.7±70.7	17.6±25.2	87.6±94.1
P		n. s.	n. s.	<0.01	<0.05

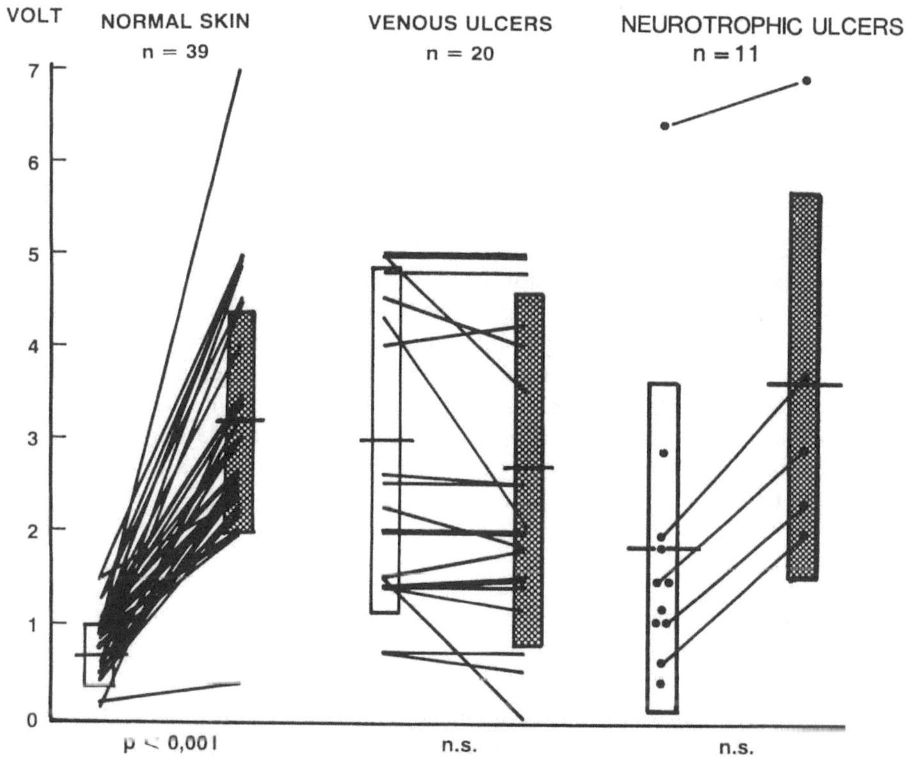

Fig. 2. Skin blood flow before *(empty blocks)* and after *(filled blocks)* direct heating. Laser-Doppler measurement: the output voltage *(V)* is proportional to the product of number and speed of blood corpuscles in subepidermal vessel plexus

Discussion

In cases of venous ulcers, the $tcPO_2$ in the vicinity of the ulcer is severely reduced. Sixty-three percent of the values obtained are below 20 mm Hg, and are thus in a range of impending amputation if arterial gangrene has been established. Not only the fact that we have never had to amputate a leg with a venous ulcer proves that the macrocirculation in these patients is intact, but also the results of all the blood flow measurements that indicate local hyperemia (xenon clearance, angioscintigraphy, angiography, and positron emission scintigraphy with $^{15}O_2$). The increased values obtained with laser-Doppler, which suggest maximum hyperemia, since they do not rise under direct heating, provide further support of this hypothesis [6].

Hyperemia of the ulcer's border, of course, does not completely rule out local ischemia of the ulcer's center. Zones with rarefication of capillaries (apart from zones with dense capillary loops) at the ulcer's border have been described both histologically and by capillary microscopy - here even by simultaneous $tcPO_2$ measurement [2]. Nevertheless, in our opinion, the reason for the local hypoxia is not so much a deficit in the oxygen transport in the region of the capillaries, but rather a permeability barrier between capillaries and tissue. Various investigations have given evidence of such a pericapillary block, which might be the result of undissolved fibrin cuffs [1, 7]. The prompt $tcPO_2$ increase after oxygen inhalation correlates well with this theory.

We also believe that neurotrophic ulcers are not primarily caused by the nutrients not being delivered, but are a result of a disturbance in the consumption of these nutrients [4]. Blood flow measurements provided evidence of shunt circulation and disturbed vessel innervation with no flow increase in the affected hyperemic areas under indirect body heating [4]. As the present results show, there is an increase in $tcPO_2$ values after oxygen inhalation as in healthy persons. Therefore, we cannot assume hypoxia as a pathophysiologically decisive factor, as we do with all other substance defects of the skin of vascular origin. Besides the lack of pain reflex and chronic traumatization, neuropeptides could play a role here, which, apart from the many mechanisms already established, possibly have a direct trophic effect on the tissue.

Summary

$TcPO_2$ measurements were carried out in 49 patients with venous ulcers of the lower leg and 16 patients with neurotrophic ulcers. Measurements were made with a TCM1 oxygen monitor (Radiometer, Copenhagen) using a electrode heated to 44 °C in the vicinity of the ulcer. The average values among the venous ulcers were 19.7 ± 18.4 mm Hg; among the neurotrophic ulcers 54.7 ± 12.6 ($P < 0.05$). (Normal values in the lower leg are 59.6 ± 11.2.) The macrocirculation was intact in all of the patients examined (Doppler sonography, acral oscillography). Local hyperemia could be determined in both groups of patients by means of xenon clearance, thermography, angioscintigraphy, and the laser-Doppler method.

It is assumed that the hypoxia of venous ulcers is primarily a result of disturbed permeability. Neurotrophic ulcers, however, apparently cannot be explained by local oxygen deficiency.

References

1. Browse, N.L., K.G.Burnand: The cause of venous ulceration. Lancet 2 (1982) 143-145
2. Franzeck, U.K., A.Huch, R.Huch, A.Bollinger: Simultane transkutane PO_2-Messung und Videomikroskopie bei Patienten mit chronisch-venöser Insuffizienz. VASA, Suppl. 9 (1982) 19
3. Nilsson, G.E., T.Tenland, P.A.Öberg: A new instrument for continuous measurement of tissue blood flow by light beeting spectroscopy. IEEE Trans. Biomed. Eng. BME-27 (1980) 12-19
4. Partsch, H.: Neuropathien vom ulzeromultilierenden Typ. Klinik, Diagnostik, Durchblutungsmessungen. VASA, Suppl. (1978) 6
5. Partsch, H.: Transkutane Messung des Sauerstoffpartialdrucks bei trophischen Hautläsionen an der unteren Extremität. In: F.Nobbe, G.Rudofsky: Probleme der Vor- und Nachsorge und der Narkoseführung bei invasiver angiologischer Diagnostik und Therapie. Pflaum-Verlag, München (1983) S.209-214
6. Partsch, H.: Hyperaemic hypoxia in venous ulceration. Br. J. Derm. 110 (1984) 249-251
7. Partsch, H.: Zur Pathogenese des venösen Ulcus cruris. Hautarzt 36 (1985) 196-202

Transcutaneous PO_2 Measurements in Dermatology

A. Ott

Introduction

There are a number of dermatological diseases in which transcutaneous (tc) PO_2 measurements have become important. Such diseases include skin ulcers, particularly among diabetics, gangrene, ulcus cruris, acrocyanosis, and scleroderma. In all diseases in which microcirculatory changes are involved and in which Doppler sonography alone is not able to provide definite information on the actual O_2 supply to the skin, $tcPO_2$ measurements have become a valuable diagnostic tool. Moreover, $tcPO_2$ measurement provides an objective method of evaluating the effect of the different modes of therapy. If amputation is necessary, the method can be used to preoperatively determine the level of amputation to ensure that the supply of oxygen is sufficient for optimum healing.

For our measurements we used the oxygen monitor TCM 1 in conjunction with the two-canal recorder TCR 2 (both manufactured by Radiometer, Copenhagen). The electrode temperature generally chosen was 45 °C.

To date we have carried out, in addition to numerous measurements in healthy volunteers, 92 measurements in 41 patients with the diseases listed above, and 76 measurements in 18 patients with psoriasis.

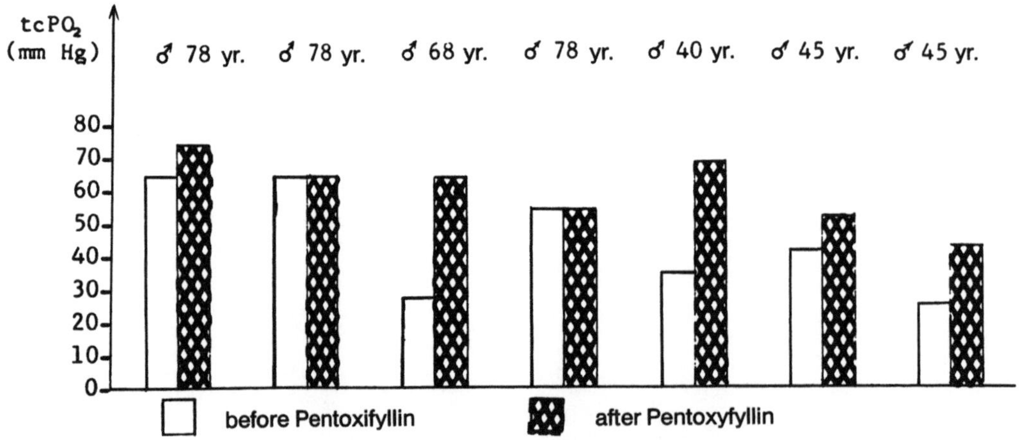

Fig. 1. $tcPO_2$ values before and after 14-day therapy with 400 mg pentoxifylline infusions in seven patients with peripheral arterial circulatory disturbances stages I and II

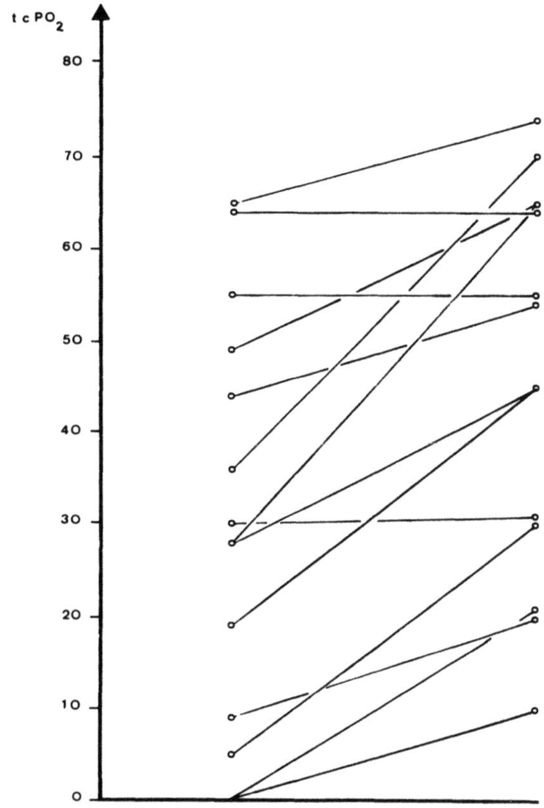

Fig. 2. $tcPO_2$ values before and after 14-day infusion treatment with pentoxifylline in 14 patients with peripheral arterial circulatory disturbances stages II-IV

Results

Seven patients with peripheral circulatory disturbances stages II and III had been treated with pentoxifylline infusions (400 mg/day) for 14 days. Before and after the therapy, $tcPO_2$ measurements were made on the forefoot (Fig. 1). Figure 2 gives an overview of such therapy control measurements in a total of 14 patients with peripheral circulatory disturbances stages II-IV. In severe cases, 1 g pentoxyfylline per day was infused.

The examples in Table 1 show that in some cases Doppler sonography alone does not provide definite information on the actual O_2 supply. On the other hand, the $tcPO_2$ measurement can have an unexpectedly good result - for instance, as in the following example, by ruling out a diminished arterial supply assumed to be the cause of the ulceration: a 62-year-old man, who had been diabetic for 25 years and had had his second to fourth toes amputated, had multiple ulcers on the feet and lower legs, which made Doppler sonographic examination only limitedly possible. The $tcPO_2$ measured on the right forefoot and right calf was 70 mm Hg, on the left calf 65 mm Hg. It was not possible to place the electrode on the left forefoot. The ulcers on the lower legs and dorsa of the feet were not caused by arterial impairments

Table 1. Patients examined

1. Male, 68 years, generalized arteriosclerosis, BP 140/80 mm Hg, varicosis, calf edema, right ulcus cruris		
	Doppler (mm Hg)	$tcPO_2$ (mm Hg)
Right foot:	ADP 80 ATP 90	28
Left foot:	ADP 0 ATP 60	65
2. Male, 40 years, diabetes mellitus, 20 cigarettes/day, mal perforant of right foot, BP 180/95 mm Hg		
	Doppler (mm Hg)	$tcPO_2$ (mm Hg)
Right foot:	ADP 180 ATP 180	35
3. Male, 45 years, intermittent claudication for 20 years, now rest pain, 40 cigarettes/day, BP 120/80 mm Hg		
	Doppler (mm Hg)	$tcPO_2$ (mm Hg)
Right foot:	ADP 160 ATP 120	Medial 42 Lateral 25
Arteriographic findings: terminal occlusion of ATP; occlusion of ATA and AP in distal third of calf; faint, corkscrew-like collaterals		

Abbreviations: ADP, arteria dorsalis pedis; ATP, arteria tibialis posterior; ATA, arteria tibialis anterior; AP, arteria peronea

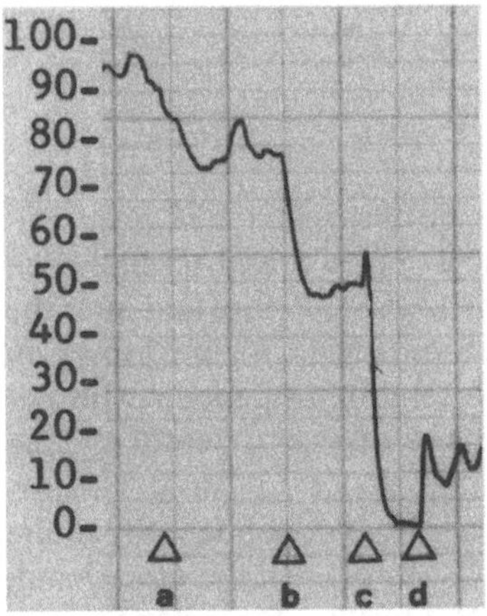

Fig. 3. Changes of the $tcPO_2$ values under pressure: pressure on the electrode positioned above the tibia with *(a)* 100 g, *(b)* 200 g, *(c)* 500 g and *(d)* 300 g

but by pressure. Neurological examination showed massive polyneuropathy with loss of distal sensitivity, including sense of posture.

The extent to which external pressure reduces the $tcPO_2$ values and thus the supply of oxygen to the skin does depends not only on the amount of pressure, but also on whether there is soft tissue or bone underneath the measuring site. In one case,

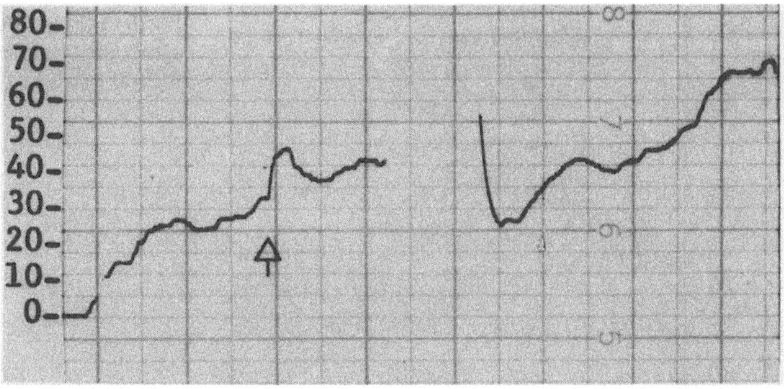

Fig. 4. *Left curve* $tcPO_2$ measured over the sacrum of paraplegic patient lying on his back in a special bed (*arrow* on patient's waking). *Right curve* Control measurement on site of the same segment without weight pressure

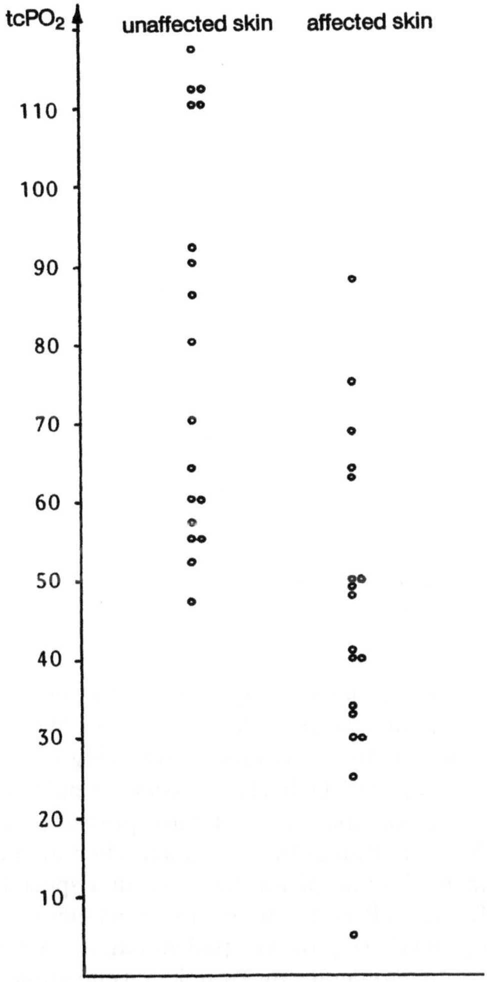

Fig. 5. $tcPO_2$ values before therapy in 18 patients with psoriasis measured on affected and nonaffected skin ($n = 18$)

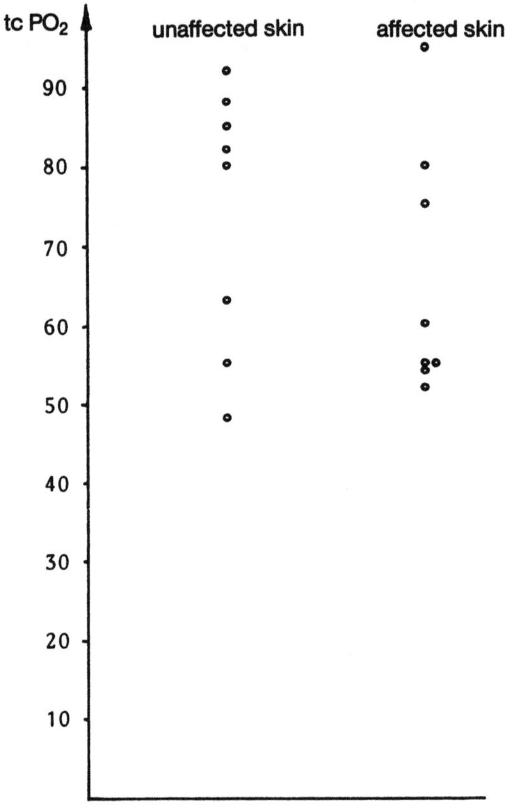

Fig. 6. $tcPO_2$ values on psoriatic and non-affected skin after 4 weeks of therapy with dithranol and SUP ($n = 8$)

Table 2. Skin $tcPO_2$ values before and after therapy (mm Hg)

	Before ($n = 18$)	After ($n = 8$)
Unaffected skin	79.4 ± 24.6	74.1 ± 16.5
Affected skin	46.3 ± 20.1	65.8 ± 15.8
Mean difference	33.1	8.3

the application of 500 g to the electrode resulted in a $tcPO_2$ decrease of only 10 mm Hg when measured on the abdomen. When measurements were made on the tibia, however, the $tcPO_2$ value of 0 mm Hg was obtained with the same amount of pressure (Fig. 3). The heel of a normal weight person in a horizontal position is subjected to more than twice as much pressure. The same applies to the conditions above the os sacrum or the trochanter when the person is lying on the side. A 68-year-old man with paraplegia had been in a special bed (Clinitron) for 8 months. We performed $tcPO_2$ measurements on his sacrum, which yielded values of between 34 and 43 mm Hg (Fig. 4). The patient had no necroses.

Investigations by Tronnier [5] encouraged us to perform $tcPO_2$ measurements in

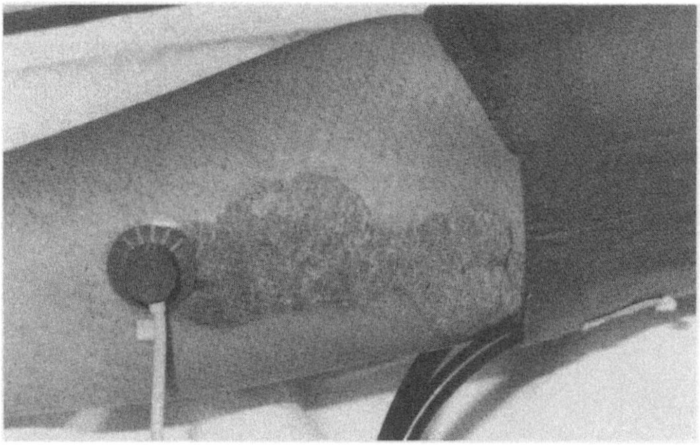

Fig. 7. Equipment for measurement of $tcPO_2$ reduction on psoriasis plaque after stopping the arterial flow

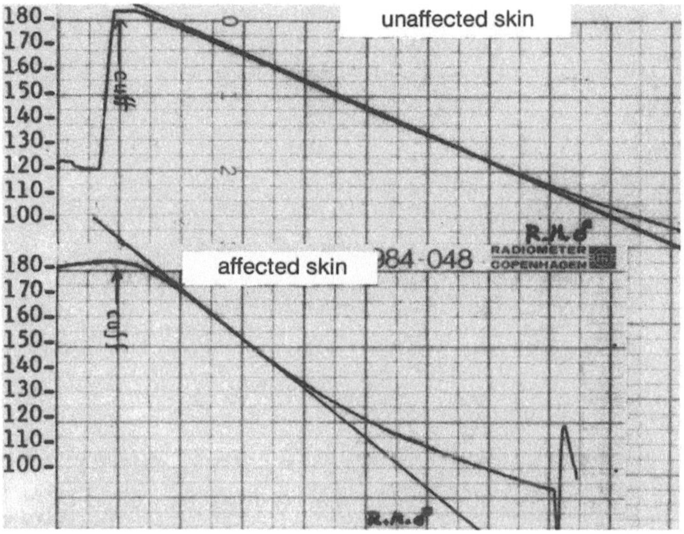

Fig. 8. $tcPO_2$ decrease on psoriatic and nonaffected skin of same patient as in Fig. 7: steep reduction on psoriatic areas. See Fig. 9

patients with psoriasis. We measured the psoriatic skin and nonaffected skin before and after the beginning of therapy with dithranol and SUP and after 4 weeks of treatment (see Table 2, Figs. 5, 6). The $tcPO_2$ values measured on the psoriatic skin are approximately 42% lower than those obtained from the healthy skin of the same patients. Under the described therapy the values obtained from the psoriatic skin increase. Reasons for the low values in the psoriatic skin could be (a) decreased O_2 supply, (b) decreased O_2 diffusion, or (c) increased O_2 consumption due to the higher rate of mitosis.

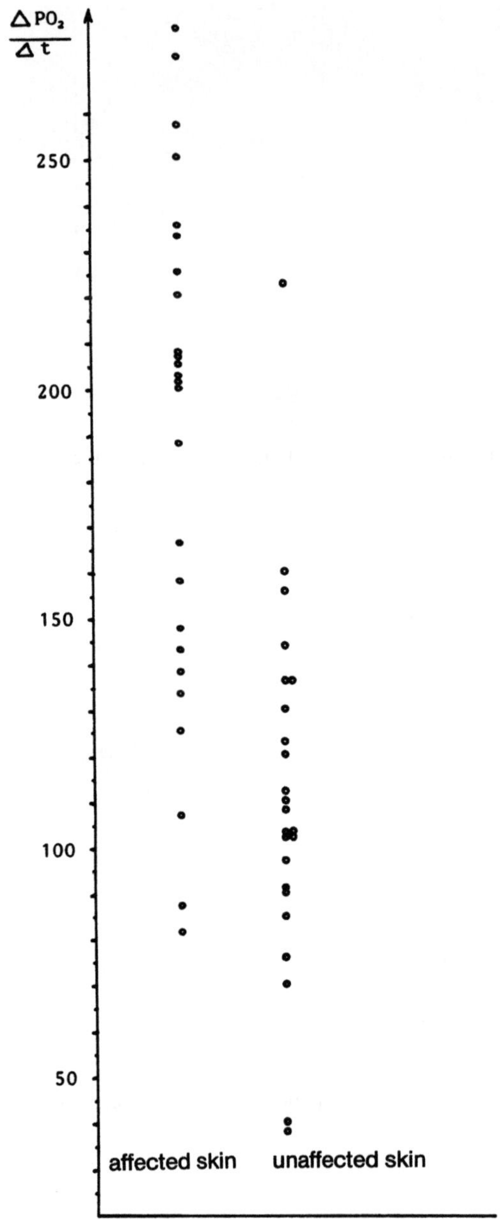

Fig. 9. O_2 consumption on psoriatic and non-affected skin ($n=25$)

We singled out the last of these three possibilities for investigation: after having stopped the arterial blood flow in one extremity with an inflated cuff (Fig. 7), we took the amount $\frac{\Delta PO_2}{\Delta t}$ as parameter for the O_2 consumption [2]. The consumption thus measured is higher on the psoriatic skin than on the nonaffected skin of the same patients (Fig. 8). The mean values ($n=24$) were 187 on psoriatic and 111 on

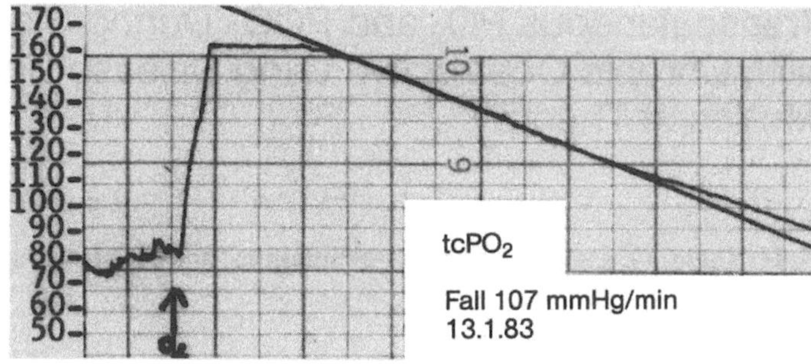

Fig. 10. $tcPO_2$ decrease on psoriatic plaque after clinically satisfactory healing

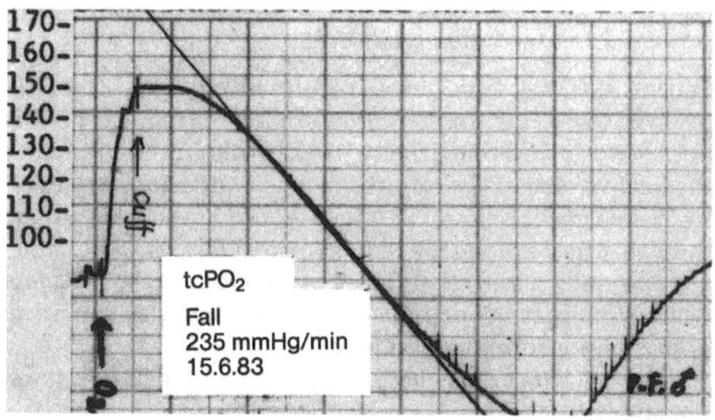

Fig. 11. $tcPO_2$ reduction on psoriatic plaque in same patients as in Fig. 10. Steep decrease after exacerbation

nonaffected skin (Fig. 9). Measurements in healthy persons with no psoriasis yielded a mean value of 94. The consumption of O_2 seems to drop when the clinical effect is good and to increase again with exacerbation of the disease (Figs. 10, 11).

References

1. Borzykowski, J., B. Krahenbuhl: Perkutane Messung der Sauerstoffspannung am Fuß zur Verlaufskontrolle der Arterienverschlußkrankheit der unteren Extremitäten. VASA 10 (1981) S. 137–140
2. Huch, R., A. Huch, D. W. Lübbers: Transcutaneous PO_2 Thieme-Verlag, Stuttgart (1981)
3. Krahenbuhl, B., J. M. Dubas: Transcutaneous oxygen pressure on the foot of normal subjects and patients suffering from arterial occlusive disease. In: A. H. M. Jageneau (ed): Noninvasive methods on cardiovascular haemodynamics. Elsevier/North-Holland, Amsterdam/New York (1981)
4. Partsch, H.: Messung des transkutanen Sauerstoffpartialdrucks im Stadium IV. In: H. Denk, G. W. Hagmüller, U. Brunner (Hrsg.): Arterielle Durchblutungsstörungen der unteren Extremitäten - Grenzzonen der Therapieentscheidung. TM-Verlag, Bad Oeynhausen (1982)
5. Tronnier, H., E.-M. Böttger, E. Hoffmann: Transkutane PO_2-Messung unter UV-Therapie von Psoriasis und Akne. Z. Hautkrankh. 54 (1979) S. 546–550

Transcutaneous PO_2 and PCO_2 During Sleep in Patients with Chronic Obstructive Lung Disease Under Different Modes of Therapy

K. H. Rühle, G. Klein, M. Schmitz-Schumann, and H. Matthys

Introduction

Patients suffering from chronic obstructive lung disease (COLD) frequently develop hypoxemia, i.e., oxygen desaturation, during sleep. This is caused by ventilation/perfusion inhomogeneity [1] or by diminished ventilation with a reduction in the tidal volume or the respiratory rate or apneic episodes (Fig. 1). It can be treated by raising the inspiratory oxygen concentration by administering 100% oxygen through a nasal prong [3]. Another possibility is stimulating the respiratory center with centrally acting respiratory analeptics. We now have a drug at our disposal which stimulates the peripheral chemoreceptors and, particularly in the case of hypoxia, increases ventilatory drive.

The following deals with the effects and side effects of these three modes of therapy.

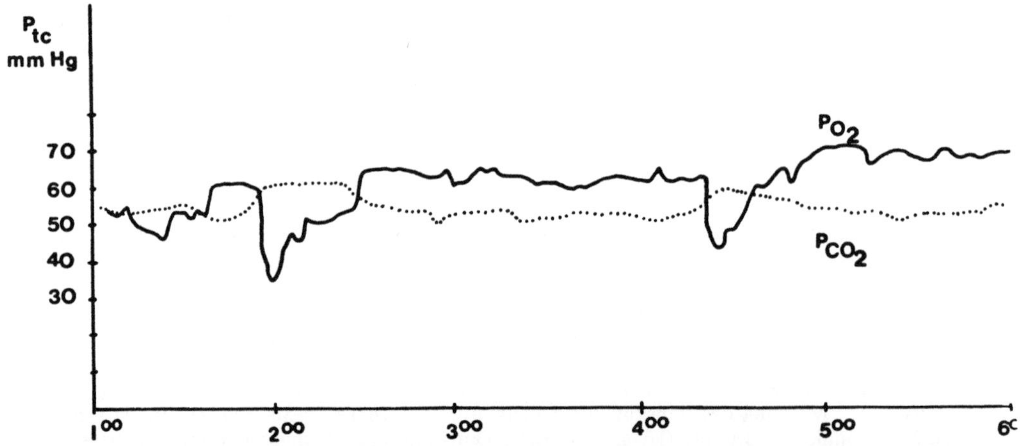

Fig. 1. Nocturnal transcutaneous blood gas measurement in a patient with chronic obstructive lung disease (measured between 1 A.M. and 6 A.M.): severe hypoxemic episode with PO_2 values below 40 mm Hg at 2 A.M. and at 4.30 A.M., each lasting about 20 min

Method

Measurements of the transcutaneous (tc) blood gas under air and oxygen breathing were carried out in a group of 14 patients (age 58 ± 15 years). Pure oxygen was applied through a nasal prong at a flow rate of 2 l/min. The electrodes for the transcutaneous measurement were fixed on the lower arm and were repositioned after 5 h measuring time. The adhesive ring for the second measurement was also fixed to the skin at the outset in order to minimize manipulation while the patient was asleep.

Monitoring was done from 8 P.M. until 6 A.M. The $tcPO_2/PO_2$ monitoring system TCM 222 (Radiometer) was used for the recordings. Partial pressures were registered with a plotter at a speed of 5 cm/h. Two-point calibration at various times during the monitoring made it possible to correct the transcutaneously measured values by analyzing the arterial blood gases from the hyperemic earlobe. In addition, a long-term ECG was recorded and evaluated with an automatic analyzer (Medilog System, Oxford Medical Systems).

In a second investigation on seven patients with global respiratory insufficiency with a PCO_2 value of 54 ± 6 mm Hg and a PO_2 of 53 ± 8 mm Hg, a respiratory analeptic agent (amiphenazole) was injected i.v. for a total period of 5 min after 20 min of 100% oxygen respiration at a flow rate of 3 l/min. The transcutaneous PO_2 and PCO_2 values were monitored for a total of 35 min after the injection.

The third group of measurements was carried out in a double blind experiment on 12 patients (age 65 ± 10 years) with severe COLD who had been given either a placebo or a peripheral chemoreceptor-stimulator (almitrine 2 × 50 mg/day). The measurements were made after the patients had taken the tablets for 15 days. The following parameters were continuously recorded polygraphically for 8 h: transcutaneous PO_2 and PCO_2, respiratory rate, heart rate, and cough frequency. In addition to the maximum and minimum PCO_2 values, the average value of these two quantities was calculated per night: every 10 min a value was read from the continuously recorded curve, and after correction a mean value from 48 values per night was calculated for each patient.

Results

Air breathing in patients with COLD frequently resulted in hypoxemia, which lasted for a considerable period of time. The mean value of the lowest PO_2 during the night was 45 mm Hg. Under oxygen respiration, however, the minimum PO_2 value of 57 mm Hg was considerably higher. The amount of time the PO_2 under air breathing remained below the critical level of 50 mm Hg was 77 min. Under oxygen breathing the critical time was only 1.9 min during the entire night (Fig. 2).

The maximum PCO_2 values under air and oxygen breathing were in the same order of magnitude (73 mm Hg air and 75 mm Hg oxygen). During air breathing, the PCO_2 was more than 10 mm Hg above the initial value for 235 min, under oxygen breathing for 360 min. The respiratory acidosis is prolonged, but the PCO_2 peaks

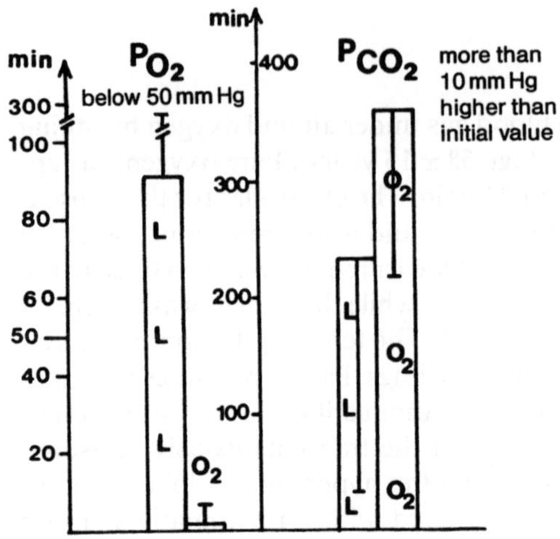

Fig. 2. Mean value of the time in which the PO_2 was below 50 mm Hg (A, air breathing; O_2, oxygen breathing)

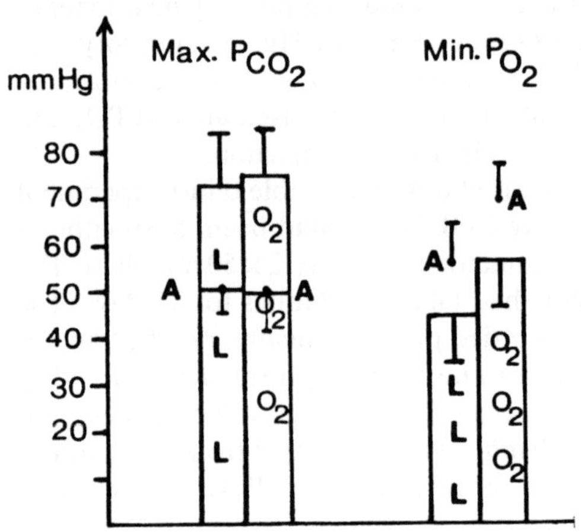

Fig. 3. Maximum PCO_2 and minimum PO_2 per night (8 P.M.–6 A.M. *L*, air breathing; O_2, oxygen breathing 2 l/min). For comparison, the initial values *(A)*; mean values and standard deviation for each

are not significantly influenced by oxygen respiration (Fig. 3). Under therapy with oxygen the incidence of ventricular ectopic beats was reduced from 141 ± 229 to 86 ± 122 per night. The application of O_2 resulted in a reduction of the ventricular ectopic beats in all patients.

The point of interest in the second investigation was whether the additional application of a centrally acting respiratory analeptic agent under oxygen therapy effects a reduction of the PCO_2 (Fig. 4). At no point during the continuous recording could a marked reduction in the arterial PCO_2 be observed. There was a tendency for the PO_2 (also measured transcutaneously) to drop.

Since central respiratory analeptic agents proved to be ineffective in the third group, we used night measurements to study the effect of almitrine, a peripheral chemoreceptor-stimulator. The administration of a almitrine resulted in an increase in the PO_2 from 58.0 ± 7.5 to 61.3 ± 8.2 mm Hg. This difference was significant in the one-tailed *t* test ($P < 0.05$). The average PO_2 during sleep tended to rise under al-

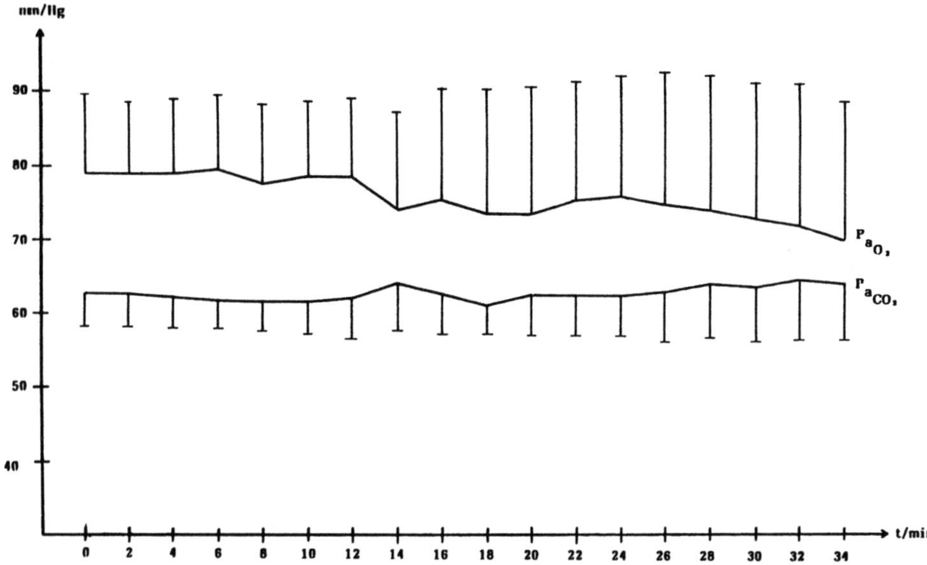

Fig. 4. Transcutaneous continuous measurement of blood gases after injection of a centrally acting respiratory analeptic agent (mean value with standard deviation); reduction of PO_2 while PCO_2 remains just about constant

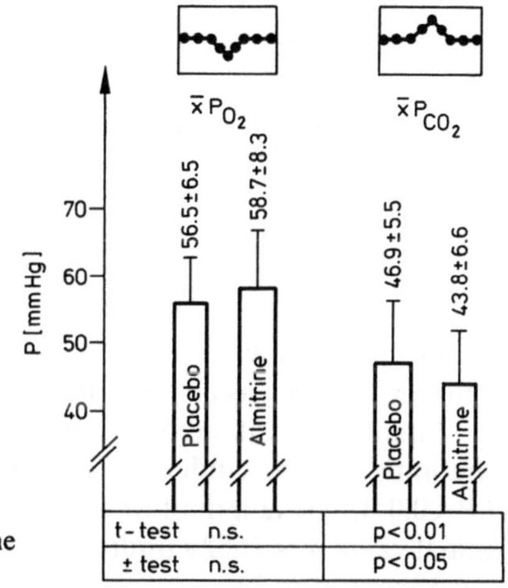

Fig. 5. Average PO_2 and PCO_2 under placebo and after Almitrine; significant reduction in PCO_2 during the course of the night

mitrine - from 56.5 to 58.7 mm Hg - but the difference was not significant. The maximum PCO_2 increase per night was considerably less under almitrine than under the placebo: it dropped from 8.2 ± 5.2 to 4.4 ± 4.3 mm Hg ($P < 0.01$) (Fig. 5).

The number of hypercapnic episodes, in which the PCO_2 increased more than 5 mm Hg in comparison with the initial value, was 6.2 ± 5.1 episodes per night under

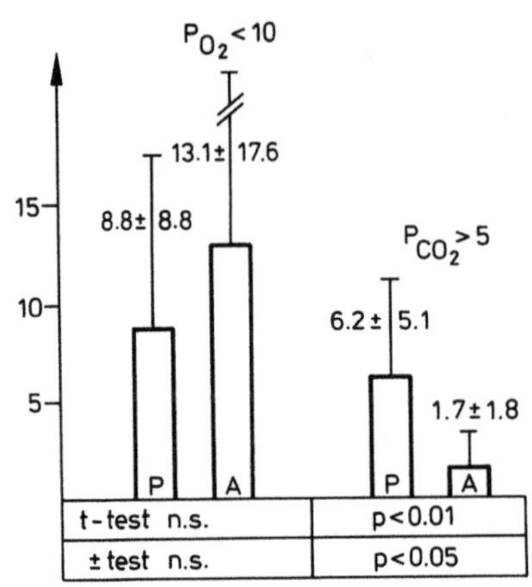

Fig. 6. Number of hypoxemic ($PO_2 > 10$ mm Hg below initial value) and hypercapnic episodes ($PCO_2 >$ 5 mm Hg above initial value) under placebo (P) and Almitrine (A)

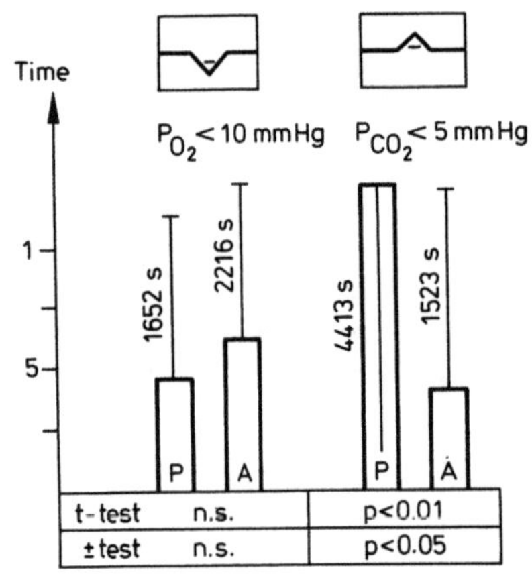

Fig. 7. Duration of hypoxemia ($PO_2 > 10$ mm Hg below initial value) and hypercapnia ($PCO_2 > 5$ mm Hg above initial value): significant reduction in the duration of the PCO_2 increase after almitrine

the placebo and was reduced to 1.7 ± 1.8 episodes per night under almitrine (Fig. 6). This difference was also significant in the one-tailed *t* test ($P < 0.01$).

Finally we evaluated the time in which the PCO_2 was more than 5 mm Hg above the initial value. Under the placebo 4413 were measured, under almitrine 1523 s ($P < 0.01$). The length of time in which the PO_2 was more than 10 mm Hg below the initial value was 1652 s under the placebo, as opposed to 2216 s under almitrine (not significant) (Fig. 7).

Discussion

As a result of the application of 2 l/min pure oxygen, the time during which the PO_2 was below 50 mm Hg was significantly reduced. This reduction in hypoxemia is responsible for the decrease in the ventricular ectopic beats, which serve as indicators of better oxygen supply of the myocardium. The increase in the PO_2 in this group did not result in a increase in the PCO_2 peaks, but in an increase in the length of time in which the PCO_2 was raised.

The danger of this therapy can be reduced by stimulating ventilation. It was evident in our attempt, however, that respiratory mechanisms in patients with severe obstruction make it impossible to stimulate alveolar ventilation with centrally acting respiratory analeptic agents. On the contrary, central respiratory stimulation in such cases usually results in a reduction of the PO_2. Nolte [4] was able to prove this phenomenon in treatments with oxygen and respiratory analeptic agents. Substances were therefore developed that would stimulate ventilation more selectively. Douglas et al. [2] showed that the O_2 saturation measured with the ear oximeter rose under almitrine from 84% to 89%. The lowest O_2 saturation value during sleep increased from 62% under the placebo to 77% after almitrine. In this investigation, however, the CO_2 partial pressure was not continuously recorded.

Conclusion

In addition to an admittedly slight increase in the PO_2 after the administration of almitrine, we were also able to show a drop in the PCO_2 and thus a stimulation of the alveolar ventilation. It therefore seems reasonable to treat patients undergoing long-term oxygen therapy in whom there is a nocturnal increase in respiratory acidosis with an additional stimulator of the peripheral chemoreceptors.

References

1. De Marco, F.J., J.W.Wynne, A.J.Block, P.G.Boysen, V.C.Taasan: Oxygen desaturation during sleep as a determinant of the blue and bloated syndrome. Chest 79 (1981) 621
2. Douglas, N.J., J.J.Connaughton, A.D.Morgan, C.M.Shapiro, N.Pauly, D.C.Flenley: Effect of Almitrine on nocturnal hypoxaemia in chronic bronchitis and emphysema, and in patients with central sleep apnoe. Bull. Europ. Physiopath. 19 (1983) 631
3. Kearley, R., J.Wynne, A.J.Block, P.G.Boysen, S.Lindsey, C.Martin: The effect of low flow oxygen on sleep-disordered breathing and oxygen desaturation. Chest 78 (1980) 682
4. Nolte, D.: Hyperoxieatmung bei Patienten mit alveolärer Hypoventilation. Pneumonol. 149 (1973) 55

Transcutaneously and Arterially Measured PO_2 in Adult Intensive Care Patients

N. Spannbrucker and F. Vogel

Introduction

The advantage of PO_2 measurement using skin surface electrodes is that it is a simple and noninvasive method which allows continuous recordings to be made. This advantage has made the procedure valuable in treating intensive care patients, and transcutaneous PO_2 measurement has become established as a routine method in neonatology [1, 2]. On the other hand, there are a number of interfering factors associated with determining the PO_2 on the surface of the skin, particularly in the case of critically ill patients, such as severe blood pressure fluctuation in hemodynamically unstable patients, reduced skin circulation due to restricted cardiac output, and skin perfusion changes induced by vasoactive drugs. These factors, along with the fact that the PO_2 permeability of the adult skin is considerably less than that of a neonate, can limit the application of transcutaneous (tc) PO_2 measurement in adult patients undergoing intensive care. A number of questions thus arise:

1. To what extent can $tcPO_2$ values provide information on the arterial blood gases?
2. Can the correlation regularly observed between higher arterial PO_2 (PaO_2) values and lower $tcPO_2$ values be compared interindividually?
3. To what extent is PO_2 measurement affected by the diminished blood flow in low output syndrome or by vasoactive drugs?
4. How does the critically ill patient with reduced skin perfusion tolerate local monitoring under the sensors?

Method

We examined 18 patients between the ages of 42 and 73 years who were undergoing internal intensive care. Fifteen of these patients had to be artificially ventilated. Seven had low cardiac output with a CI of 2–2.4 l/m^2 per min. In 32 series of measurements lasting 3–4 h, eight $tcPO_2$ values from each series were compared with the arterial values determined at the same time. The $tcPO_2$ measurement was carried out with Kontron ($n = 176$) and Radiometer ($n = 80$) sensors. The temperature of the sensors was 45 °C in every case. The sensor was located in the infraclavicular region. The arterial samples taken at the same time were processed in a blood gas analyzer ABL II (Radiometer).

Results

The 45 °C hyperthermia applied for 3-4 h did not cause any skin damage, except for some redness of the skin surrounding the sensor, which completely disappeared within 24 h. No blister formation was observed.

To determine the total correlation of the transcutaneously and arterially measured PO_2, 256 $tcPO_2$ values and the corresponding arterial values of all of the patients were included. In this interindividual comparison there was no satisfactory correlation for $tcPO_2$ and PaO_2 values, with a correlation coefficient of $r = 0.6$. The standard estimation error $s_{yx} = 26$ and the prediction range of ± 50 mm Hg for the PaO_2 value show that interindividual comparison for the purpose of estimating arterial values is not accurate enough to be applied in practice. The reason for this poor correlation is assumed to be a so-called individual skin factor [3], i.e., each individual's skin has its own specific oxygen permeability. This hypothesis would indicate that an intraindividual correlation of $tcPO_2$ and PaO_2 values - in other words, of several series of measurements made on one person - would be more reliable.

We therefore tried to calculate such an intraindividual correlation of $tcPO_2$ and PaO_2 values in patients in whom several series of measurements had been carried out. The results showed that this correlation in one single person based on several measuring series with correlation coefficients of $r = 0.6-0.7$ and a prediction range for the PaO_2 from ± 30 to ± 50 mm Hg was by no means better than in the interindividual comparison.

The only way to achieve a correlation of transcutaneous and arterial values that can be used in practice is to determine a correlation within each of the series of measurements. Here, the correlation coefficients are considerably better: in 27 of the 32 measuring series the correlation coefficients were above 0.8, with a prediction range for the PaO_2 from ± 5 to ± 15 mm Hg.

In the interindividual comparison, it could not be established to what extent reduced skin perfusion during low output syndrome or the application of vasoactive drugs influenced the $tcPO_2$ measurements. The separate correlation of measured pairs from patients with low CI or from patients under catecholamine therapy revealed no essential change in comparison with the total correlation. The difference between transcutaneously and arterially measured PO_2, or the quotient of $tcPO_2/PaO_2$, which was 0.4-0.8 in patients with stable circulation, was not very different in patients with reduced perfusion from that in the other patients: in the seven patients with reduced cardiac output, the correlation of $tcPO_2$ and PaO_2 was with one exception, between 0.8 and 0.9, with a prediction range for the PaO_2 from ± 8 to ± 12 mm Hg.

Regarding the practical application of $tcPO_2$ measurement, continuous monitoring has proved advantageous, particularly for the ventilated patients. Changes in the inspiratory oxygen concentration were immediately shown by movements in the $tcPO_2$ curve. Inadequate respiration patterns could be corrected and optimized by controlling the continuous PO_2 curve. Insufficient oxygenation in ventilated patients caused by cramps or counterbreathing brought on characteristic changes in the $tcPO_2$ profile. The disconnection of the respiratory system in the presence of a suction maneuver could be recognized by a sudden and deep drop in the PO_2 curve with a decline in the $tcPO_2$ of 50-60 mm Hg - something not observed in the usual arterial measurements. This was not an artificial "skin phenomenon." Such changes in the oxygen curve are known from continuous invasive PO_2 measurements [3]. The fact that this process is also observed in transcutaneous measurements shows the high standard of the method.

Summary

Transcutaneous PO_2 measurement in adults can provide only conditional information on the arterial PO_2. The electrode must be recalibrated for every single measuring period, since in each region of skin of any individual, even within a limited area, the PO_2 permeability fluctuates considerably. We did not observe any influence of cardiac output on the method up to a CI of $2\,l/m^2$ per min. Neither did catecholamine have any effect on the method when there was sufficient cardiac output. One of the advantages of transcutaneous measurement is that continuous monitoring makes it possible to detect any changes that occur. In connection with transcutaneous PCO_2 measurement, this method is especially valuable in patients being artificially ventilated. The need to measure arterial blood gas is not replaced, but the frequency of the arterial measurements can be markedly reduced.

References

1. Huch, R., D. W. Lübbers, A. Huch: Reliability of transcutaneous monitoring of arterial PO_2 in newborn infants. Arch. Dis. Child 49 (1974) 213
2. Versmold, H. T., O. Linderkamp, M. Holzmann, I. Strohhacker, K. Riegel: Transcutaneous monitoring of PO_2 in newborn infants: Where are the limits? Influence of blood pressure, blood volume, blood flow, viskosity and acid base state. Birth Defects 4 (1979) 285
3. Goeckenjan, G.: Kontinuierliche Messung des arteriellen Sauerstoffpartialdrucks. Anästhes. Intensivmed. 137 (1981)

Transcutaneous PO_2 and Local Perfusion in Healthy Adults, Artificially Ventilated Patients with Stable Circulation, and Patients in Cardiogenic Shock

R. Lemke, D. Klaus, and D. W. Lübbers

Introduction

Transcutaneous (tc) PO_2 measurement is possible because oxygen from the skin capillaries can diffuse through the intact skin. The heated electrode produces maximum local dilatation of the vessels directly under the sensor. When circulation is stable with optimum peripheral perfusion pressure, maximum dilatation results in maximum blood flow increase at the measuring site. Under such conditions, the $tcPO_2$ is close to the arterial PO_2 (PaO_2) or the quotient $tcPO_2/PaO_2$ is close to 1 [1]. Various studies have shown that during shock the $tcPO_2/PaO_2$ quotient drops considerably, according to the severity of the shock. When normal circulatory conditions are restored, the $tcPO_2/PaO_2$ value returns to normal values close to 1 [2, 3]. This phenomenon is probably brought on by the decrease in the peripheral blood flow caused by the reduced perfusion pressure, vasoconstriction, and shunt diffusion of O_2 from the venous capillary limb to the arterial limb [4].

The objective of our investigation was to find out whether the PO_2 increase measured with a commercial $tcPO_2$ electrode after perfusion has been stopped can provide information on the extent of local blood flow at the site of measurement, and whether the parameter we determined has any connection with the known relation $tcPO_2/PaO_2$.

Method and Patients

The following parameters were measured: (a) $tcPO_2$ on the forearm, (b) PaO_2 from the radial artery or from the hyperemic ear lobe, and (c) the rate of the $tcPO_2$ reincrease ($tcPO_2$/min). For the measurements, a blood pressure cuff on the upper arm was inflated until suprasystolic values were reached and circulation was cut off until the $tcPO_2$ had dropped to 0 mm Hg. After the occlusive pressure was released, the $tcPO_2$ rose to the initial value at various rates. The $\Delta tcPO_2$ was determined by applying a tangent and calculated as a percentage of the initial value (reincrease rate percentage) (Fig. 1). For this measurement it is necessary to use a $tcPO_2$ electrode with a 90% response time, which is lower than the reincrease rate.

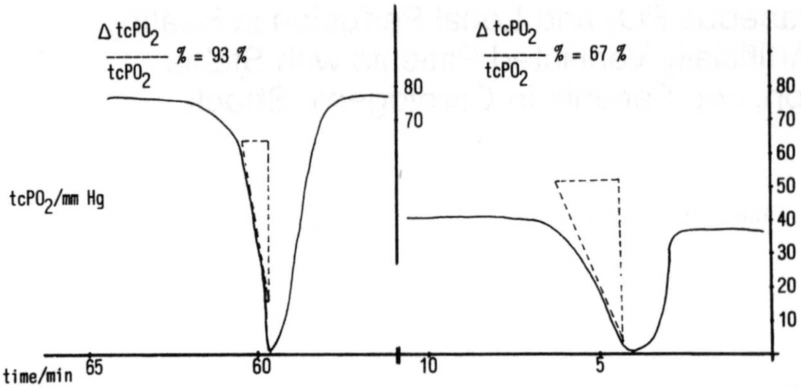

Fig. 1. Different reincrease rate percentages: (*r*) low $tcPO_2$, low local flow, low reincrease rate; (*l*) high $tcPO_2$, good local flow, high reincrease rate

Table 1. Oxygen pressure parameters in the four groups of patients

Group (*n*)	$tcpO_2$ (mm Hg)	apO_2 (mm Hg)	$tcpO_2$/ apO_2	Δ $tcpO_2/tcpO_2$ %
Stable circulation (28)	74,0 ± 30,6	79,0 ± 26,8	0,93 ± 0,13	87,9 ± 10,1
Stable circulation, artificial ventilation (14)	87,4 ± 43,7	93,3 ± 41,9	0,95 ± 0,19	84,3 ± 19,1
Shock, no artificial ventilation (5)	42,0 ± 8,7	55,1 ± 11,7	0,78 ± 0,18	66,4 ± 18,2
Shock, artificial ventilation (5)	36,4 ± 13,8	81,8 ± 47,9	0,51 ± 0,20	43,5 ± 19,1

The following groups of patients were included in the investigation:

28 patients with stable circulation
14 artificially ventilated patients with stable circulation
5 patients in slight shock, not artificially ventilated
5 artificially ventilated patients in severe shock

Results and Discussion

The results obtained in the four groups of intensive care patients using the method described are compiled in Table 1. The patients with stable circulation and the artificially ventilated patients with stable circulation had a $tcPO_2/PaO_2$ quotient close to 1; that means that local hyperemia produced optimum blood flow conditions. This finding correlates with the data given in the literature [2]. In these patients, the $tcPO_2$ had reached 80%–90% of the original value within 1 min after the circulation

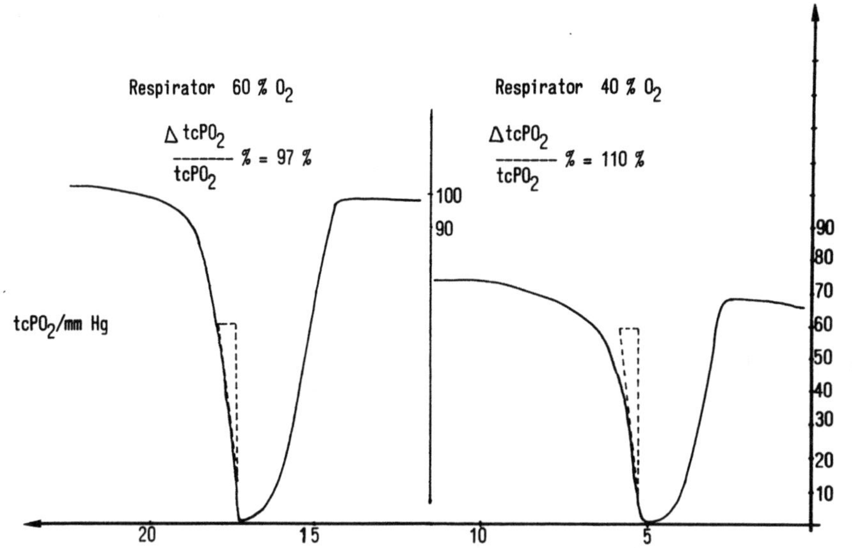

Fig. 2. Stable ventilated patient: high reincrease rate with different $tcPO_2$ values

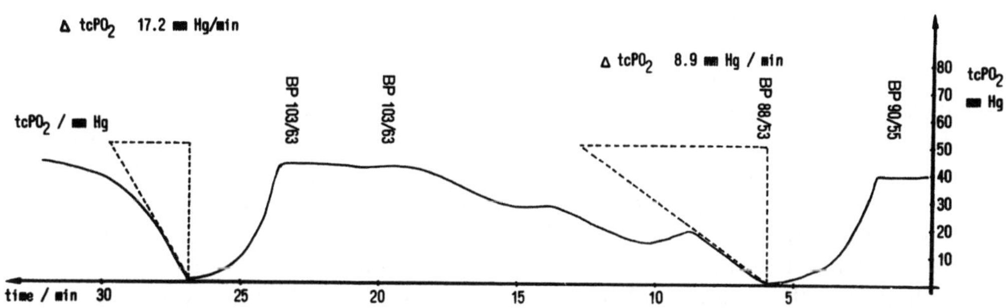

Fig. 3. Shock patient: reincrease rate in the presence of low BP (*r*) 22.2% ($\Delta tcPO_2 = 8.9$ mm Hg/min), and of higher BP 37.4% ($\Delta tcPO_2 = 17.2$ mm Hg/min)

was restored. The $tcPO_2/PaO_2$ value in patients in shock dropped considerably, depending on the severity of the shock. In these patients the reincrease rate percentage was also clearly reduced. From this we conclude that the reincrease rate percentage after suprasystolic blockage, particularly for low $tcPO_2$ values, can be used as a noninvasive parameter to estimate the local flow.

If further studies confirm these results, this parameter can be used to distinguish between, for example, a low $tcPO_2$ measurement caused by low blood O_2 and one due to reduced local flow because of shock, without the arterial PO_2 being known. This is illustrated in Fig. 2 in an example of an artificially ventilated patient with stable circulation. The $tcPO_2$ value varies depending on the FIO_2; the reincrease rate percentage is not reduced when the $tcPO_2$ value is low, but is infact even higher (110% vs 97%). By contrast, a patient with shock (Fig. 3) shows a $\Delta tcPO_2$ of

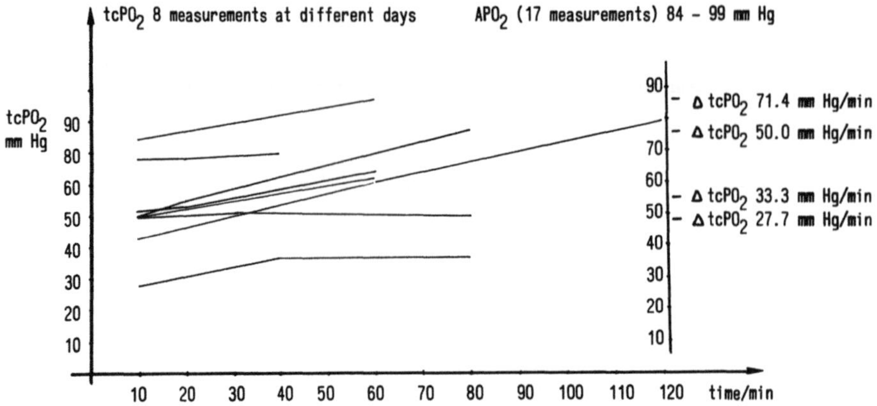

Fig. 4. Various measurements on one volunteer on different days: initial values vary, some being very low, with continuous $tcPO_2$ increase for up to 2 h

8.9 mm Hg/min, dependent on the blood pressure together with a low respiratory rate (RR) value. This corresponds to a reincrease rate of 22.2%. When the RR values rise temporarily, the $\Delta tcPO_2$ increases to 17.2 mm Hg/min, which corresponds to a reincrease rate of 37.4%.

There is a group of healthy volunteers and patients with stable circulation who do not respond to local vasodilatation with a maximum blood flow increase (so-called slow or nonresponders). This reaction might be caused by insufficient blood inflow from the larger vessels, possibly due to vegetative counterregulation mechanisms of the skin. In this group, the $tcPO_2$ as well as the quotient $tcPO_2/PaO_2$ are low. The reincrease rate percentage is also low. In terms of clinical applicability, the values measured in these patients with stable circulation and in healthy volunteers were similar to those measured in shock patients. Figure 1 shows, on the right, such a volunteer, with low initial $tcPO_2$ and a $\Delta tcPO_2$ of 25 mm Hg/min, corresponding to a reincrease rate of 67%. On another day (left) the $tcPO_2$ was high (85 mm Hg), with the PaO_2 unchanged and the reincrease rate was 93%.

Figure 4 shows that the duration of the monitoring is of major importance. Measurements lasting various amounts of time were carried out on one volunteer on different days. Apart from the different initial values, there was a $tcPO_2$ increase with no change in the PaO_2 for up to 2 h. Accordingly, the reincrease rate percentage varied from 45% to 85% (27.2 to 71.4 mm Hg/min). It would be important to identify such patients right away or to find a way to achieve optimum hyperemia in this group as well. In the first experiments, in which 1% Scandicaine (0.5 ml) was injected under the $tcPO_2$ electrode, the $tcPO_2$ increased from 56 to 79 mm Hg within 12 min (arterial PO_2 approximately 85 mm Hg, $n=4$). Further experiments are necessary to show whether this reaction can be used for diagnostic purposes.

References

1. Huch, R., A. Huch, D. W. Lübbers: Transcutaneous PO_2. Thieme-Verlag/Stratton, Stuttgart/New York (1981)
2. Schober, J. G., K. Stübing: Arterial and transcutaneous PcO_2 monitoring in cardiogenic shock. In: R. Huch, A. Huch (eds): Continuous blood gas monitoring. M. Dekker New York/Basel (1983)
3. Tremper, K. K., W. C. Shoemaker: Transcutaneous oxygen monitoring of critically ill adults, with and without low flow shock. Crit. Care Med. 9 (1981) 706-709
4. Lübbers, D. W.: Transcutaneous PO_2 in shock (in print)

Transcutaneous PO_2 Measurement at 37 °C in Children with Diabetes

U. Ewald, G. Rooth, and T. Tuvemo

Introduction

Transcutaneous oxygen pressure measurements ($tcPO_2$) are used widely in the fields of perinatology and anesthesiology [1]. Noninvasive continuous monitoring of arterial oxygen pressure (PaO_2) requires maximal hyperemia of the skin under the electrode. This is achieved with an electrode temperature of 44°-45 °C. If the circulation in the skin is impaired, i. e., the patient is in shock with disturbed central and/or peripheral circulation, a discrepancy between $tcPO_2$ and PaO_2 occurs [2]. The influence of the circulation on $tcPO_2$ becomes more evident when the normal regulation of the vessel tone is not disturbed by heating. Absence of heating also permits the study of physiological vascular reactivity. With an electrode temperature of 37 °C, we obtained reproducible measurements of the changes in blood flow occurring during postocclusive reactive hyperemia [3, 4].

There is a close relationship between $tcPO_2$ at 37 °C and the tissue PO_2 in the epidermis as the measuring probe is separated from the living epidermis only by the electrolyte layer (5-10 μm), the Teflon membrane (25 μm), and the dead stratum corneum (10-15 μm). No oxygen consumption takes place in these layers and the oxygen consumption of the electrode (proportional to PO_2) counteracts the diffusional resistance. The changes in $tcPO_2$ therefore mirror variations in tissue PO_2 which only depend on blood flow changes in the capillary loops.

We have developed a model based on the general laws of gas. Using this model we have been able to calculate blood flow during reactive hyperemia. For the calculation we used data from the literature on the thickness of the skin layers, on hemoglobin concentration, on diffusibility and solubility constants, and on the metabolic rate of the skin [5]. As we have recently demonstrated that children with diabetes type I have reduced reactive hyperemia [6, 7], the objective of this investigation was to calculate skin blood flow in diabetic and healthy children using the $tcPO_2$ method at 37 °C.

Methods and Results

Reactive hyperemia was induced by a 4-min cuff occlusion of the forearm circulation. A conventional $tcPO_2$ electrode at 37 °C was attached to the forearm skin and the resulting $tcPO_2$ values were used for calculation of skin blood flow. Experiments were performed in ten children with type I diabetes mellitus and in ten healthy children matched as to age and sex. Figure 1 gives a representative example of $tcPO_2$ recordings at 37 °C of the postocclusive reactive hyperemia in a healthy child and in a child with diabetes mellitus.

Figure 2 shows the distribution of the postocclusive peak $tcPO_2$ values. The maximal peaks noted in children with diabetes were considerably lower than those in control children ($P<0.001$; Student's *t* test). There was also a significant differ-

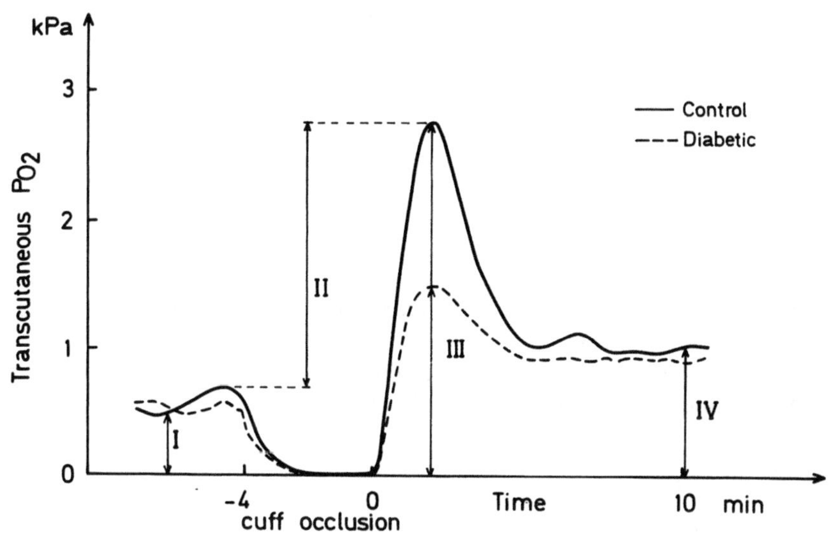

Fig. 1. $tcPO_2$ of the forearm at an electrode temperature of 37 °C during and after arterial occlusion in a diabetic and a nondiabetic child

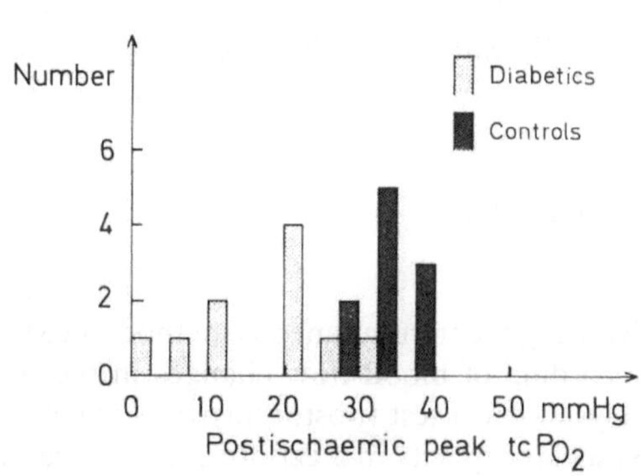

Fig. 2. Distribution of postocclusive peak $tcPO_2$ values in ten diabetic children and ten matched controls

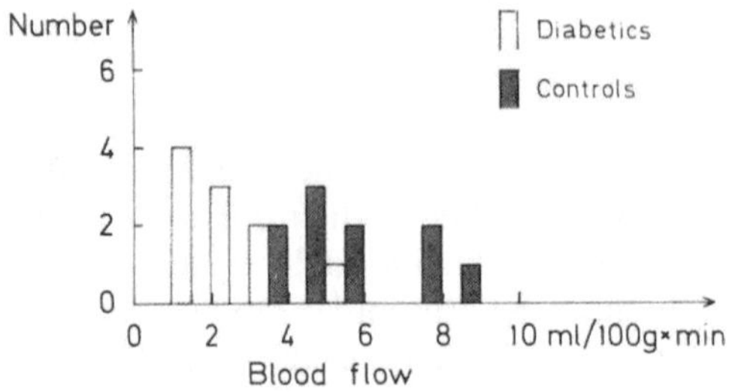

Fig. 3. Calculated peak values of blood flow during reactive hyperemia in the children of Fig. 2

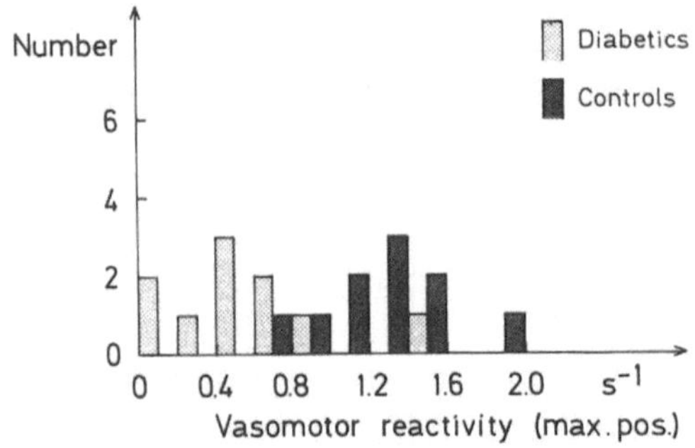

Fig. 4. Calculated maximal rate of vasodilation (VMR) in the groups of Fig. 2

ence between healthy children and those with diabetes with respect to the distribution of the calculated values for the maximal blood flow during postocclusive reactive hyperemia (Fig. 3).

As a measure for the velocity of the vasodilative phase, vasomotor reactivity (VMR) value can be calculated as follows:

$$\mathrm{VMR} = \frac{1}{Q(t)} \times \frac{dQ(t)}{dt}$$

where $Q(t)$ is blood flow per unit time.

Figure 4 shows the distribution of the maximal positive VMR values in the ten pairs. In the children with diabetes the vasodilation was much slower than in the healthy children.

From Figure 5 it is obvious that children with diabetes exhibited both the slowest VMR and the smallest postocclusive $tcPO_2$ increase.

Discussion

By using the transcutaneous method of oxygen measurement at 37 °C a noninvasive recording of blood flow changes in the dermis is possible. Using a standardized physiological test (postocclusive reactive hyperemia), a measure of the functional vascular capacity in a certain skin area can be obtained.

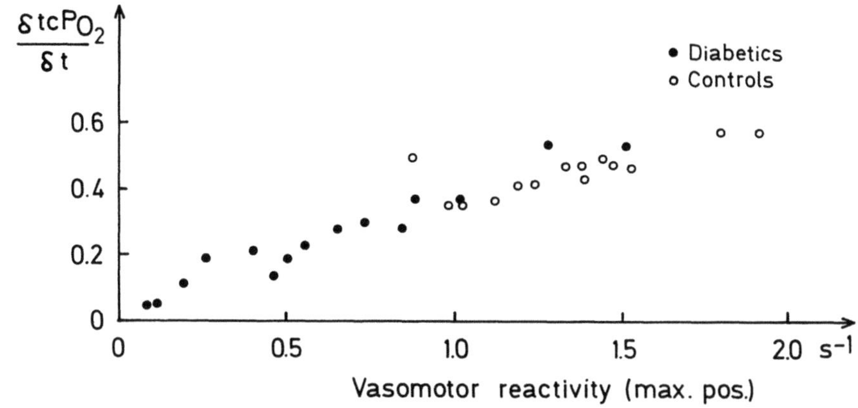

Fig. 5. Relation between maximal VMR and $tcPO_2$ increase in the children of Fig. 2

By applying this technique, a functional vascular disturbance was identified in diabetic children with only a few years duration of disease. Although these children had no micro- or macroangiopathy, we found that they exhibited slower vasodilation and impaired vascular reactivity. This confirms the hypothesis that functional disturbances precede the later structural lesions [8]. The inability to respond adequately to tissue hypoxia shown here at an early stage of diabetes mellitus in very young persons might well have pathogenetic significance in the initiation and development of diabetic angiopathy. The advantage of the described method is that it gives reproducible estimates of the vascular capacity in a local skin area in response to a physiological stimulus. Since it is a noninvasive method, it can be used for repeated measurements even in children.

References

1. Huch, R., A. Huch, D. W. Lübbers: Transcutaneous PO_2. Thieme-Stratton, New York (1981)
2. Tremper, K. K., K. Waxman, R. Bowman, W. C. Shoemaker: Continuous transcutaneous oxygen monitoring during respiratory failure, cardiac decompensation, cardiac arrest, and CPR. Crit Care Med 8 (1980) 377–381
3. Ewald, U.: Evaluation of the transcutaneous oxygen method used at 37 °C for measurement of reactive hyperaemia in the skin. Clin Physiol 4 (1985) 413–423
4. Ewald, U., G. Rooth, T. Tuvemo: Postischaemic hyperaemia studied with a transcutaneous oxygen electrode used at 33–37 °C. Scand J Clin Lab Invest 41 (1981) 641–645
5. Caligara, F., G. P. Mezzi, U. Ewald, T. Tuvemo, G. Rooth: Skin blood flow, vasomotor reactivity and tissue PO_2 calculated from non-steady state transcutaneous PO_2 measurements. Int J Microcirc (submitted 1986)
6. Ewald, U., T. Tuvemo, G. Rooth: Early reduction of vascular reactivity in diabetic children detected by transcutaneous oxygen electrode. Lancet 1 (1981) 1287–1288
7. Ewald, U., T. Tuvemo: Reduced vascular reactivity in diabetic children and its relation to diabetic control. Acta Paediatr Scand 74 (1985) 77–84
8. Altura, B. M., S. Halevy, P. D. M. Turlapaty: Vascular smooth muscle in diabetes and its influence on the reactivity of blood vessels. In: Davis E (ed) The microcirculation in diabetes. Karger, Basel (1979) pp. 118–150

The Influence of the Extent of the Heated Skin Area on Transcutaneous Oxygen Pressure

F. Fallenstein, R. Lemke, J. M. Steinacker, R. E. Wodick, A. Huch, R. Huch, and D. W. Lübbers

Introduction

Transcutaneous (tc) blood gas measurement [1] can be considered the standard method in the field of neonatology worldwide [2, 3]. In the beginning stages, when commercial $tcPO_2$ monitors first came on to the market, various studies were carried out which were devoted, among other things, to establishing the ideal electrode temperature. All of these investigations confirmed that the higher the electrode temperature, the smaller the absolute difference between $tcPO_2$ and arterially measured PO_2 (PaO_2), i.e., the closer the correlation between PaO_2 and $tcPO_2$ [4]. The problem, however, is that higher temperatures raise the risk of damaging the skin, from severe redness up to first degree burns, with the duration of continuous electrode application playing an important role.

With regard to the problem of electrode temperature, a compromise has to be found between the wish to achieve a correlation with the central PO_2 using the highest temperatures possible, and the desirability of leaving the oxygen electrode in the same place for as long as possible. In the Department of Neonatology of the Women's Hospital in Zürich, an electrode temperature of 44 °C with an uninterrupted measuring period of a maximum of 4 h has been used successfully for a number of years. When the electrode is repositioned, the response time usually takes less than 10 min, which means that there is hardly any loss of measuring time.

The transcutaneous PO_2 method applied in adults has not been as successful as in newborns, because in many cases the hyperemia induced by the external heating of the area to be measured is not sufficient. Consequently, the transcutaneously measured values are sometimes far below the arterial level. Another factor in this connection is the dependence of the $tcPO_2$ on changes in local skin blood flow: since the hyperemia achieved does not result in an oversupply of oxygen beneath the measuring area, flow changes due to oxygen consumption in the surrounding tissue (and possibly in the polarographic system of the $tcPO_2$ electrode) contribute to the results of the recordings.

Approaches to Improving Hyperemia

The problems described above cannot be eliminated by further increase of the electrode temperature: 45 °C is the maximum temperature of a heated electrode with a diameter of 5–6 mm that can be tolerated – both subjectively and objectively. Higher temperatures are found unpleasant ("biting" or "pinching") after just a short time and cause skin irritations that can last for more than 24 h.

There are various approaches to the problem of achieving better hyperemia. One of them is based on the principle of diathermy [5]: an electrode in the form of a high-frequency antenna emits shortwave electromagnetic energy through the surface of the skin. The high-frequency energy is converted into heat, first in the vital skin layers and then in the capillary area. Since the electrode itself remains "cold," the heat gradient is reversed in comparison with the directly heated electrode, so that the skin is subjected to less thermal irritation. Pilot studies showed that this method could be very promising. The problems of controlling the penetration depth and the temperature in the heated tissue layer, however, have yet to be adequately solved.

Another method is the so-called "stripping technique" [6, 7], by which the stratum corneum of the skin is removed layer by layer. The effect can be seen both in a reduced O_2 diffusional resistance and in increased heat conduction between the electrode and the capillary region. The mean $tcPO_2$ levels are 11 °C higher. It must be noted here that these methods are right at the borderline between noninvasive and invasive.

A further approach is based on the idea that by enlarging the heated skin surface, an increase in blood flow and thus an improvement of the $tcPO_2$ measurement could be achieved. In the following, three studies are described, which were carried out in Dortmund, Ulm and Zürich.

Methods and Results

The Dortmund Study

The Dortmund study group led by Lemke and Lübbers [7] used a ring-shaped accessory heating element with a diameter of 30 mm in the form of a copper shield. The device was placed concentrically over a $tcPO_2$ electrode on the skin surface. The accessory heater was electrically controlled and set at temperatures between 37° and 42 °C. The arterial PO_2 was determined each time, and the $tcPO_2$ levels noted with and without the additional heating (partly with several successive raised temperatures). The study was carried out on 27 patients with stable circulation (measuring done on the arm) and 16 patients with arterial blood flow disturbances of the lower extremities (measuring done on the outside of the lower leg).

Out of 27 measurements made in the first group, 3 showed a continuous $tcPO_2$ increase, which was very probably caused by a parallel increase in the PaO_2. One

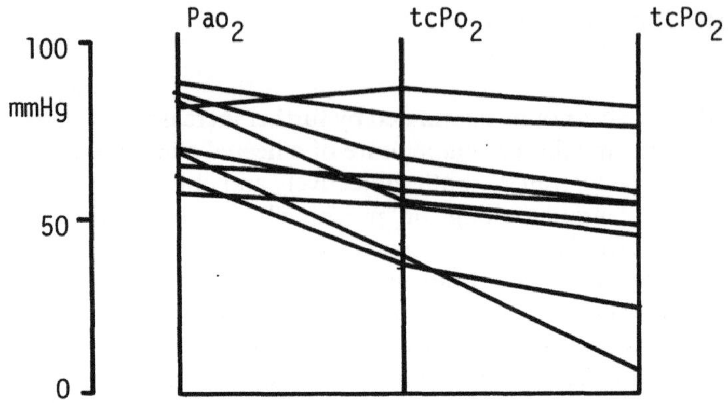

Fig. 1. Volunteers with stable circulation (Dortmund): $tcPO_2$ drops *(right)* after the electrode has been additionally heated

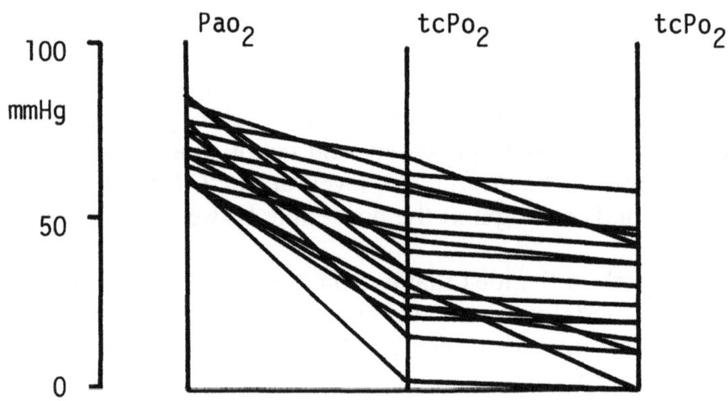

Fig. 2. Patients with arterial circulatory disturbances (Dortmund): $tcPO_2$ drops *(right)* after the electrode is additionally heated

measurement revealed an increase which was possibly due to the effect of the heating cover. In 14 measurements, no changes were observed. In 9 measurements (Fig. 1), the $tcPO_2$ dropped drastically after the accessory heating was turned on. It can be seen that in the cases where the $tcPO_2$ drops most, the initial $tcPO_2$ already deviates considerably from the arterial value.

In the second group of patients with circulatory disturbances, one person was measured 3 times, making a total of 18 measurements (Fig. 2). With one exception, all of the measurements show an appreciable drop in the $tcPO_2$ after the heating cover is turned on. The difference between the initial $tcPO_2$ and PaO_2 is also greater in this group than in the group of patients with stable circulation.

The tentative conclusion of this study is that accessory heating offers no advantage for the $tcPO_2$ method under normal circulatory conditions. If the circulatory conditions are already restricted, enlarging the heated skin area apparently even has an adverse effect on the local skin circulation.

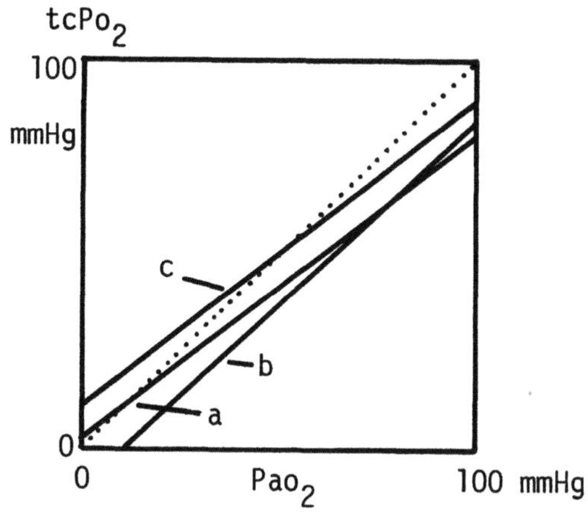

Fig. 3. Regressions between blood PO_2 measured invasively and transcutaneously with two electrodes (Ulm). *a)* Electrode I without accessory heating ($n=86$): $tcPO_2=0.79$ mm Hg; $PaO_2=+3.7$ mm Hg ($r=0.74$). *b)* Electrode I with accessory heating ($n=83$): $tcPO_2=0.94$ mm Hg; $PaO_2=-9.5$ mm Hg ($r=0.83$). *c)* Electrode II ($n=87$): $tcPO_2=0.82$ mm Hg; $PaO_2=+10$ mm Hg ($r=0.59$)

The Ulm Study

For the series of investigations carried out by the Ulm group led by Steinacker and Wodick [8], two $tcPO_2$ electrodes (I and II) were used, both of which were heated to 45 °C. Electrode I was embedded in a large copper plate, 63 mm in diameter, which could be electronically heated to 40 °C. Measurements were carried out on ten adult volunteers and the following parameters were established: the $tcPO_2$ with electrode I without accessory heating, with electrode I with accessory heating, and with electrode II, and the PaO_2 from blood samples taken regularly from the radial artery. During the measurements, each volunteer carried out ergometric exercise which was gradually stepped up until the point of physical exhaustion.

The results of this study are characterized by the regression between the PO_2 values of the three transcutaneous measurements and their respective arterial values (Fig. 3). A better correlation between $tcPO_2$ and PaO_2 was achieved with the accessory heating; however, the average level within the physiologic range was not altered by the accessory heating.

The Zürich Study

The data obtained in the Dortmund and Ulm studies from measurements on volunteers with stable circulation do not contradict each other. This induced Fallenstein et al. [6] to perform a further investigation to find out what influence a larger area of heated skin has on the $tcPO_2$ measurement. Measurements were made with two $tcPO_2$ electrodes, one on the left and one on the right forearm. One of the two electrodes was equipped with a special integrated heatable metal ring 23 mm in diame-

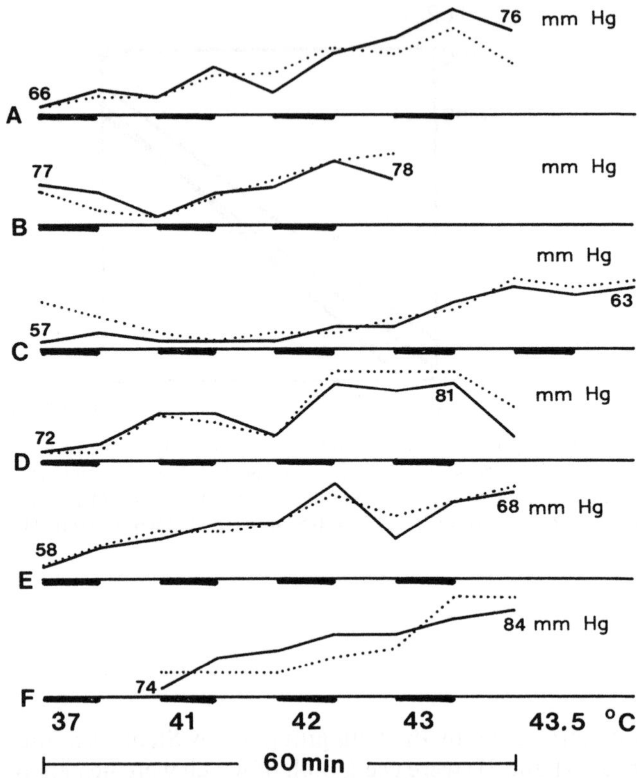

Fig. 4. Six volunteers at rest: $tcPO_2$ with one normal (...) and one additionally heated (—) electrode at various temperatures of additional heat (Zürich)

ter instead of the normal screw ring. This electronically regulated accessory heating element could be set at temperatures up to 45 °C. The temperature of both electrodes was 45 °C. Measurements were made on six healthy adult volunteers. The volunteers were at rest during the measurements in order to ensure stable physiologic conditions. The accessory heating element was not switched on during the response time up to the point when stable initial level was reached (20 min). There was then an 8 min cycle with accessory heating at 37 °C, followed by 7 min with the accessory heating switched off. These cycles were repeated at temperatures of 41°, 42°, 43° and 43.5 °C. The experiment was in each case discontinued when the volunteer could no longer bear the increasing heat.

The development of $tcPO_2$ levels measured by the additionally heated electrode and the reference electrode is shown in Fig. 4. In all cases, a continuous increase was registered, which is reflected in the curves of both electrodes. There is no divergency in the behavior of the additionally heated electrode compared with that of the reference electrode.

Our explanation for the continuous increase in the $tcPO_2$ level in all six measurements is that absolute physical resting conditions and thus steady breathing

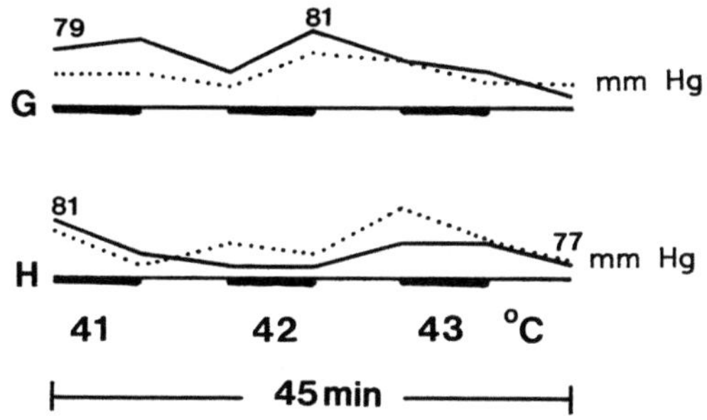

Fig. 5. Two volunteers under ergometric stress of 40 W: $tcPO_2$ with one normal (...) and one additionally heated (–) electrode at various temperatures of additional heat (Zurich)

produce a higher central PO_2 level. To test this hypothesis, we carried out two additional measurements, this time during constant exercise with 40 W ergometric power. The other measurement conditions were the same as before (Fig. 5). As expected, there was no increase in the PO_2.

Discussion

The results of three investigations carried out independently of one another indicate that enlarging the heated area when using $tcPO_2$ electrodes is not suitable for bringing the $tcPO_2$ values measured in adults closer to the PaO_2 level. The Dortmund studies show that in the presence of arterial circulatory disturbances the $tcPO_2$ values are lowered even further by the additional heating. We are convinced that the phenomena observed are of a physiologic origin. Two models are feasible:

1. The skin capillaries are already dilated as far as possible by the relatively small heating area of the normal $tcPO_2$ electrode and so cannot be widened any further even if the heating area is enlarged. The difference between the $tcPO_2$ and the PaO_2 is a result of high diffusional resistance and of the oxygen consumption in the measured area. If the arterial circulation is already impaired, the remaining flow reserves distribute themselves onto the enlarged heating area with maximally dilated skin capillaries, which results in a decrease in the local specific circulation. This could be considered a steal phenomenon, with respect to circulation.
2. The specific skin circulation, i.e., the flow in relation to the tissue volume, increases with the enlargement of the heating area within the region heated. However, increase in oxygen consumption caused by increased metabolism is associated with this hyperthermia. These factors with their opposite effects can compensate for each other locally in the area of the polarographic electrode. This, then, could be considered a metabolic steal phenomenon. In the case of general circulatory insufficiency, the metabolic components may prevail, which would explain the observations of the Dortmund group.

References

1. Huch, R., A. Huch, D. W. Lübbers: Transcutaneous PO_2. Thieme-Verlag/Stratton, New York (1981)
2. Huch, A., R. Huch, J. F. Lucey (eds.): Continuous Transcutaneous Blood Gas Monitoring. R. Liss, New York (1979)
3. Huch, R., A. Huch (eds.): Continuous Transcutaneous Blood Gas Monitoring. M. Dekker, New York (1983)
4. Loefgren, O., L. Jacobson: The Influence of Different Electrode Temperatures on the Recorded Transcutaneous PO_2 Level. Pediatrics 64 (1979) 892-897
5. Huch, A., K. Meinzer, D. Seiler, F. Fallenstein, D. W. Lübbers, R. Huch: A Modified $tcPO_2$ Electrode: A New Concept Using Diathermy. In [2] S. 103-107
6. Fallenstein, F., W. Nef, A. Huch, R. Huch: Effect of Stripping on the Level and Variability of Transcutaneous PO_2. In [3] S. 161-167
7. Lemke, R., D. Klaus, D. W. Lübbers: Experiences with the commercially available $tcPO_2$ electrode in adults. In [3], S. 143-151
8. Steinacker, J. M., R. E. Wodick: Transcutaneous Measurement of Arterial PO_2 in Adults: Design of an Improved Electrode. In [2] S. 133-141

Animal Experiments

Tissue PO_2 of the Surface of the Kidney Under Hemodilution

W. Hartel

Introduction

There have been numerous publications dealing with the hemodynamics and the metabolic and respiratory changes in hemorrhagic shock, but due to technical reasons the important question, namely that of tissue oxygen tension, has not been conclusively answered. During the last few years, however, platinum electrodes have been developed, with which the O_2 pressure on the surface of the organ can be determined. In the trial here discussed, therefore, it was decided to apply tissue surface electrodes during hemorrhagic shock, since blood flow and blood gas measurements provide only indirect conclusions about the peripheral O_2 supply.

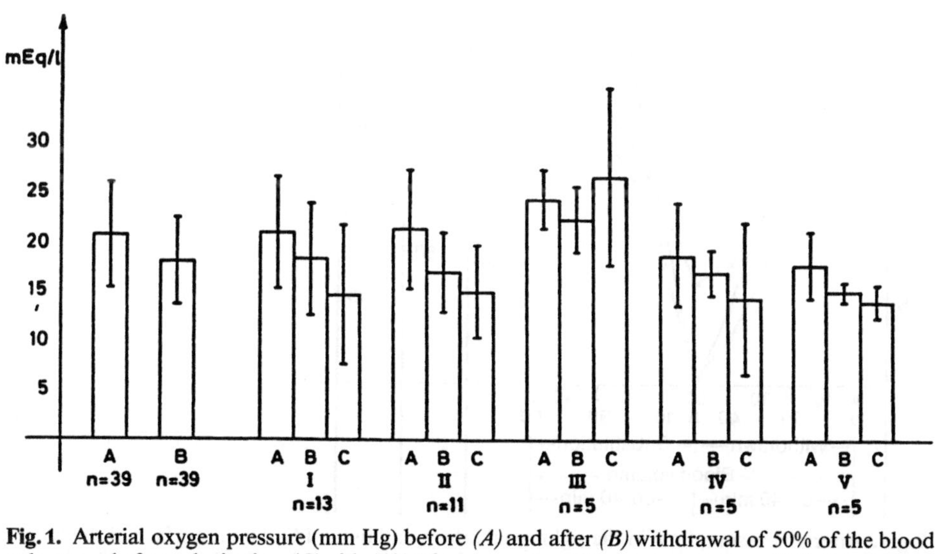

Fig. 1. Arterial oxygen pressure (mm Hg) before *(A)* and after *(B)* withdrawal of 50% of the blood volume and after substitution *(C)* with 10% Rheomacrodex *(II)*, tutofusin lactate *(III)*, 5% albumin (Behring Werke) *(IV)*, and whole blood *(V)*

Materials and Methods

Under constant respiratory conditions, tissue O_2 tension was measured on the kidneys in 39 dogs and on the liver in 12 dogs under light Nembutal anesthesia. At the same time, the arterial and venous pressure, the flow in the supplying arteries, and the blood gas were registered continuously. Hemorrhagic shock was induced by the withdrawal of 50% of the blood volume within approximately 40 min. The arterial pressure was 30–40 mm Hg at the peak of shock. Volume substitution was carried out with the following substitutes at a ratio of 1:1 - 10% dextran 40 (Rheomacrodex) in 0.9% NaCl, 5% human albumin, and whole blood. The blood lost was substituted with isotonic electrolyte solution (tutofusin lactate) at a ratio of 1:3.

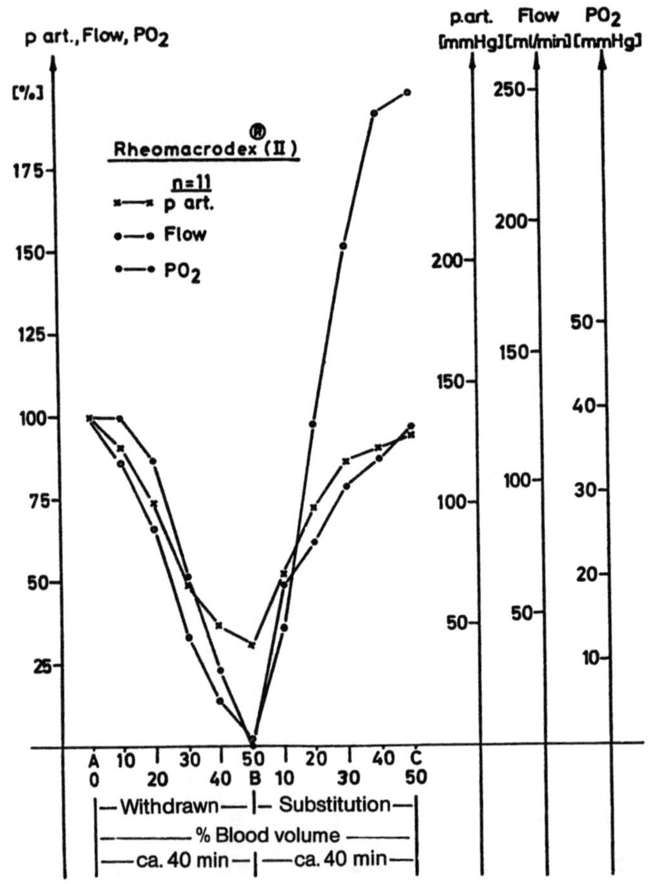

Fig. 2. Arterial pressure, flow in the right renal artery, and tissue PO_2 on the surface of the kidney after withdrawal of 50% of the blood volume and full substitution with 10% Rheomacrodex in 0.9% NaCl

To polarographically determine the tissue O_2 on the surface of the kidney, a Teflon-coated multiwire platinum electrode developed by Clark was used, which consisted of five platinum wires with a diameter of 20 μm each. The oxygen pressure from the five different pressure fields was pooled and registered continuously.

Results

Figure 1 shows the average arterial PO_2 values of the individual substitution groups. Even before the blood volume had been withdrawn the mean PO_2 had dropped to 83.7 mm Hg. The major cause of this could be multiple atelectasis areas as a result of secretion, which, despite an initial application of ⅛ mg Atropine, could not be prevented. After blood withdrawal, the average O_2 pressure fell to 74.8 mm Hg. Limited uptake of oxygen during shock can be the result of an occlusion of the lung capillaries due to microthrombosis. According to [5], this and the declining pulmonary pressure may cause the dead space ventilation, usually at 30%, to rise to 80%. After volume substitution, the average PO_2 value can be balanced out. In the dextran and electrolyte group, however, a deficit remained (though insignificant for O_2 saturation).

Figures 1-4 show proportional and absolute values of the arterial pressure and flow in the renal artery, along with the O_2 pressure on the surface of the kidney in the different substitution groups. In the phase of blood withdrawal, all of the animals had the following in common: an average O_2 pressure of 44 mm Hg was mea-

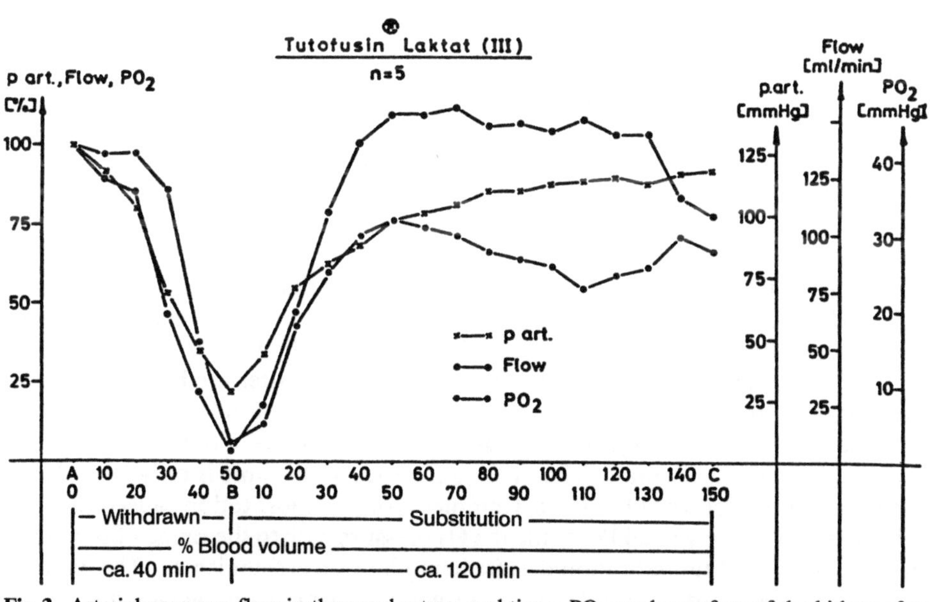

Fig. 3. Arterial pressure, flow in the renal artery, and tissue PO_2 on the surface of the kidney after withdrawal of 50% of the blood volume and threefold substitution with tutofusin lactate

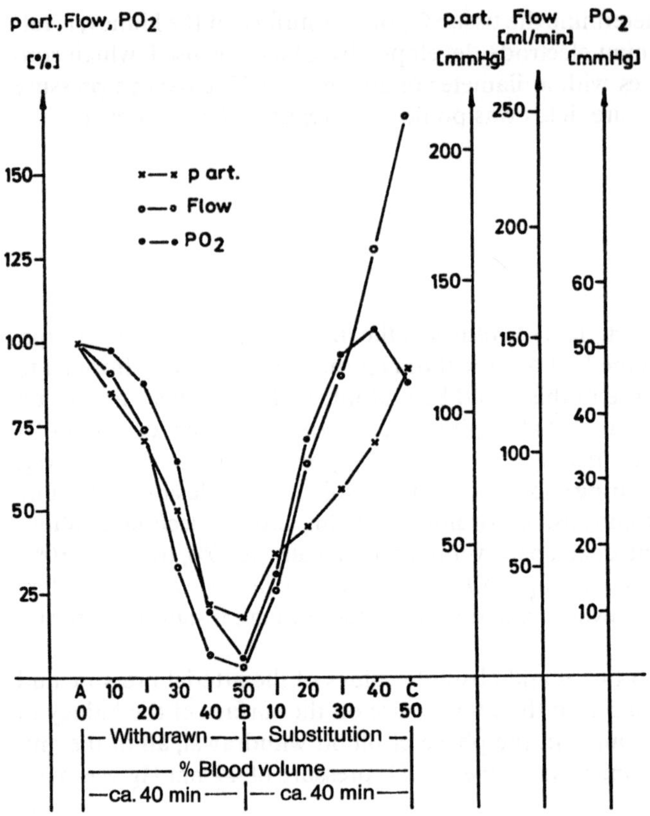

Fig. 4. Arterial pressure, flow in the renal artery, and tissue PO_2 on the surface of the kidney after withdrawal of 50% of the blood volume and full substitution with 5% albumin (Behring Werke)

sured on the surface of the kidney (highest value 66, lowest value 23). Up to 20% withdrawal, the PO_2 declined only slowly as a result of restricted oxidative processes, according to the findings of Kramer [4] and Deetjen [1]: the O_2 pressure of all the animals (Fig. 2) was still approximately 90% at this time, although the flow had already dropped to 55%. After withdrawal of 50% of the blood volume, the PO_2 on the surface of the kidney fell to 8–23 mm Hg, which for the kidney is critical.

Figure 2 demonstrates the rapid recovery of the flow after the blood volume is replaced with 100% Rheomacrodex. After 20% volume substitution, about 100% of the initial flow is reached again. The PO_2 value acts the same, although the proportional O_2 increase lags behind. The av-O_2 difference at this time is only 2 vol. %, an expression of the delayed O_2 release. Complete volume substitution caused the flow to increase excessively to 200%, while the O_2 pressure rose to the range of the initial level. The degree of O_2 restitution despite hemodilution is astonishing. The increase in flow in the renal artery could have a causal relationship with the widening of the previously constricted vessel regions. The excessive flow must be considered to be

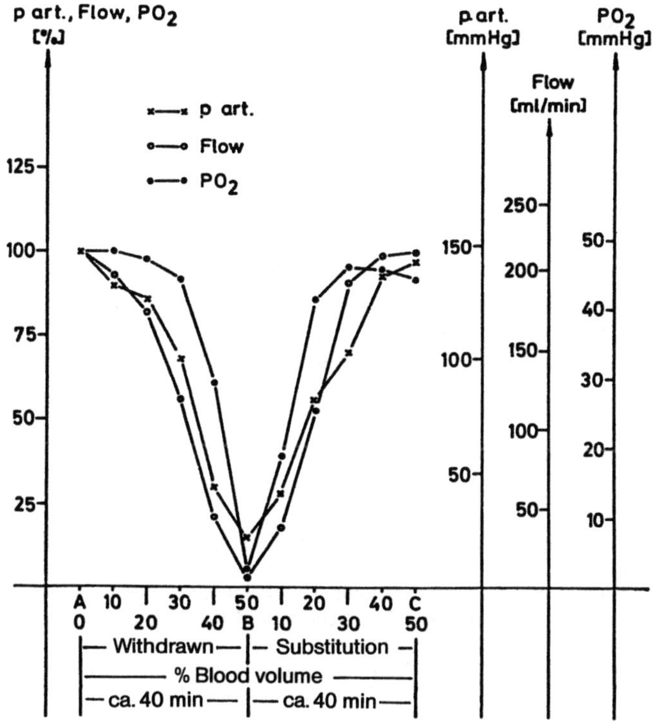

Fig. 5. Arterial pressure, flow in the renal artery, and tissue PO_2 on the surface of the kidney after withdrawal of 50% of the blood volume and full substitution with whole blood

due to the preservation of renal autoregulation. Reduction of the peripheral resistance due to hemodilution and damage to the smooth muscles as a result of shock-induced acidosis both probably contribute to this to an equal extent.

Figure 3 shows the same conditions for the crystalloid solution, which brings the flow back to the initial level, but cannot raise the tissue O_2 to above 76.2%. With increasing hemodilution, a decreasing trend appears for both parameters. While the volume is being replenished with 5% human albumin (Fig. 4), a slight lead of the tissue O_2 over the flow can be seen, in contrast to the Rheomacrodex group. Bartel and associates [3] pointed out that the O_2 affinity of hemoglobin increases under artificial colloid administration. For unknown reasons, this impedes the O_2 supply to the tissue. A possible explanation could be the higher PO_2 values on the surface of the kidney in the human albumin group.

The lead of the tissue PO_2 becomes more clear under retransfusion (Fig. 5): with a 20% volume substitution, 50% of the flow and 80% of the tissue PO_2 is restored. This is not surprising, since hemodilution is taking place. Furthermore, the lack of excessive flow and the small amplitude in the flow registration should be noted. This indicates a low cardiac volume load.

Discussion

The following conclusions can be drawn from the investigations:

1. The measurement of the tissue O_2 is a valuable method of evaluating microcirculation during hemorrhagic shock. In contrast to flow, the nutritive peripheral blood supply can be reliably assessed.
2. In the earlier phase of hemorrhagic shock, the PO_2 on the surface of the kidney decreases more slowly than the flow.
3. There are indications that volume substitution with albumin or blood restores the tissue PO_2 more quickly than that with artificial colloids.

References

1. Deetjen, P.: Normal and critical oxygen supply of the kidney. In: D. W. Lübbers, U. C. Luft, G. Thews, E. Witzleb (eds.): Oxygen Transport in Blood and Tissue. Thieme-Verlag, Stuttgart (1968)
2. Hartel, W.: Tierexperimentelle Untersuchungen über Veränderungen des Gewebssauerstoffdrucks im hämorrhagischen Schock und nach Volumenersatz. Habil.-Schrift, Univ. Frankfurt (1970)
3. Knorpp, K., C. Bauer, A. M. Rathschlag-Schaefer, H. Bartels: Einfluß von Plasmaersatzlösungen auf die Atemfunktion des Blutes beim hämorrhagischen Schock des wachen Kaninchens. Arzneim.-Forsch./Drug. Res. 6 (1970) 853
4. Kramer, K.: Renal failure in shock. In: K. D. Bock (ed.): Ciba Foundation Symposion on Shock: Pathogenesis and Therapy. Springer-Verlag, Berlin/Heidelberg/New York (1962)
5. Ulmer, W. T.: Die Lungenfunktion. Thieme-Verlag, Stuttgart (1970)

Response of Tissue PO_2 of the Subcutis and Skin Muscles During Hemodilution with Dextran 60 or Ringer's Lactate

W. Funk, V. Baldinger, and K. Messmer

Introduction

The use of crystalloid or colloidal solutions in primary fluid substitution therapy for hypovolemia has long been the subject of sometimes rather heated discussion, and there is still no agreement today on the indication and mode of action of the two substitution solutions [6]. In particular, little is known about the effect of hemodilution brought about by crystalloids or colloids at the microcirculatory level. This induced us to carry out a comparative study of the microangiodynamics and of the local tissue oxygenation under hemodilution with colloidal or crystalloid solutions.

Materials and Methods

As a standardized experimental model, an acute, limited hemodilution to a hematocrit of 30% was selected, which was achieved by exchanging whole blood with either dextran 60 or Ringer's lactate solution. An aluminum chamber was implanted in a dorsal skin fold of 63 hamsters (body weight 65–80 g) anesthetized with pentobarbital [2]. In addition, permanent catheters (PE 10) were inserted into the right jugular vein and the right carotid artery. Forty-eight hours later, at the earliest, the awake hamsters were immobilized in a Lucite tube for examination. Following a control measurement, the blood of the animals was diluted by an exchange of blood with dextran 60 at a ratio of 1:1 and with Ringer's lactate at a ratio of 1:4. The surface PO_2 was measured continuously during the exchange. Further measurements were performed at a hematocrit of 30% as well as 30 min and 60 min after this value had been reached. Since for Ringer's lactate a rapid equilibration takes place between the intravasal and interstitial space, additional amounts of the solution were infused after the exchange to maintain the central venous pressure at control level.

Aortic pressure, central venous pressure, and heart rate, taken as parameters of the macrohemodynamics, were registered continuously. For surface oxygen pressure measurements, a micromanipulator was used to place the multiwire surface electrode (Kessler and Lübbers [4]) on 25 preparations superfused with NaCl solution (0.9%, 30 °C, pH 7.4). The correct positioning of the electrode was controlled by the PO_2 values obtained, which rapidly drop to 0 when the preparation is compressed and show uniform values corresponding to the PO_2 of the superfusion solution when the electrode is too far away from the preparation. It was also checked that the measured values did not change as a result of inflation with ambient air or nitrogen. The results are shown in histograms; the Kolmogoroff-Smirnoff Test [7] was used for the statistical evaluation.

In another experimental series ($n=20$), intravital microscopy was performed in preparations covered with a glass slip. The contrast of the blood vessels was intensified with an i.v. injection of fluorescein isothiocyanate (FITC)-dextran (MW 150000, 100 mg/kg) and the picture recorded by means of moonlight video camera. The blood cell velocities in the capillaries were then determined by the dual window cross correlation technique [1]. The number of capillaries perfused with erythrocytes which branched off from one terminal arteriole (functional capillary density) was counted. The Wilcoxon rank sum test was used for statistical evaluation.

Results

The results of our investigation are summed up in Table 1. In neither group could a change in central blood pressure or heart rate during the entire course of the experiment be detected. To keep the central venous pressure constant in the Ringer's lactate group, considerable amounts of the solution were necessary during the first hour following hemodilution: an average of seven times the amount of blood withdrawn had to be infused. The colloidosmotic pressure in these animals dropped significantly in comparison with the control group and the animals treated with dextran. Their body weight increased by an average of 13%.

Figure 1 shows on the right the changes in the surface PO_2 in the animals of the Ringer's lactate group during the course of the experiment. Initially, the histogram shows a physiologic distribution with a mean value of 20.1 mm Hg. The data were obtained in ten experiments and contain 1111 single measurements. The supply of oxygen to the tissue is not influenced immediately after hemodilution; there is only

Table 1. Colloidosmotic pressure (COP), surface oxygen pressure (PO_2), functional capillary density, and capillary blood cell velocity (vEry)

		COP (cm H_2O)		$tcPO_2$ (mm Hg)		Capillary density (*n*)		vEry (mm/s)	
		Contr.	60′	Before	60′	Before	60′	Before	60′
Dextran 60	Mean	26.4	28.4	19.1	16.6	9	7	0.37	0.29
	SD	4.3	4.7	—	—	3	4	0.22	0.16
Ringer's solution	Mean	26.4	13.1*	20.1	7.8*	12	3*	0.33	0.03*
	SD	4.3	2.5	—	—	6	3	0.11	0.23

* $P<0.01$

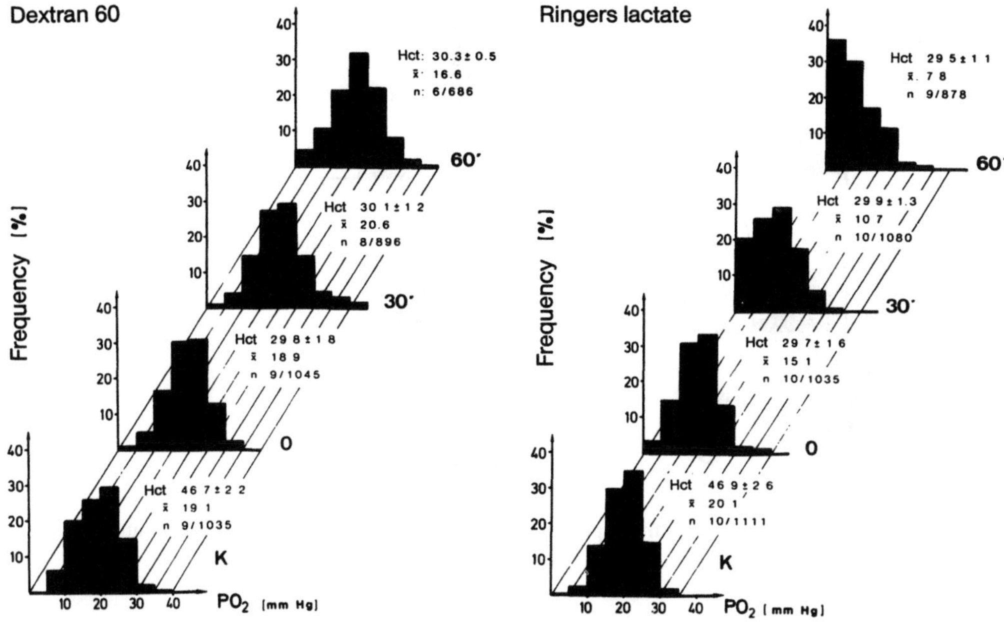

Fig. 1. Changes in the tissue surface PO_2 during hemodilution, before the exchange, immediately after *(O)*, and 30 min and 60 min later. *Left*, during substitution with dextran; *right*, with Ringer's lactate

a slight shift to the left in the histogram. After 30 min, however, and even more after 60 min, a significant shift to the left ($P < 0.01$) appears: 36% of the values are now in the hypoxic range between 0 and 5 mm Hg, the mean value has dropped to 7.8 mm Hg. In contrast, there is only a slight change in the tissue oxygen supply in the dextran group, as is demonstrated by the unchanged shape and position of the histogram. The mean value declines from 19.1 to 16.6 mm Hg; the distribution, however, remains homogeneous.

These changes are concurrent with alterations in the blood cell velocity in the capillaries (Fig. 2): hemodilution with dextran 60 does not cause any significant change in the frequency distribution of the blood cell velocities. In contrast, 60 min after hemodilution, the histogram of the Ringer's lactate group is divided into two larger classes: 59 out of 89 capillaries show a blood cell velocity below 0.1 mm/s, which indicates capillary stasis. The blood cell velocity in the few capillaries perfused further, on the other hand, remained in the range of the control values. This can be interpreted as a formation of functional arteriovenous shunts.

The homogeneity of the tissue perfusion can also be demonstrated by the number of perfused capillaries (cf. Table 1). This parameter changes only slightly under hemodilution with dextran 60. In contrast, in the Ringer's lactate group, significantly fewer capillaries ($P < 0.001$) in comparison with the initial value were perfused 60 min after hemodilution.

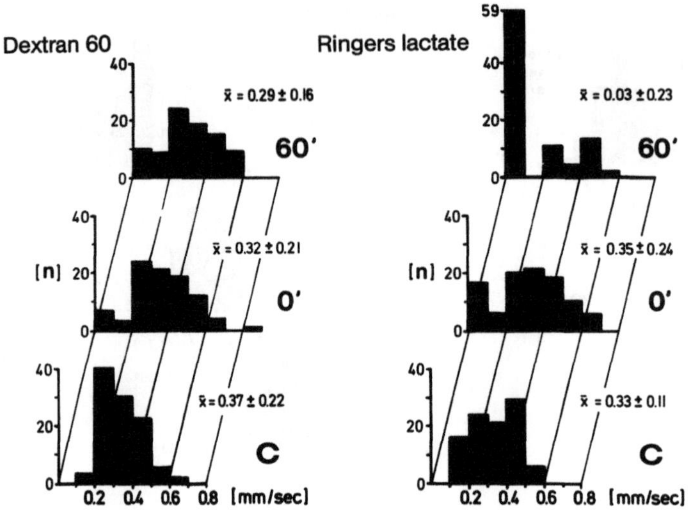

Fig. 2. Frequency distribution of the capillary blood cell velocity *(C)* before, *(O)* immediately after, and 60 min after hemodilution

Discussion

From our results we conclude that volume substitution therapy with dextran 60 does not disturb the function of the microcirculation, i.e., tissue oxygenation remains adequate. Crystalloid solutions, however, strongly interfere with the microcirculation of the subcutis and the skin muscle of the hamster: severe tissue hypoxia results, which is accompanied by interstitial edema.

The maldistribution of the capillary flow, which is manifested in the velocity as well as the PO_2 histograms, is likely to be caused by the altered fluid balance across the capillary wall. The quick filtration of Ringer's lactate might result in hemoconcentration and loss of fluid, particularly in the distal portion of the longer capillaries. At the same time, an increase in interstitial pressure is caused by an edema resulting from an overloading of the lymphatic system. These two mechanisms jeopardize capillary perfusion and tissue oxygenation.

Our results correlate well with those of Heughan et al. [3], who observed a hindrance of oxygen transport to the cutis during an excessive infusion of saline solution. Recently, Larsson and Ware [5] showed, in a model with isotonic fluid load in the rat, that a similarly unfavorable situation with edema formation can be expected with a high tissue compliance in other organs, namely, the stomach, the small and large intestine, and the skeletal muscle.

References

1. Arfors, K. E., D. Bergquist, M. Intaglietta, G. Westergreen: Measurement of blood flow velocity in the microcirculation. Ups. J. Med. Sci. 80 (1975) 27–33

2. Endrich, B., K. Asaishi, A. Götz, K. Messmer: Technical report - A new chamber technique for microvascular studies in unanesthetized hamsters. Res. exp. Med. 177 (1980) 125-134
3. Heughan, C., J. Niinikoski, T. K. Hunt: Effect of excessive infusion of saline solution on oxygen transport to tissue. Surg. Gynec. Obstet. 135 (1972) 257-265
4. Kessler, M., D. W. Lübbers: Aufbau und Anwendungsmöglichkeiten verschiedener pO_2-Elektroden. Pflüg. Arch. 291 (1966) 32-42
5. Larsson, M., J. Ware: Effects of isotonic fluid load on plasma water and extracellular fluid volumes in the rat. Eur. Surg. Res. 15 (1983) 262-267
6. Meßmer, K.: Kristalloide oder Kolloide in der Volumenersatztherapie. Beitr. Intensiv-Notfallmed. 2, Karger-Verlag, Basel (1984) S. 105-112
7. Ödman, S., N. Lund: Data acquisition and information processing in MDO oxygen electrode measurement of tissue oxygen pressure. Acta Anaesth. Scand. 24 (1980) pp. 161-165

The Effects of Bleeding and Hemodilution on Muscular PO_2 in the Dog

W. Fleckenstein, T. Kersting, A. Schaffler, R. Heinrich, K. Reinhart, and C. Weiss

Introduction

Changes in the oxygen pressure in the skeletal muscle of the dog following experimentally induced changes in the circulating intravasal volume were studied with a newly developed method [4, 5, 6, 7] for fast registration of tissue PO_2 histograms (based on 200 single local PO_2 values obtained within 6 min). The objective of the experiments was to study the relationship between muscular PO_2 on the one hand, and systemic O_2 delivery, systemic O_2 consumption, cardiac output, and systemic vascular resistance, on the other.

Materials and Methods

Since PO_2 microprobes are fragile and surface PO_2 probes require that the tissue to be measured be surgically exposed, we used polarographic needle probes with unbreakable stainless steel shafts (hypodermic needle type, diameter 350 μm) to measure muscular PO_2. The inevitable tissue displacement caused by the volume of an inserted needle probe – especially one 350 μm thick – results in a reduction of blood flow in the vicinity of the probe. The tissue cells continue to consume O_2, which causes the tissue PO_2 to decrease. Taking this chain of events into account, a procedure for tissue PO_2 measurement was developed that allowed to determine the local PO_2 before the effect of the reduced blood flow on the local PO_2 had time to significantly alter the PO_2 at the tip of the probe [6]. For this purpose, the probe was moved forward stepwise in the tissue and the PO_2 was determined during the "standstill" periods between the steps.

The technical prerequisites for this type of PO_2 measurement were (a) a PO_2-sensitive probe with a sufficiently short response time ($t\,90 = 500$ ms) and (b) a mechanical driving device providing for sufficiently fast axial movements and exact positioning of the probes in the tissue.

The response time (t 90) of the probes used for this study was < 500 ms at 37 °C. The diameter of the probe shaft was 350 μm. The sharply ground tip of the probe contained a membranized, polarographic, recessed, gold microcathode in the form of a ring with a diameter of 15 µm. A commercially available Ag/AgCl self-adhesive ECG electrode served as polarographic anode, which was placed on the underbelly skin of the dogs. After a maximum of 1.5 h of continuous tissue contact, the probes were recalibrated in a sterile calibration chamber. The following technical data were obtained from the probes in 0.9% NaCl solution at 37 °C and at a polarization potential of -750 mV: O_2 sensitivity 15-30 pA/mm Hg PO_2; probe current in O_2-free solution: 50 pA; no detectable effect of pH on the polarographic current signal in the pH range of pH 5.5 to 10.5; temperature effect: 1.9%-2.2% per °C; drift: 2% per h; usable time after first polarization: 30 h; stirring effect: about 4% of the oxygen-dependent probe current.

For the tissue PO_2 measurement the probe was guided into the muscle (m. gracilis, m. semimembranosus, m. semitendinosus) through the lumen of a previously inserted Abbocath cannula (no. 18) to the longitudinal axis of the fibers at an angle of between 15° and 30°. Within the muscle, the stepwise movement of the probe consisted of a rapid forward movement of 1.12 mm, which was followed by an equally rapid backward movement of 0.32 mm. This meant an effective forward step of 0.8 mm. Moving the probe in steps this way minimized the mechanical effects of the adhesive forces between the probe surface and the surrounding tissue. Following this technique, the probe was moved a maximum distance of 20-30 mm into the muscle, depending on the muscle thickness. One second after each movement, a single local PO_2 value was measured. To obtain the 200 single PO_2 values for a PO_2 histogram, the probe was immediately drawn back after it had reached the predetermined maximum depth and was inserted again in steps at a slightly different angle.

A few seconds after the 200 single PO_2 values had been measured, which required 6 min, the PO_2 histograms were displayed. For each histogram a new Abbocath cannula was inserted. The sampling of the 200 PO_2 values for each tissue PO_2 histogram was carried out at 15 min intervals.

The five 1-year-old beagles used for the study weighed from 10 kg to 12.5 kg. After being relaxed with pancuronium (1 mg/h), they were intubated and anesthetized with N_2O ($FIN_2O = 0.75$) and with buprenorphine (0.1 mg/h). The normocapnic respiration of the animals ($FIO_2 = 0.25$) was adjusted in such a way that the arterial PO_2 was between 120 and 130 mm Hg. Using a Swan-Ganz catheter equipped with a thermosensor, cardiac output was determined by thermodilution of cooled (0 °C) physiologic saline (cardiac output computer, Edwards Lab, AHS-Corp, USA). An arterial and a venous catheter were inserted to continuously monitor the arterial and venous blood pressures and to obtain samples of arterial and venous blood to determine the pH, base excess, PO_2, PCO_2, and hematocrit.

The first PO_2 measurements were performed 1.5 h after anesthesia had begun. While the 200 PO_2 values for the histogram were being recorded, cardiac output, arterial and mixed venous PO_2, PCO_2, pH, base excess, and hematocrit were determined. To change the circulating volume, the nonsplenectomized dogs were either bled or had a physiologic saline or hydroxyethyl starch solution administered intravenously. Both solutions infused were warmed up to 37 °C.

The hemodilution and bleeding experiments were carried out for each dog on two consecutive days. On the 1st day, after the reference values were recorded (120 min), 1000 ml physiologic saline (up to 10% of body wt.), and subsequently

(120 min later) 500 ml colloidosmotically hypertonic hydroxyethyl starch (up to 5% of body wt.) were infused (HAES-sterile 10%, Fresenius, Oberursel, FRG, mean molecular weight 200000). On the 2nd day, after the reference values were recorded (60 min), the dogs were bled in 3 steps (1st step: 2.5%-3.7% of body wt.; 2nd step: 3.8%-5% of body wt.; 3rd step: 5.1%-6.3% of body wt.). Afterwards (60 min after the last step), 500 ml hydroxyethyl starch solution were infused. Finally (60 min later), the dogs received their own heparinized blood.

Results

Figure 1 contains the sequential PO_2 histograms from the pooled data of all the dogs, which were obtained in the hemodilution experiments of the 1st day. It can be seen that during the fast infusion (within 5 min) of 1000 ml physiologic saline solution, the mean muscular PO_2 rose, at a constant mean arterial pressure, from 31.9 mm Hg (reference value) to 41.9 mm Hg within the first 30 min. Within the following 90 min, however, the mean muscular PO_2 dropped to below the reference values and an increased number of hypoxic values occurred. One hundred and twenty minutes after the infusion of physiologic saline, the hematocrit (Hct), the cardiac index (CI, related to kg body wt.), and the systemic vascular resistance (SVR) reached the reference values measured before the saline infusion had been started (mean values from five dogs).

Under the subsequent infusion of 500 ml colloidosmotically hypertonic hydroxyethyl starch solution (carried out 120 min after the infusion of physiologic sa-

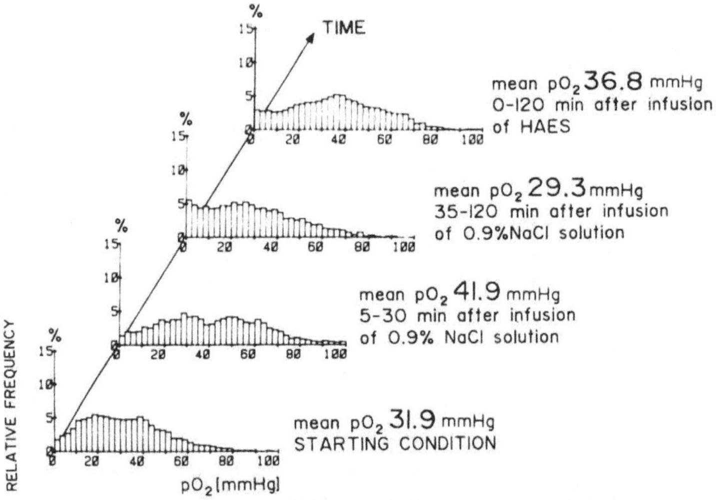

Fig. 1. PO_2 histograms: pooled data from five dogs before *(bottom)* and after infusion of 1000 ml of 0.9% NaCl solution *(2nd and 3rd histograms from bottom),* followed by infusion of 500 ml hydroxyethyl starch solution *(top). Ordinates,* relative frequency of PO_2 classes; *abscissae,* PO_2 (mm Hg)

line), the mean muscular PO_2 rose continuously until the end of the period of observation (Fig. 1, uppermost histogram; C in Fig. 2) and attained values higher than the reference values. The number of hypoxic values dropped significantly. The systemic oxygen delivery and the systemic oxygen consumption also increased (mean values from five dogs), whereas the systemic oxygen extraction ratio and arterial blood pressure remained unchanged after the infusion of hydroxyethyl starch. During the hemodilution experiments of the 1st day (Fig. 3), the systemic oxygen delivery was not closely related to the systemic oxygen consumption.

During the bleeding experiments of the 2nd day, the mean muscular PO_2 dropped after each step of bleeding (Fig. 4; E, F, G in Fig. 5). The histograms revealed a shift to the left (Fig. 4). In the 2nd and 3rd stages of bleeding, practically no tissue PO_2 values above 40 mm Hg were observed. The lowest mean muscular PO_2 was reached 10–15 min after the end of each bleeding (Fig. 4; E, F, G in Fig. 5). Afterwards, the mean muscular PO_2 values increased slightly and stabilized at a lower level approximately 30 min later.

Hydroxyethyl starch (500 ml) was infused 60 min after the 3rd bleeding. The resulting hemodilution led to an average hemoglobin content of 47 g/l of blood (SD ± 20.5).

In spite of these low Hb values, the mean muscular PO_2 reached the reference values (measured before bleeding) within 90 min after the infusion of hydroxyethyl starch (Fig. 4; H in Fig. 5), while the CI values were twice as high as the reference values. The final retransfusion of the withdrawn blood led to an increase in the mean muscular PO_2 to above the reference level (Fig. 4, uppermost histogram; I in Fig. 5).

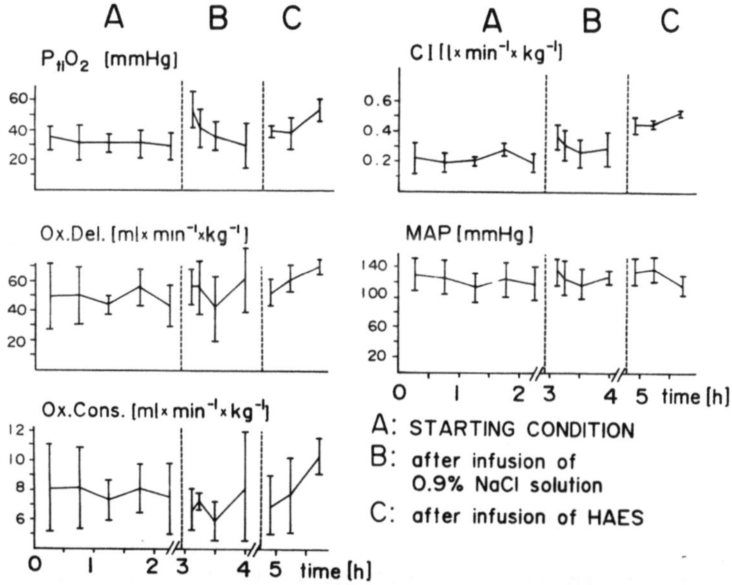

Fig. 2. Pooled data of five dogs. *Ordinates*, mean tissue PO_2 *($PtiO_2$)*, cardiac index *(CI)*, systemic oxygen delivery *(Ox. Del.)*, mean arterial blood pressure *(MAP)*, and systemic oxygen consumption *(Ox. Cons.)*; *abscissae*, time (h)

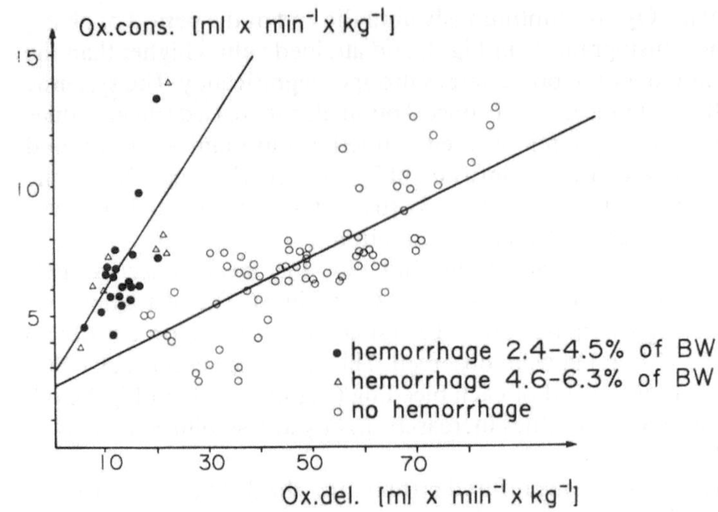

Fig. 3. Systemic oxygen consumption plotted against oxygen delivery data from bleeding experiments *(open circles);* bleeding experiments (*closed circles* and *triangles*). *Ordinate,* systemic oxygen consumption *(Ox. Cons.); abscissa,* oxygen delivery *(Ox. Del.)*

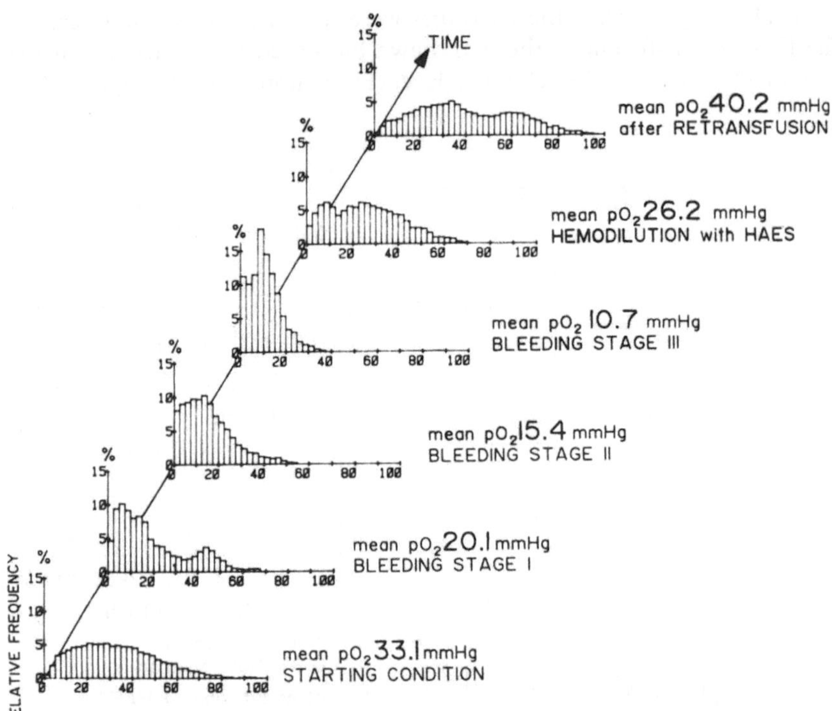

Fig. 4. PO_2 histograms: pooled data from five dogs before *(bottom)* and after each bleeding *(2nd, 3rd, 4th from bottom),* followed by infusion of 500 ml of hydroxyethyl starch solution *(5th from bottom),* and subsequent retransfusion of withdrawn blood *(top). Ordinates,* relative frequency of PO_2 classes; *abscissae,* PO_2 (mm Hg)

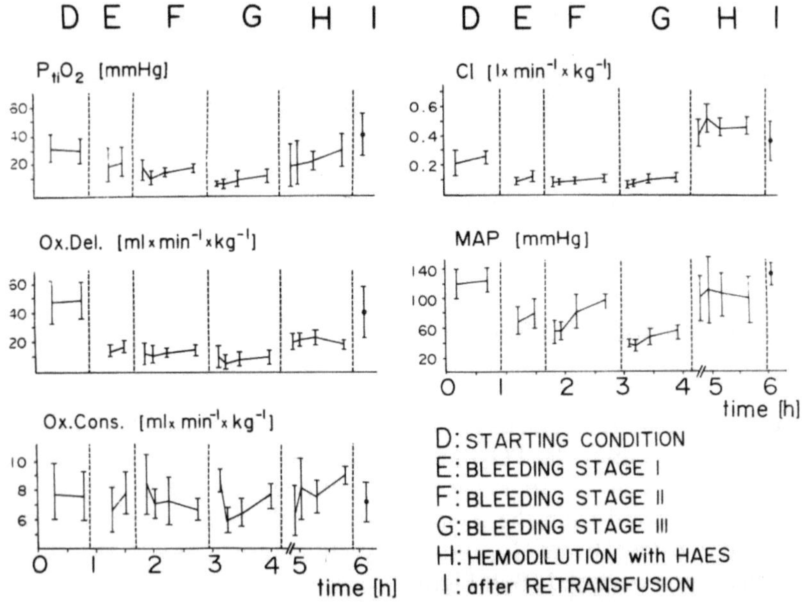

Fig. 5. Pooled data of five dogs subjected to bleeding, hemodilution, and retransfusion. *Ordinates,* mean tissue PO_2 *($PtiO_2$),* cardiac index *(CI),* systemic oxygen delivery *(Ox. Del.),* mean arterial blood pressure *(MAP),* and systemic oxygen consumption *(Ox. Cons.); abscissae,* time (h)

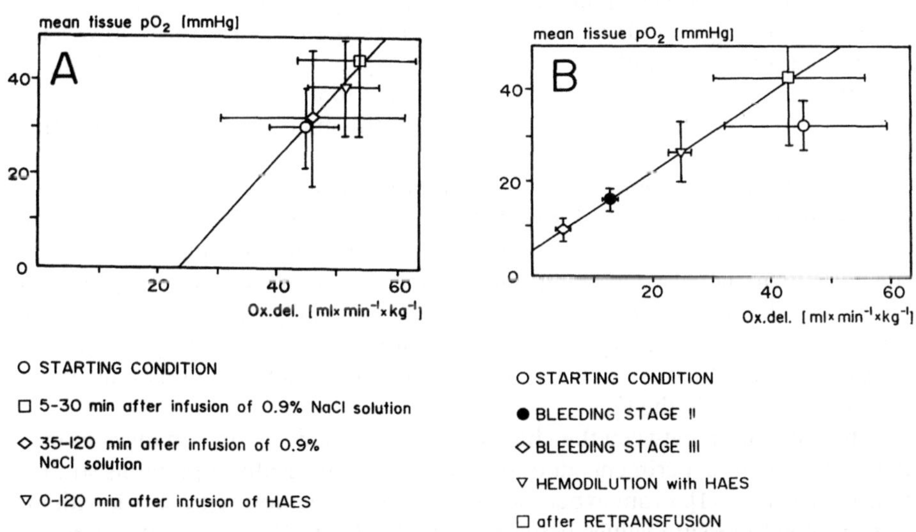

Fig. 6. Mean tissue PO_2 values plotted against mean systemic oxygen delivery (values from five dogs): *(A)* infusion experiments on 1st day; *(B)* bleeding, hemodilution, and retransfusion experiments on 2nd day

As Fig. 6 shows, a linear relation existed between the systemic oxygen delivery and the mean muscular tissue PO_2 during the bleeding experiments. Such a relation did not exist between the tissue PO_2 and the systemic oxygen consumption; neither was a linear relation between the tissue PO_2 and the SVR observed.

The systemic oxygen consumption remained essentially unaffected by the bleeding, although the oxygen extraction ratio increased from 14% to a maximum of 76% (starting condition, stage III of bleeding, respectively).

Discussion and Conclusions

During hemodilution (1st day), and even more pronouncedly during bleeding (2nd day), the mean muscular PO_2 of the skeletal muscle followed the changes in the systemic oxygen delivery. In no case, however, could the levels of systemic oxygen delivery be inferred from the mean muscular PO_2 unless the mean PO_2 values decreased to below 20 mm Hg, and hypoxic tissue zones as well as a shift to the left of the histogram occurred (B in Fig. 6).

The decrease of the mean muscular PO_2 at reduced systemic oxygen delivery in the bleeding experiments (2nd day) was less pronounced than the increase in the mean muscular PO_2 at increased oxygen delivery in the hemodilution experiments (1st day). This is demonstrated by the different slopes of the regression lines A and B in Fig. 6. One possible explanation for this finding is a decrease of muscular oxygen uptake during periods of low perfusion pressure [10, 13]. Since, however, our experiments did not include measurements of oxygen uptake of the muscle tissue, it cannot be concluded whether this finding is due to a decreased muscular oxygen uptake or to a redistribution of blood flow in the microcirculation of the muscle studied.

Two hours after the end of the infusion of physiologic saline (8%–10% of body wt.), an increased incidence of low PO_2 values was observed (Fig. 1, 3rd histogram from the bottom). The microcirculatory disturbance could be due to an interstitial edema or to an increased vascular muscle tone. The number of low PO_2 values was reduced by the subsequent infusion of colloidosmotically hypertonic hydroxyethyl starch. Another contributory factor is an abnormally high level of muscle perfusion (so-called "luxury perfusion"). The lowered blood viscosity [3] at this stage induced a twofold increase in the CI.

The correlation between the changes in oxygen delivery and changes in muscle PO_2 found in our experiments does not agree with the results of previous studies by Sinagowitz et al. [11]. From experiments with dogs in hemorrhagic shock with mean arterial blood pressure values between 40 and 110 mm Hg, those authors reported mean muscular PO_2 values of just above 0 mm Hg already occurring in the early stages of hemorrhagic shock. As the method of measuring tissue PO_2 used by Sinagowitz et al. (i.e., using multiwire surface electrodes) is different from the method we used, the results cannot be directly compared. Several other authors [9, 12, 13, 14], however, have reported a reduction but not total cessation of the oxygen consumption in the hindlimb of the dog, even at arterial perfusion pressures in the range of 40 mm Hg. It is conceivable that the weight of a multiwire surface electrode could suffice to compress microvessels in the vicinity of the probe at critically low perfusion pressures.

Apparently, in order to sustain continuing oxidative metabolism even in hemorrhagic shock, functionally significant PO_2 gradients are maintained within the muscle. The fewer capillaries perfused under conditions of low arterial pressure [1, 8, 9, 14], the more pronounced these PO_2 gradients must be. Our results support this interpretation.

References

1. Äppelgren, L.: Perfusion and diffusion in shock. Acta Physiol. Scand. 378 (1972) 5-72
2. Cherniack, N.S., N.H. Edelmann, A.P. Fishman: O_2 exchange across spleen during asphyxia. Am. J. Physiol. 219 (1970) 1585-1589
3. Ehrly, A.M., H. Landgraf, K. Saeger-Lorenz, S. Hasse: Verbesserung der Fließeigenschaften des Blutes nach Infusion von niedermolekularer Hydroxyäthylstärke (Expafusin) bei gesunden Probanden. Infusionstherapie 6 (1979) 331-336
4. Fleckenstein, W., C. Weiss: Evaluation of pO_2-histograms obtained by hypodermic needle electrodes. In: W. Bleifeld, D. Harder, H.K. Leetz, M. Schaldach (eds.): Proceedings World Congress on Medical Physics and Biomedical Engineering, 7.14, Kinzel, Göttingen, FRG (1982)
5. Fleckenstein, W.: Ein neues Gewebe-pO_2-Meßverfahren zum Nachweis von Mikrozirkulationsstörungen. Focus MHL 1 (1984) 74-84
6. Fleckenstein, W., C. Weiss: A comparison of pO_2-histograms from rabbit hindlimb muscles obtained by simultaneous measurements with hypodermic needle electrodes and with surface electrodes. Adv. Exp. Med. Biol. 169 (1984) 447-455
7. Fleckenstein, W., R. Heinrich, T. Kersting, H. Schomerus, C. Weiss: A new method for the bedside recording of tissue pO_2-histograms. Verh. Dtsch. Ges. Inn. Med. 90 (1984) 439-443
8. Grünewald, W., W. Sowa: Capillary structures and O_2 supply to tissue. Rev. Physiol. Biochem. Pharmacol. 77 (1977) 149-209
9. Honig, C.R., J.L. Frierson, C.N. Nelson: O_2 transport and VO_2 in resting muscle: significance for tissue-capillary exchange. Am. J. Physiol. 220 (1971) 357-363
10. Pappenheimer, J.R.: Blood flow, arterial oxygen saturation, and oxygen consumption in the isolated perfused hindlimb of the dog. J. Physiol. 99 (1941) 283-303
11. Sinagowitz, E., H. Rahmer, R. Rink, L. Görnandt, M. Kessler: Local oxygen supply in intra-abdominal organs and in skeletal muscle during hemorrhagic shock. Adv. Exp. Med. Biol. 37 A (1973) 505-511
12. Stainsby, W.N., A.B. Otis: Blood flow, blood oxygen tension, oxygen uptake, and oxygen transport in skeletal muscle. Am. J. Physiol. 206 (1964) 858-866
13. Stainsby, W.N., C.R. Lambert: Determinants of oxygen uptake in skeletal muscle. Exerc. Sport Sci. Rev. 7 (1979) 125-151
14. Whalen, W.J., D. Burek, C. Thuning, B.E. Kanoy, W.N. Duran: Tissue pO_2, VO_2, venous pO_2 and perfusion pressure in resting dog gracilis muscle perfused at constant flow. Adv. Exp. Med. Biol. 75 (1976) 639-655

Oxypherol – A New Way to Preserve Organs?

O. Ruland, H. U. Spiegel, J. Hauss, and K. Schönleben

Introduction

Despite the good organization of transplantation services in Europe, only 50% of the demand for kidney transplants in the Federal Republic of Germany is presently being met. This problem, as we all know, is extremely complex. A crucial factor is the amount of time that elapses before the transplant can be removed from the potential donor because of the examinations necessary to determine the time of death. Several times, among our own patients, it has happened that a potential kidney could no longer be explanted because, despite the application of drugs, hypotonic circulation could not be prevented.

Our hypothesis is that such situations can be avoided by isolated in vivo perfusion of the kidney with a suitable perfusion agent. The arterial and venous flow of

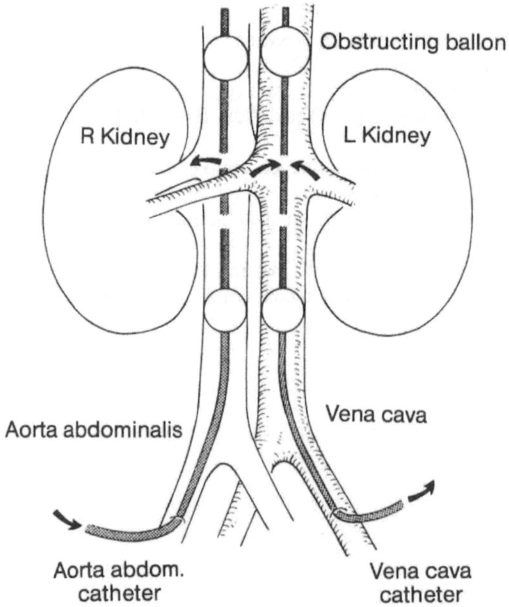

Fig. 1. Diagram for isolated renal perfusion experiment

the kidneys can be excluded from the systemic circulatory system with two balloon catheters and the kidneys can be perfused with a suitable medium via these catheters if sufficient perfusion pressure is maintained (Fig. 1). We therefore looked for a suitable perfusion agent, which, in our opinion, had to fulfill two conditions: (a) it must be able to transport oxygen and deliver it to the tissue, and (b) it must possess a certain degree of temperature stability, since it is warmed by the body.

Materials and Methods

For the perfusion medium we chose Oxypherol. This is fluosol 43 to which, following Yokoyama's recommendation, modified Ringer's solution, glucose, and human albumin is added.

The fluid ready for use is a perfluorotributylamine emulsion, which, when equilibrated with oxygen, physically dissolves the oxygen and delivers it to the tissue. It contains 124 mmol sodium and 11 mmol potassium per liter.

Preliminary experiments were first carried out on the explanted kidneys. Ten isolated dog kidneys were perfused with Oxypherol in a modified Gambro machine for 10 h at 22 °C. As a control, five dog kidneys were perfused under the same experimental conditions with Euro-Collins, since this electrolyte solution imitates the intracellular ionic content (10 mmol/l, Na, 115 mmol/l, K). This group should not be ascribed any particular importance, since the solution used cannot be compared with the oxygen-transporting liquid.

During the course of the experiment, we constantly recorded the perfusion pressure, the pH in the equilibrating vessel, the pH of the renal tissue, the PO_2 on the arterial side, and the local tissue PO_2.

The local tissue PO_2 measurements were carried out with the platinum multiwire surface electrode developed by Kessler and Lübbers, which measures PO_2 with eight platinum wires at eight different places at the same time. We considered the parameters mentioned to be essential, since we expected to be able to obtain information on the functioning of the organ at the microcirculatory level.

Other measurements to assess vitality included gas analyses and pH measurements in the perfusate, measurement of urine production, measurements of electrolytes and osmolarity in the perfusate and urine, enzyme determination in the perfusate and urine, and examinations with light and electron microscopes (Fig. 2).

Results

The gas analyses show high PO_2 values in accordance with the equilibration with 95% O_2 and 5% CO_2 on both the arterial and venous sides. The pH remained practically constant during perfusion with Oxypherol (Fig. 3).

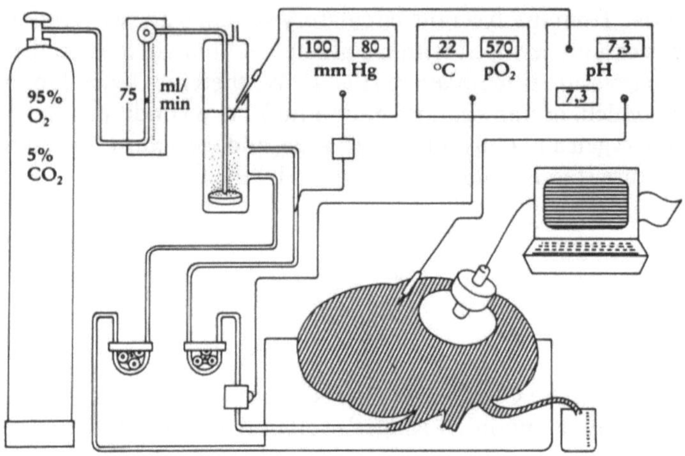

Fig. 2. Schematic presentation of experimental setup and registered values for perfusion of explanted kidney of the dog

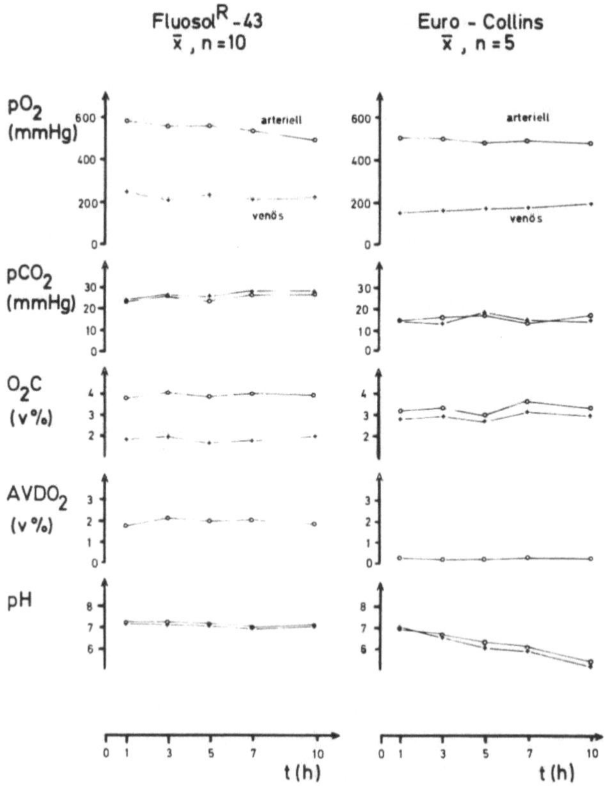

Fig. 3. Results of gas analyses and pH measurements

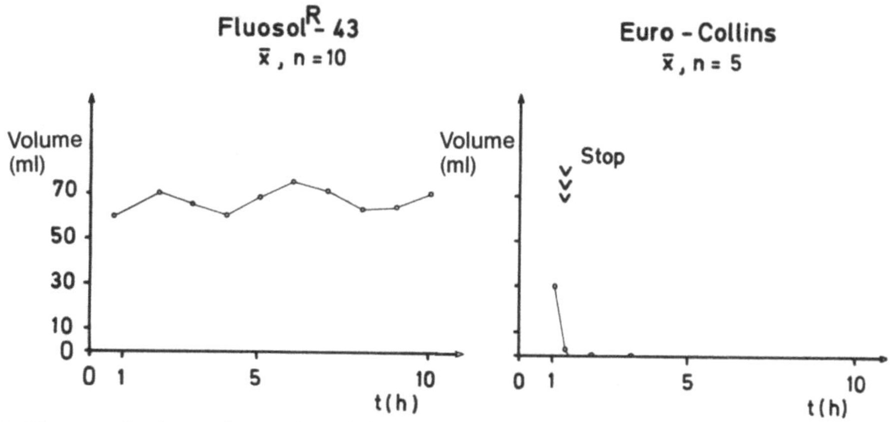

Fig. 4. Urine production under renal perfusion

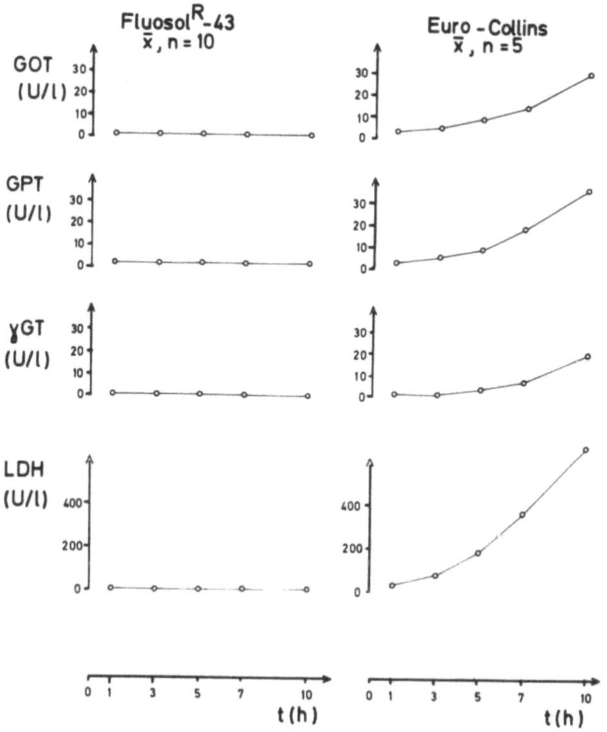

Fig. 5. Enzyme values in the urine under both perfusion solutions

Urine production of the Oxypherol (fluosol) group was 50–70 ml/h (Fig. 4); the electrolyte values fell within the physiologic range.

The determination of enzymes in the urine (Fig. 5) yielded values of only 2–3 units/l in the Oxypherol group. The amount of enzymes in the perfusate – including lactic dehydrogenase – was also only 3–4 units/l (Fig. 6). Even after 10 h of perfusion no increase could be registered.

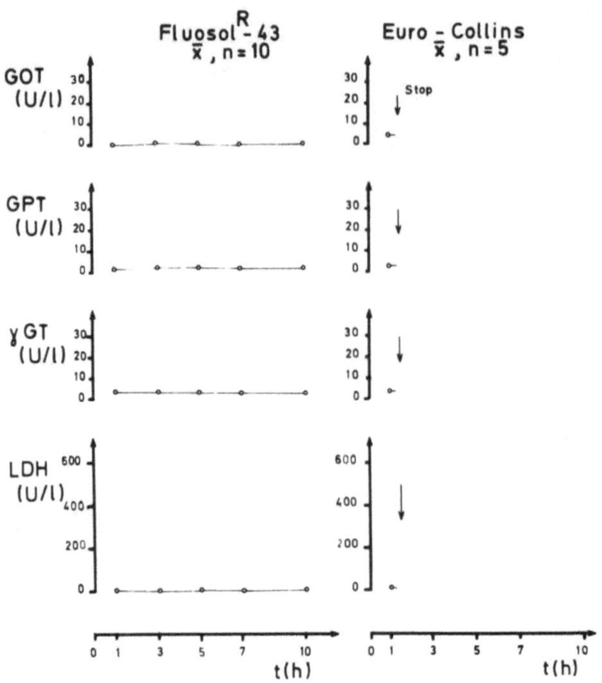

Fig. 6. Enzyme values in perfusate

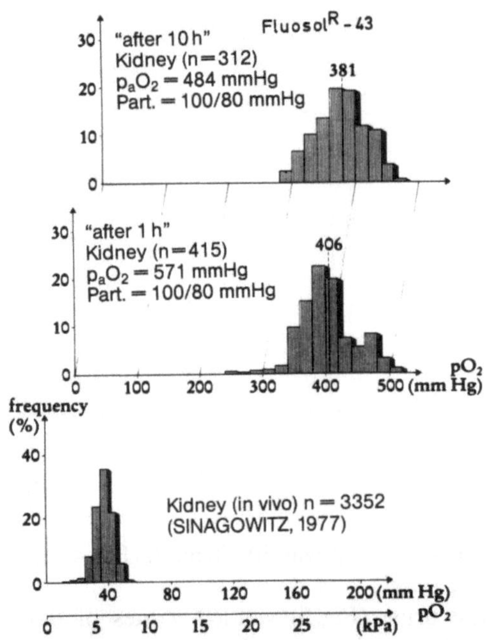

Fig. 7. Tissue partial oxygen pressure histograms

Investigations with the light and electron microscopes showed in the Oxypherol group normal mitochondria and intact microvilli structures in the tubus cells, which, of course, react very sensitively to hypoxia.

The histograms obtained after 1 h and 10 h of perfusion, which give the level and the statistical distribution of the local PO_2 in the tissue, are at a higher level under fluosol perfusion, due to the almost 100% equilibration of the perfusate with oxygen. The histograms show, however, a constant normal configuration after 1 h and 10 h. In particular, we were unable to establish any deterioration of oxygen supply, which would have shown up by a shift to the left or in a splitting of the configuration of the histograms, or by the appearance of anoxic values or values close to anoxia.

Summary

On the basis of the results of our experiments, we feel justified in saying that Oxypherol meets the requirements for an in vivo perfusion medium. This could mean a new clinical perspective for the preservation of organs, especially if other investigations come to the same results. The measurement of local oxygen pressure, in our opinion, is an important and reliable parameter for evaluating the quality of organ perfusion.

References

1. Belzer, F.O., J.H. Southard: The future of kidney preservation. Transplantation 3 (1980) 161-165
2. Grundmann, R., H. Pichlmayr: Organkonservierung. In: R. Pichlmayr (Hrsg.): Transplantationschirurgie. Springer-Verlag, Berlin (1981) S. 123-169
3. Kessler, M., D.W. Lübbers: Aufbau und Anwendungsmöglichkeiten verschiedener pO_2-Elektroden. Pflüg. Arch. 291 (1966) 82
4. Liebau, G., R. Grundmann, H. Pichlmayr: Vitalitätskriterien isolierter perfundierter Hundenieren. Arch. Klin. Chir. 329 (1971) 976
5. Shindo, K. et al.: Experimental studies on kidney preservation by perfusion with fluorochemical (FC-43) emulsion at room temperature. Proc. IVth Int. Symp. on perfluorochemical blood substitutes, Kyoto 1978

PO_2 Distribution in Muscle and Renal Cortex of Rats with Chronic Liver Changes

R. Heinrich, S. Dette, N. Grein, M. Günderoth-Palmowski, W. Grauer, W. Fleckenstein, and H. Schomerus

Introduction

In man, cirrhosis with portosystemic shunting as a consequence of portal hypertension produces changes of systemic circulation, i.e. hyperdynamic circulatory syndrome, which are characterized by decreased peripheral vascular resistance as well as increased cardiac output, systolic ejection rate, and intrapulmonary venous admixture [4, 11, 12]. In addition, changes of renal perfusion, i.e., hyperperfusion of the medulla with diminished perfusion of the cortex, have been demonstrated. Functional renal failure as a consequence of decompensated cirrhosis (i.e., so-called hepatorenal syndrome) have been partially attributed to these intrarenal perfusion changes [8].

Using well-defined animal models, we investigated the extent to which chronic liver changes due to (a) portocaval anastomosis (PCA) or (b) thioacetamide (TAA)-induced cirrhosis with consecutive portal hypertension cause changes of the PO_2 distribution in skeletal muscle or on the renal cortex. Since rats without cirrhosis but with portocaval end-to-side anastomosis (PCA) also develop changes in intrarenal perfusion resembling those of cirrhotics, in addition to a hypercirculatory syndrome, this animal model (rats with PCA) is well suited for investigating shunt-dependent perfusion changes [5, 10].

In the course of time, TAA administration in rats leads to changes resembling those of human cirrhosis, but without any demonstrable drug-induced muscle or kidney changes [1]. Both animal models were compared to clarify whether the PCA-related PO_2 distribution changes in muscle we found in animals without pre-existing cirrhosis, are also present in cirrhotic animals with consecutive spontaneous shunting [6]. We then attempted to define the temporal relation between creation of a PCA or development of cirrhosis and presumed oxygen distribution changes in muscle and renal cortex.

Research supported by the Deutsche Forschungsgemeinschaft (He 1293-1)

Methods

In the PCA Group, 32 Sprague-Dawley rats with a portocaval shunt created according to the procedure described by Lee and Fischer [9] and 24 sham-operated animals were studied. In the TAA Group, to induce cirrhosis, 500 ml TAA per liter drinking water were administered to 16 Sprague-Dawley rats for a total of 126 days. As is generally known, a prestage of cirrhosis (direct hepatotoxic damage with fatty degeneration of liver cells) is present after 40 to 50 days and hepatic fibrosis or cirrhosis with portal hypertension, after the 90th day of administration [1, 2]. Each group was compared with a corresponding group composed of 15 age-matched intact rats.

PO_2-pressure was measured with a platinum multiwire surface electrode according to Kessler and Lübbers [7]. The PO_2 histogram for the PCA group was recorded from the surface of the abdominal musculature between the 7th and 245th postoperative day and on the renal cortex on the 7th and 70th postoperative day, respectively. Analogous to the assumed stage of cirrhosis, PO_2 in the TAA group was measured between the 42nd and 64th as well as between the 112th and 126th day of administration. Each examined animal is represented on the pooled histogram by 2 histograms, individually recorded for each side, right and left abdominal musculature or renal cortex. The measurements were performed during the steady state of continuous ether anesthesia.

Results

The pooled histogram on the right of Fig. 1 shows muscle PO_2 of rats with PCA at different postoperative points in time. 7 days after creation of the PCA, mean PO_2 rose to almost 55 mm Hg with inhomogeneous distribution of PO_2 values. In the further postoperative course, mean PO_2 fell to just below 32 mm Hg on the 245th postoperative day. Concurrent with the mean PO_2 drop, the intermediate bell-shaped configuration of the histogram was lost again.

By contrast, the pooled histogram on the left of Fig. 1 shows PO_2 from the abdominal musculature of the sham-operated animals at the same postoperative points in time. Mean muscle PO_2 is not essentially influenced for 203 days after the sham operation. 245 days after the sham operation, mean PO_2, however, fell to 25 mm Hg.

The pooled histogram of local PO_2 distribution on the renal cortex of intact as well as sham-operated rats on the 7th and 70th postoperative day showed a bell-shaped, homogeneous distribution with a mean PO_2 of 40 mm Hg. No changes of mean PO_2 or PO_2 distribution on the renal cortex were detectable on the 7th or 70th day after creation of a PCA.

Muscle PO_2 distribution of the rats in the TAA group compared with that of the intact control animals is shown on the left of Fig. 2. Mean PO_2 and the bell-shaped configuration of the histogram are not essentially influenced up to 42 days after the

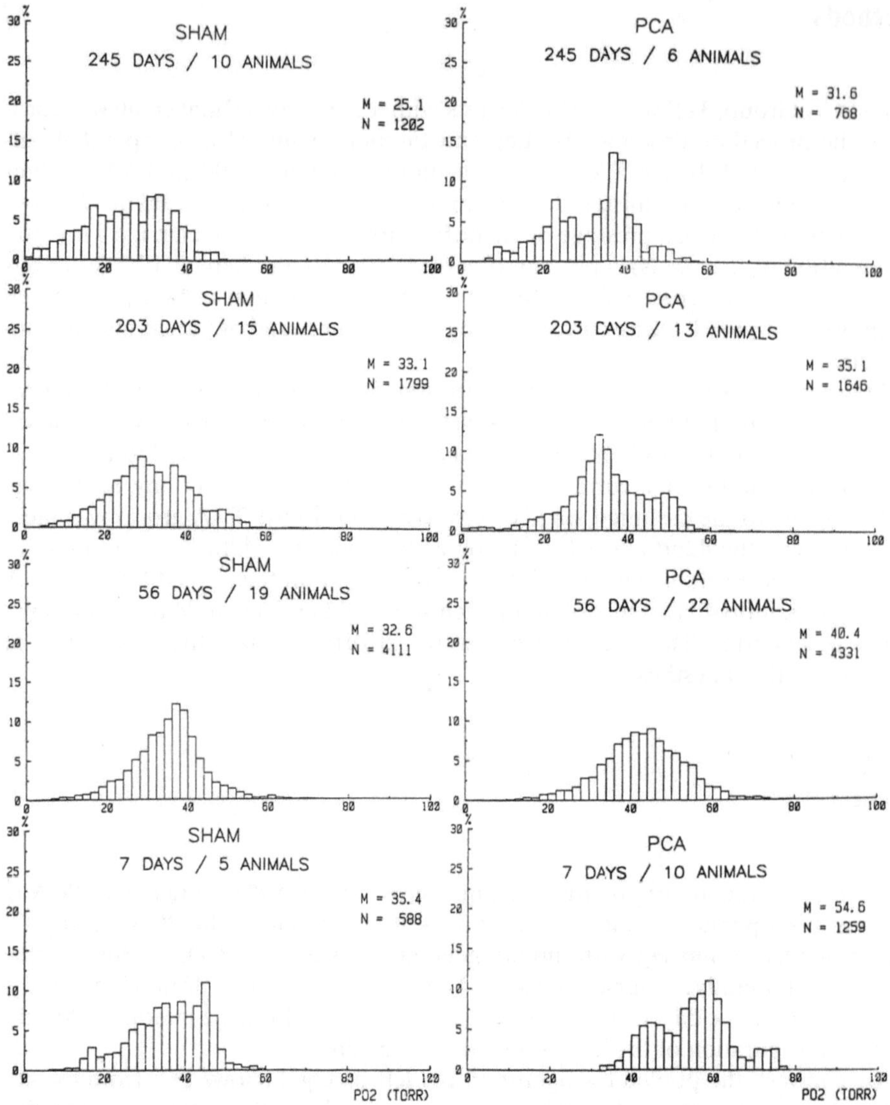

Fig. 1. PO_2 pooled histograms from muscle at different postoperative points in time (SHAM = sham-operated rats, PCA = rats with portocaval anastomosis, M = mean tissue PO_2, n = number of measuring points)

start of TAA administration. The distribution of all histograms was normal. Mean PO_2 rose to almost 40 mm Hg with a relative loss of PO_2 values below 15 mm Hg and a pronounced shift to the right 112 days after the start of TAA administration. This rise was significant on the unpaired t test ($p < 0.05$, n = number of animals examined).

The PO_2 distribution on the renal cortex of the rats in the TAA group was compared with that in intact control animals on the right of Fig. 2. Compared to the con-

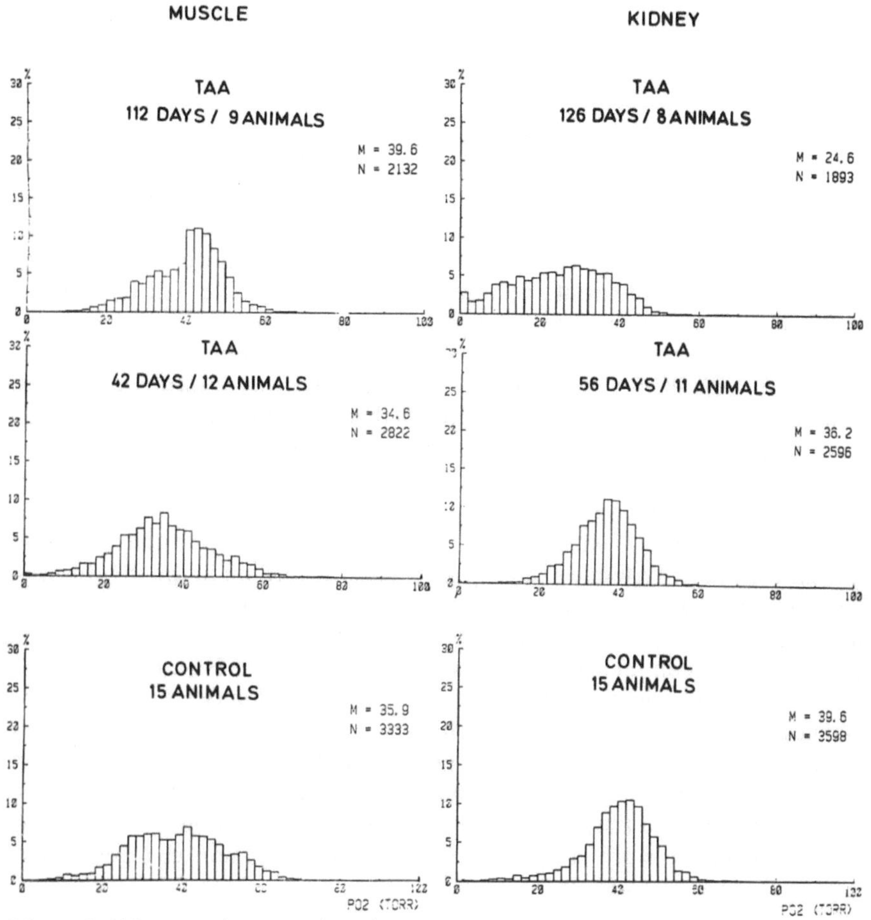

Fig. 2. PO_2 pooled histogram from muscle and on renal cortex of rats with thioacetamide (TAA)-induced liver changes and intact control animals (abbreviations see Fig. 1)

trol animals, the pooled histogram shows no significant changes in form or mean PO_2 (40 mm Hg) up to 56 days after the start of TAA administration. Mean PO_2 on the renal cortex first fell slightly below 25 mm Hg 126 days after the start of TAA treatment. Concurrently, the distribution of the PO_2 classes is inhomogeneous with a marked increase of PO_2 values below 20 mm Hg and virtually constant distribution over all PO_2 classes between 0 and 40 mm Hg. On the unpaired t test, the mean PO_2 of this pooled histogram is significantly lower ($p < 0.01$) than that of the control group. 9 of the 16 histograms (56%) show definite signs of maldistribution; 3 (19%) were normally distributed. The changed form of the pooled histogram therefore is influenced by the individual histograms.

Discussion

It is well known that, in rats, a portocaval anastomosis, including the associated liver changes which are not of a cirrhotic nature, leads to a temporarily limited, consecutive hypercirculatory syndrome [10]. The altered hemodynamics can be assumed to be an essential cause of the postoperative course of mean muscle PO_2. With increasing time after creation of the PCA, in accordance with the oxygen histogram, signs of maldistribution, however, are also probable. The reason why muscle PO_2 declined, compared to the preliminary measurements, in the sham-operated rats 245 days after surgery, but not in the intact control animals, has not been clarified. In rats with PCA, Grün et al. demonstrated changes of intrarenal perfusion up to diminished relative cortical perfusion [5] that resembled cirrhosis-induced perfusion changes in man [3, 8]. We detected no influence of PO_2 distribution or mean PO_2 on the renal cortex. This tends to indicate that the known intensification of total renal perfusion compensates for a possible hypoperfusion of the renal cortex. Drug-induced cirrhosis with TAA in rats with consecutive portal hypertension and spontaneous portosystemic shunting also leads, according to the PO_2 histogram, to hyperperfusion of the musculature. This hyperperfusion could also be the consequence of a hyperdynamic circulatory state. By contrast, mean PO_2 on the renal cortex declines with an increase of PO_2 values below 20 mm Hg and signs of changed distribution. There is no known histologic correlate for the appearance of these changes which would indicate that, in rats, TAA-induced cirrhosis reduces renal cortical perfusion without increasing overall renal perfusion. In rats with spontaneous portosystemic shunting as a consequence of TAA cirrhosis, the perfusion changes assumed on the basis of the PO_2 histograms from muscle or on the renal surface, therefore, more closely resemble known perfusion changes of cirrhotics than do those of rats with PCA.

References

1. Bachmann, E., J. Lindner, K. Grasedyck, R. Eurich: Rattenleberkollagen bei protrahierter Thioacetamidintoxikation. Arzneim. Forsch./Drug Res. 28 (1960) 2260
2. Brodehl, J.: Thioacetamid in der Leberforschung. Klin. Wschr. 39 (1961) 956
3. Conn, H. O.: Rational approach to the hepatorenal syndrom. Gastroenterol. 65 (1973) 321
4. Gordon, M. J., L. R. M. Del Guercio: Late effects of portal systemic shunting procedures on cardiorespiratory dynamics in man. Ann. Surg. 176 (1972) 672
5. Grün, M., H. Liehr, H. Thiel, U. Rasenack: Effekt einer Endotoxinämie auf die renale und intrarenale Hämodynamik bei Ratten mit und ohne portokavale Anastomose. Z. Gastroenterol. 76 (1976) 285
6. Heinrich, R., W. Grauer, H. Schomerus, M. Günderoth, M. Kessler, J. Höper: Measurements of local pO_2 in resting skeletal muscle of rats with portocaval anastomosis (PCA) under normoxic conditions. In: Oxygen Transport to Tissue VI Plenum. Publishing Corporation 1985
7. Kessler, M., D. W. Lübbers: Aufbau und Anwendungsmöglichkeiten verschiedener pO_2-Elektroden. Pflüg. Arch. 291 (1966) 82
8. Kew Mc, P. W. Brunt, R. R. Varma, K. J. Hourigan, H. S. Williams, S. Sherlock: Renal and intrarenal blood-flow in cirrhosis of the liver. Lancet II (1971) 504

9. Lee, S. H., B. Fisher: Portocaval shunt in the rat. Surgery 50 (1961) 6668
10. Liehr, H., M. Grün, H. Thiel: Hepatic blood flow and cardiac output after portocaval anastomosis in the rat. Acta Hepato-Gastroenterol. 23 (1976) 31
11. Liehr, H., M. Grün, H. Thiel: Systemische Zirkulation bei portaler Hypertension. Verh. Ges. Inn. Med. 82 (1976) 159
12. Schomerus, H., I. Buchta, H. Arndt: Pulmonary function studies on oxygen transfer in patients with liver cirrhosis and different degree of portosystemic encephalopathy. Respiration 32 (1976) 1

Endotoxin-Induced Shock in the Pregnant Miniature Pig – Changes in Macro- and Microcirculation

J. Hauss, E. H. Schmidt, H. U. Spiegel, B. Flötotto, W. Holzgreve, F. K. Beller, and H. Bünte

Introduction

In 1935, Apitz induced a generalized Shwartzman phenomenon (bilateral renal cortex necrosis, hemorrhagic diathesis, and shock) in rabbits by giving them two injections of endotoxin at a 24-h interval. He observed that one animal had already died after the first injection. This animal was pregnant [1]. Based on this observation, numerous investigations were carried out which proved that animals in the last third of their gestation period generally require a lower dose of endotoxin to develop shock than nonpregnant control animals. The results all indicated that pregnancy has a predisposing effect for generalized Shwartzman phenomenon and for shock.

In the present animal experimental study, an attempt was made to analyze the hemodynamic and microcirculatory changes that occur after the application of endotoxin during pregnancy. Previous investigations had been carried out concerning the effect of endotoxin on the nonpregnant organism [2].

Materials and Methods

The experimental arrangement is illustrated in Fig. 1. The experiments were carried out on a total of ten pregnant Göttinger mini pigs (mean weight: 52 kg). All of the animals were in the last third of their 110-day gestation period.

One hour before the beginning of the experiment, the animals were premedicated with i.m. injections of 0.5 mg atropine and 120–200 mg azaperone (Stresnil). They were anesthetized with 2.5–10.0 mg/kg b.w. metomidate hydrochloride (Hypnodil), which was slowly injected intravenously. The animals were tracheotomized, intubated, and ventilated with a volume- and pressure-controlled respirator (Servoventilator 900 C, Siemens-Elema), with an FIO_2 of 0.3. Catheters were inserted into the pulmonary artery, the aortic arch, and the vena cava. After median laparotomy, the uterus was exposed with an approximately 4-cm long incision while the placenta was kept intact. The umbilical cord of the fetus was drawn out so that the pulse rate could be measured and blood could be taken from the umbilical vein. The multiwire surface electrode was placed on the anterior surface of the uterus

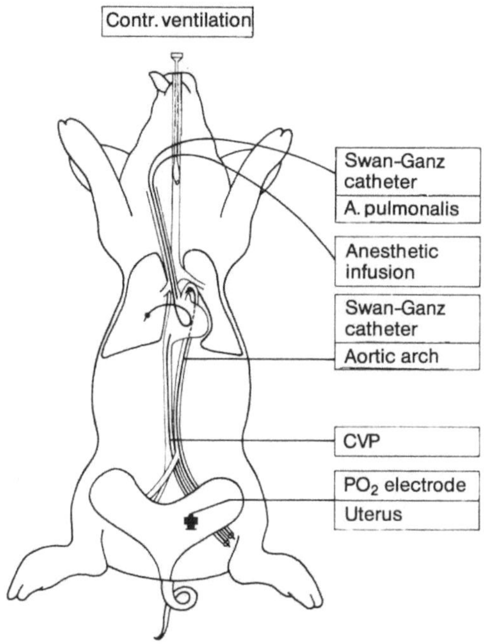

Fig. 1. Experimental protocol. *CVP,* central venous pressure

(Fig. 2) and on the surface of the liver. To suppress contractions, which would have made PO_2 measurements on the uterus impossible, a mixture of 2 mg fenoterol hydrobromide (Partusisten) and 125 mg verapamil hydrochloride (Isoptin) in 500 ml 0.9% NaCl solution was infused with a perfusor. Labor suppressing agents were discontinued during the time the PO_2 histogram was recorded. In seven animals, the infusion of 0.2 mg/kg b.w./h endotoxin (*E. coli* B6 lipopolysaccharide, Diffco Lab., Detroit, Michigan, USA) was dissolved for 5 h and diluted in 50 ml/kg b.w. Ringer's solution for 12 h or until the animal died. Three pigs served as control animals. The course of the experiment is illustrated in Fig. 3.

Results

In all of the animals the heart rate remained constant (Fig. 4). Cardiac index, stroke index, and central venous pressure decreased significantly. The systolic arterial blood pressure rose in the 1st h from 95 to 102 mm Hg (Fig. 5). Afterwards, a drop in blood pressure to 49 mm Hg after 3 h was observed ($P \leq 0.05$). The total peripheral resistance in the 1st h showed a sharp increase from 800 to 1140 dyn × s × cm^{-5} and dropped to 590 dyn × s × cm^{-5} in the following 2 h.

The systolic pulmonary arterial pressure at the beginning of the experiment was 25 mm Hg (Fig. 6). After 1 h, a value of 30 mm Hg was measured, and at the end of the experiment a pressure reduction of 19 mm Hg. The systolic pressure of the pul-

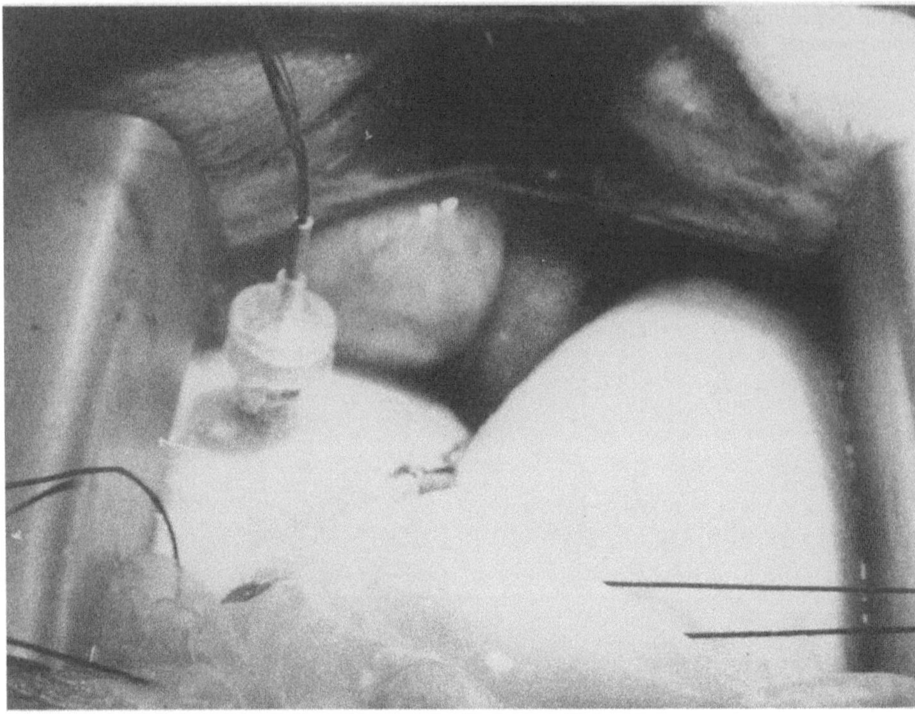

Fig. 2. Site of operation, electrode placed on the anterior surface of the uterus

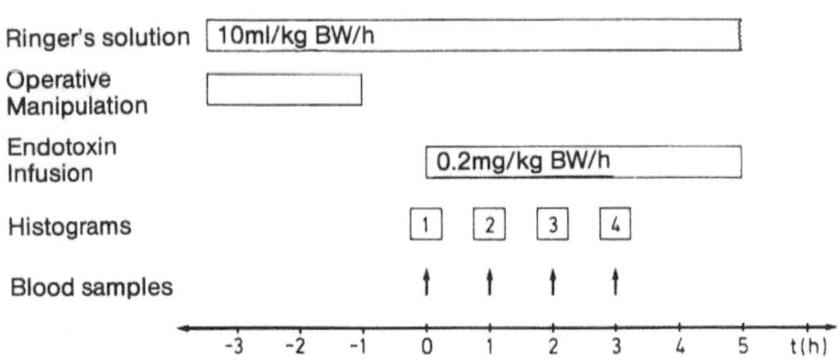

Fig. 3. Time course of the experiment

monary artery among the control animals rose continuously from 20 to 26 mm Hg. The course of the diastolic and mean pressure of the pulmonary artery was the same. The resistance of the pulmonary artery was constantly higher after the application of endotoxin. During the experiment, the heart output decreased from 7.5 to 3.6 l/min ($P \leq 0.05$; Fig. 7). The stroke output dropped continuously from an initial 85.3 ml; 49.7 ml were measured after 3 h ($P \leq 0.05$). The stroke output of the control animals diminished slightly at the beginning of the experiment and then remained constant at about 72 ml. The pulmonary-capillary wedge pressure dropped from 6.5 to 3.8 mm Hg during the experiment. In the control animals, the pressure increased

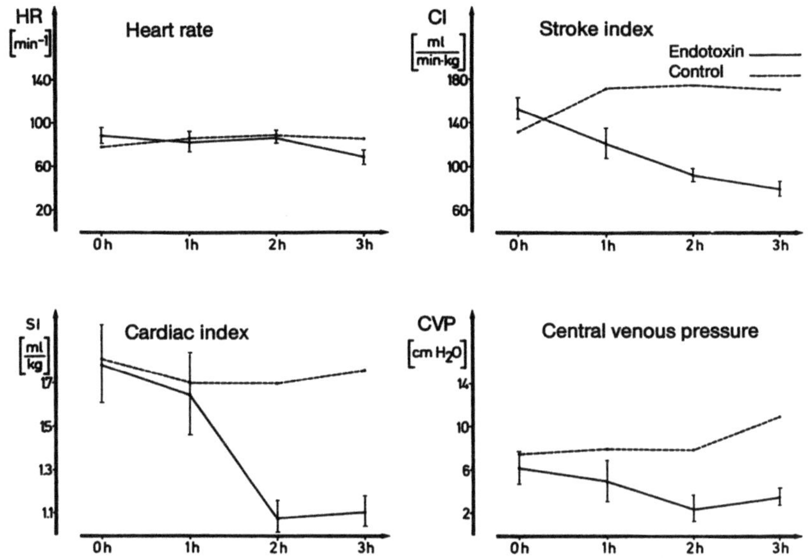

Fig. 4. Registered values of heart rate, cardiac index, stroke index, and central venous pressure

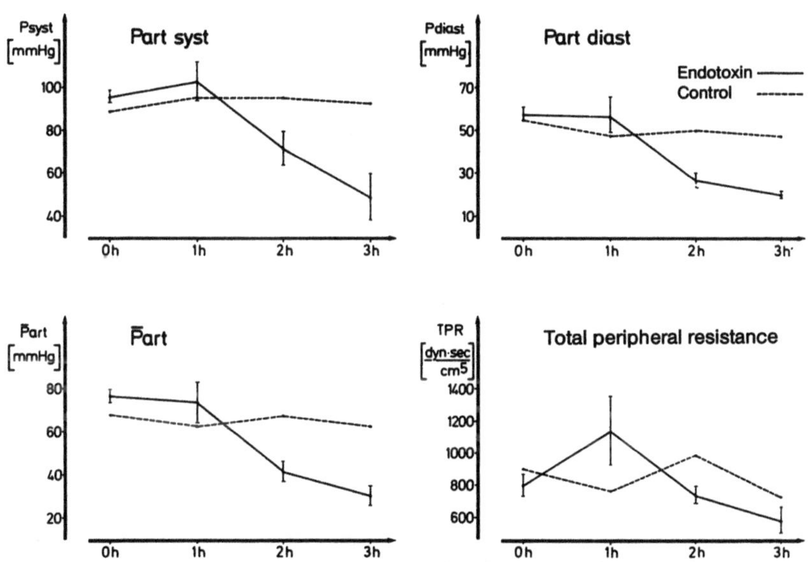

Fig. 5. Systolic (Part syst), diastolic (Part diast), mean arterial blood pressure (P̄art) and total peripheral resistance

to 10.0 mm Hg at the onset of the experiment and then remained constant. Oxygen consumption decreased from 419 to 95 ml/min.

The arterial partial oxygen pressure (PaO_2) of the endotoxin animals diminished from 102.5 to 61.6 mm Hg ($P \leq 0.05$; Fig. 8). The values among the control animals

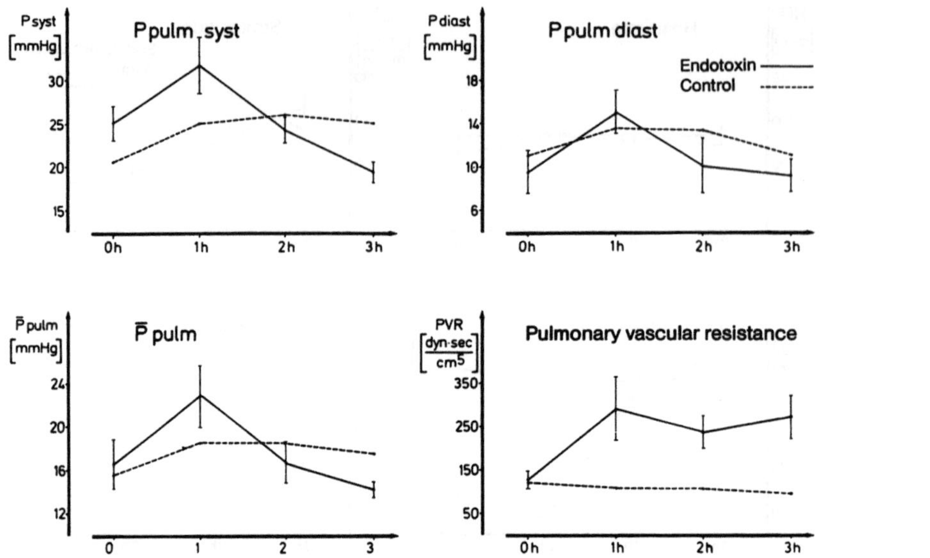

Fig. 6. Systolic (P pulm syst), diastolic (P pulm diast), mean pulmonary arterial pressure ($\overline{\text{P}}$ pulm), and pulmonary vessel resistance

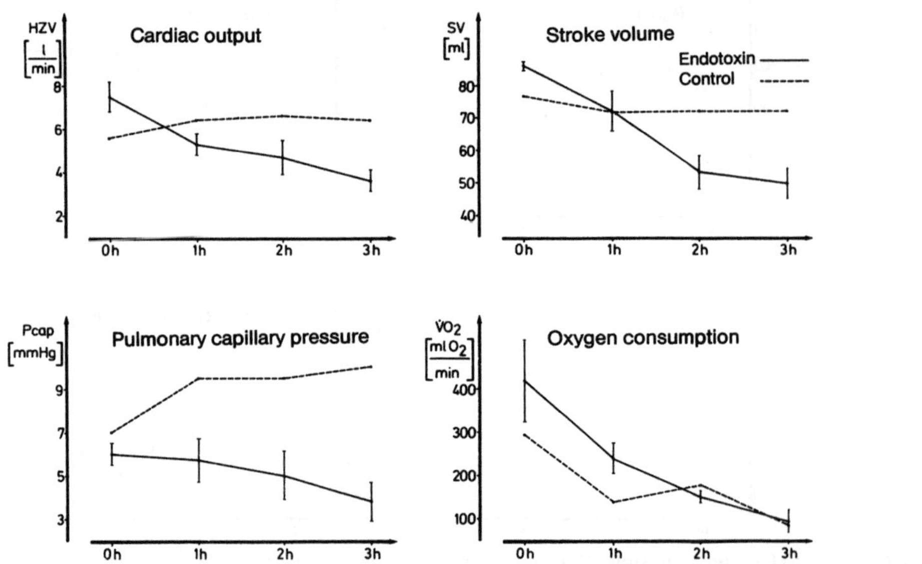

Fig. 7. Cardiac output, stroke volume, lung capillary pressure, and oxygen consumption

remained almost constant (between 82 and 91 mm Hg). The partial carbon dioxide pressure ($PaCO_2$) rose from 32.0 to 37.7 mm Hg within 2 h of the beginning of the experiment; after 3 h it was 36.8 mm Hg. The $PaCO_2$ of the control animals remained just about constant at 31 mm Hg after an initial increase. The arterial pH value decreased almost linearly from 7.41 to 7.05 during the experiment ($P \leq 0.05$).

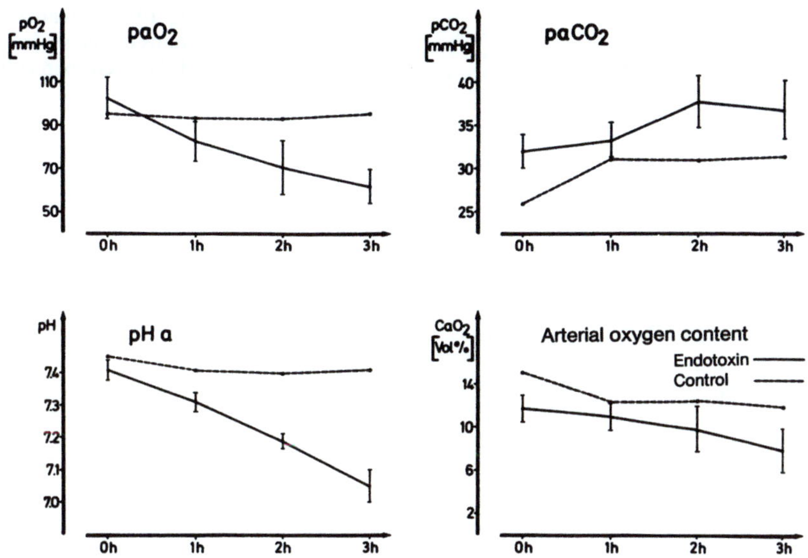

Fig. 8. Arterial partial oxygen pressure (PaO_2), arterial partial carbon dioxide pressure ($PaCO_2$), arterial pH, and arterial oxygen content

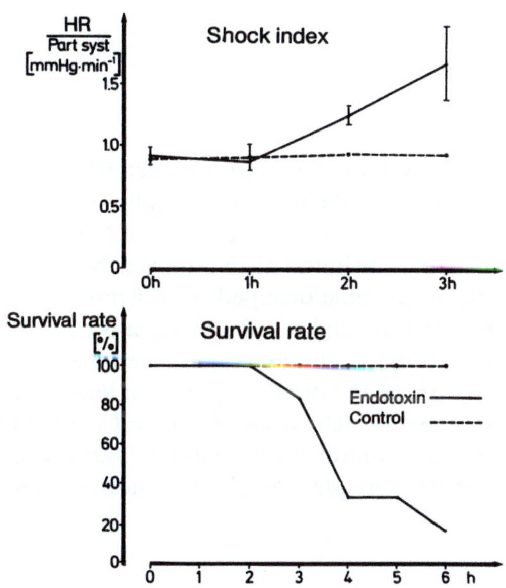

Fig. 9. Time-dependent changes in shock index and survival rate

In contrast, the pH values of the control animals, with the exception of a slight reduction at the beginning of the experiment, remained constant at values around 7.4. The arterial oxygen content diminished from 11.7 to 7.8 ml/100 ml ($P \leq 0.05$) during the experiment.

The shock index remained constant at 0.9 in the 1st h of the experiment and then rose linearly to 1.76 after 3 h ($P \leq 0.05$; Fig. 9). In comparison, the shock indices of

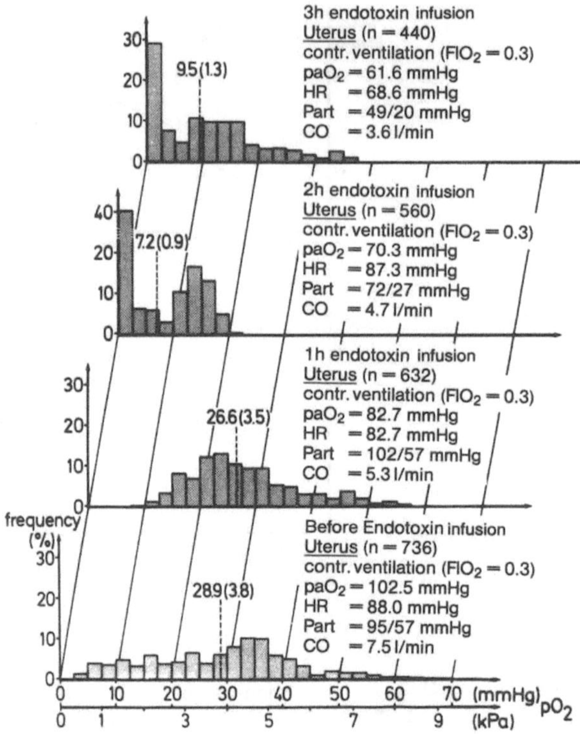

Fig. 10. Summarized PO_2 histograms on the surface of the uterus of the endotoxin animals

the control animals were between 0.88 and 0.93. The survival rate decreased from the 2nd h on; the mean survival time among the endotoxin animals was 3.4 h.

To evaluate the tissue PO_2, the PO_2 histograms of six animals were summarized (Fig. 10). The initial histogram was broad-based with a mean value of 28.9 mm Hg. The mean value dropped to 26.6 mm Hg 1 h after the beginning of endotoxin administration, and to 7.2 mm Hg after 2 h. Forty percent of all the values measured were in the smallest class, between 0 and 2.5 mm Hg. The PO_2 histogram recorded after 3 h of endotoxin application had a mean value of 9.5 mm Hg; again the peak of frequency fell within the lowest class. Figure 11 shows the PO_2 histograms of a pregnant control animal without endotoxin. The configuration of all of the PO_2 histograms were physiologic. The mean values were between 44.3 and 50.4 mm Hg.

Discussion

The results of PO_2 measurements in tissue following the administration of endotoxin have varied. Kaelin [3] and Casey [4] proved a reduction in the tissue PO_2 in the liver of the rat. After 6 h of endotoxin application 88% of the measured values were below 5 mm Hg, and the mean value had dropped to about 10%. Rink [5] observed

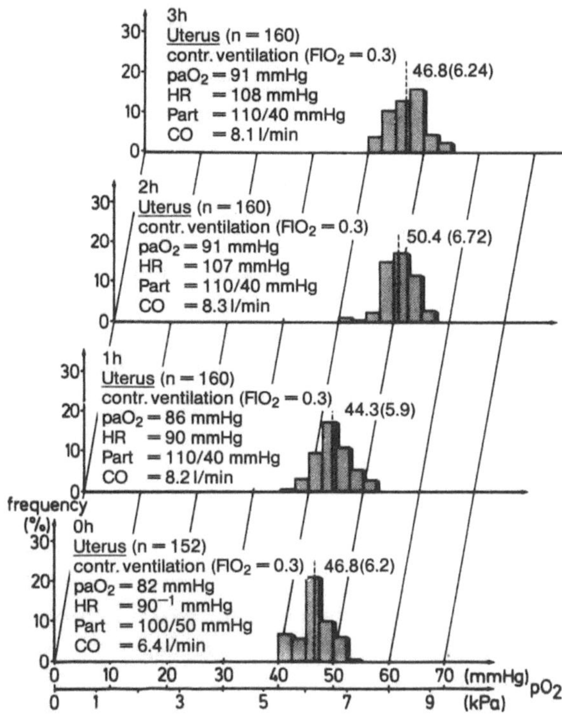

Fig. 11. PO_2 histograms on the surface of the uterus of a control animal

marked differences between the PO_2 histograms of the surviving animals and of the nonsurviving animals. In the former group, the mean tissue PO_2 fell from 25 to 9 mm Hg, while a PO_2 of 3 mm Hg was measured among the non surviving animals in shock. Schosser [6] and Mela [7] also observed a reduction of the mean tissue PO_2 of the liver of about 70% and 50% respectively. The mean tissue PO_2 pressure of the skeletal muscle and subcutaneous tissue in the dog also decreased from 45 mm Hg to 26 and 16 mm Hg, respectively. Correspondingly, Hiller [8] reported a drop to 45% of the control value in the oral submucosa of the dog.

The common result of all of the investigations mentioned above and of our own measurements on the uterus of the pregnant mini pig is that the tissue PO_2 of the organ studied decreases after the application of endotoxin. This reduction is independent of the hemodynamic reaction to endotoxin, for it occurs along with both the hypodynamic and the hyperdynamic reactions. The authors cited above give no data on the time of tissue hypoxia development. We therefore cannot judge whether there was a phase in the 1st h in which just about normal PO_2 pressure was followed by sudden decompensation, as was the case in our investigations. A comparison with the results obtained from measurements on the nonpregnant Göttinger mini pig [2] shows that in correlation with the clinical picture in humans, endotoxin shock during pregnancy occurs suddenly. In this connection, it is striking that the physiologic changes in pregnancy are in many ways identical to those in the early phase of endotoxin shock: the heart rate decreases by 10%–15%, the cardiac output increases by a third, and the peripheral vascular resistance decreases, partly as a result of diminished resistance of the precapillary arterioles, and partly as a result of the opening of the precapillary shunts. Respiratory output increases by 40%. Under reduced bicarbonate concentration, there is a normal pH value representing a compensated respiratory alkalosis. While the PO_2 pressure remains constant, the oxygen consumption increases, caused by the excess consumption of the uterus and the fetus, and the arteriovenous difference drops.

Water retention seems to play a significant role: additional water - approximately 10% of the body weight - is absorbed. This is distributed among all three compartments, whereby the largest amount is retained in the interstitial space. Schmidt [2] consistently observed an initial hyperdynamic reaction of the circulatory system in his investigations on the nonpregnant Göttinger miniature pig. The comparison shows that the physiologic changes during pregnancy apparently prevent such hyperdynamic compensation.

Summary

The influence of pregnancy on the pathophysiology of septic shock was studied after the infusion of endotoxin in Göttinger mini pigs. Seven anesthetized and ventilated animals received 0.2 mg/kg b.w./h *E. coli* endotoxin in the last 2 weeks of pregnancy. The following reactions were observed:

1. Circulatory system: a reduction in arterial and pulmonary pressure, central venous pressure, cardiac output, stroke volume, cardiac index; an increase in the pulmonary vessel resistance; in the ECG an ST reduction and T reversal,
2. Acid-base balance: reductions in the pH, the bicarbonate concentration, and the excess of bases,
3. Blood gases: a decrease in the arterial and mixed-venous PO_2, oxygen saturation, oxygen content, oxygen transport, and oxygen consumption, as well as an increase in the PCO_2,
4. Tissue PO_2: a clear shift to the left in the PO_2 histograms of the uterus, a reduction in the mean value, a marked increase in the anoxic measured values and those close to anoxia.

A comparison with the results of the nonpregnant Göttinger miniature pig shows - in correlation with the clinical picture in humans - that endotoxin shock in pregnancy occurs suddenly.

Cardiac and pulmonary insufficiency seem to be in the foreground of pathophysiologic changes. The reduced oxygen supply leads to metabolic acidosis. During the 2nd h of the experiment, the supply of oxygen to the tissue deteriorated immensely.

The results suggest that during pregnancy, hyperdynamic compensation such as is observed in the nonpregnant animal is prevented by physiologic changes. This seem to be responsible for the severity and the higher lethality of endotoxin shock during pregnancy.

References

1. Apitz, K.: A study of the Generalized Shwartzman Phenomenon. J. Immunol. 29 (1935) 255-266
2. Schmidt, E. H.: Pathophysiologische Veränderungen beim Göttinger Miniaturschwein nach Infusion von Endotoxin. Habil.schrift (1978)
3. Kaelin, C. R., R. D. Rink: Effect of Phentolamin on Hepatic pO_2 in Endotoxinemia. Can. J. Physiol. 58 (1980) 281-286
4. Casey, J. P., B. L. Short, R. D. Rink: The Effect of Methylprednisolone on Hepatic Oxygen Supply and Plasma Lactate and Glucose in Endotoxinemia. Circ. Shock 6 (1979) 245-253
5. Rink, R. D., B. L. Short, C. Pennington: Hepatic Oxygen Supply and Plasma Lactate and Glucose in Endotoxin Shock. Circ. Shock 5 (1978) 105-113
6. Schosser, R., H. Sakoda, H. Grafelmann, K. Messmer: Cardiac Output, Regional Blood Flow Distribution, Metabolism and Hepatic Oxygen Supply in Rats during Hyperdynamic Endotoxin Shock. Circ. Shock 6 (1979) 181
7. Mela, L., L. D. Miller, L. V. Bacalzo, K. Olofson, R. R. White: Role of Intracellular Variations of Lysosomal Enzyme Activity and Oxygen Tension in Mitochondrial Impairment in Endotoxemia and Hemorrhage in the Rat. Ann. Surg. 178 (1973) 727-735
8. Hiller, C., R. Bone, D. Hill, F. Wilson: Mixed Venous Oxygen Tension Compared to Tissue Oxygen Tension as a Measure of Oxygen Delivery in Endotoxin Shock. Clin. Res. 27, A399 (1979)

Effect of Somatostatin on the Oxygen Transport to the Splanchnic Organs

O. Ruland, H. U. Spiegel, A. Holzgreve, and J. Hauss

Introduction

In the literature it has been reported that somatostatin reduces the blood flow in the visceral vessels [1]. This effect is of therapeutic value. In view of the fact that somatostatin is applied in patients with diseases of the visceral organs (bleeding ulcer, cirrhosis of the liver, pancreatitis), it appears particularly important to know whether this compound, apart from having an effect on the larger vessels in this area, also has an effect - either direct or indirect - on the oxygen supply and the microcirculation in the tissue.

Among the various ways of investigating microcirculation indirectly, the measurement of the local tissue oxygen pressure (PO_2) using the multiwire surface electrode (Kessler and Lübbers) provides a "window" through which the microcirculation can be watched. The local PO_2 in the tissue is measured through an electrode with eight measuring wires; the electrode is placed on the exposed surface of the organ. Besides continuous measurement, a so-called PO_2 histogram can be recorded, which shows the statistical distribution of oxygen pressure in the tissue.

Materials and Methods

In four dogs weighing 25 kg each and under standardized anesthesia (controlled ventilation, high dose piritramide basic anesthesia), PO_2 histograms of the liver, the pancreas, and the small intestine were established. The dogs then received a continuous i.v. infusion of somatostatin 0.33 μg/min b.w. (0.5 mg/h) until the end of the experiment. During this infusion, another histogram for each of the three organs was recorded without any alteration of anesthesia and ventilation. During the entire duration of the experiment, the dogs were given i.v. 300 ml/h Ringer's lactate as a volume substitute. From the pulmonary and circulatory data obtained with Swan-Ganz catheters placed in a pulmonary artery and the aorta ascendens, as well as a central venous catheter in the v. cava, the complete cardiorespiratory status could be worked out for the respective histogram (Fig. 1). The important circulation and respiration parameters at the time of the histogram recording are shown on the right side of the diagram.

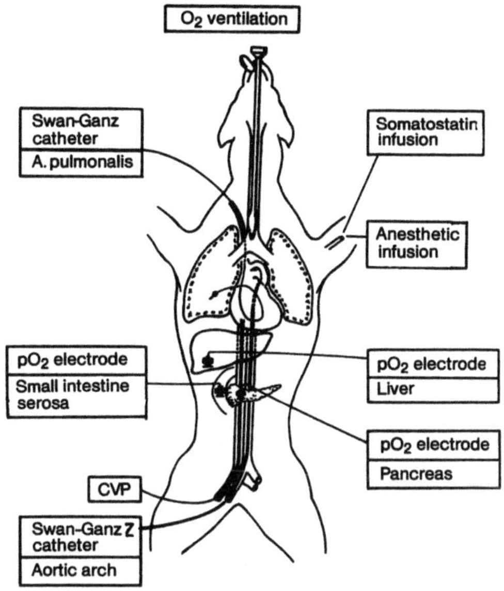

Fig. 1. Schematic presentation of experimental setup

Results

A look at the values of the experimental parameters reveals that during the controlled ventilation, which, once set, was not changed again, the arterial PO_2 and PCO_2, the pH, and the Hb remained almost constant before and under somatostatin infusion (Fig. 2). Only the arteriovenous difference in oxygen ($avDO_2$) increased slightly.

The circulation data obtained and computed before and after somatostatin infusion presents a somewhat different picture. The heart rate (HR) and the cardiac output of the heart (CO as right/left measurement) remained almost constant. The arterial mean pressure ($\overline{p}_{art}$) rose slightly under somatostatin, which correlated with the increase in the systemic vascular resistance (SVR) when the cardiac output remained constant (Fig. 3).

The summarized PO_2 histograms of the small intestine (Fig. 4) do not differ significantly from one another. There is no indication of a deterioration of the local oxygen supply under somatostatin, which would be shown here by a shift to the left of the histogram or by the occurrence or accumulation of anoxic or almost anoxic values. Under otherwise identical experimental conditions, the summarized PO_2 histograms of the pancreas during somatostatin infusion showed no deviation from the summarized histogram without somatostatin (Fig. 5). A comparison of the summarized PO_2 histograms of the liver also reveals no differences in the level and distribution of oxygen pressure in the tissue (Fig. 6).

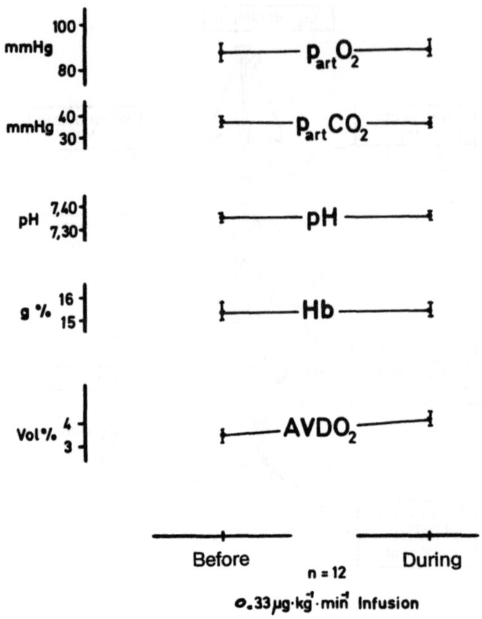

Fig. 2. Oxygen supply and measured values under somatostatin

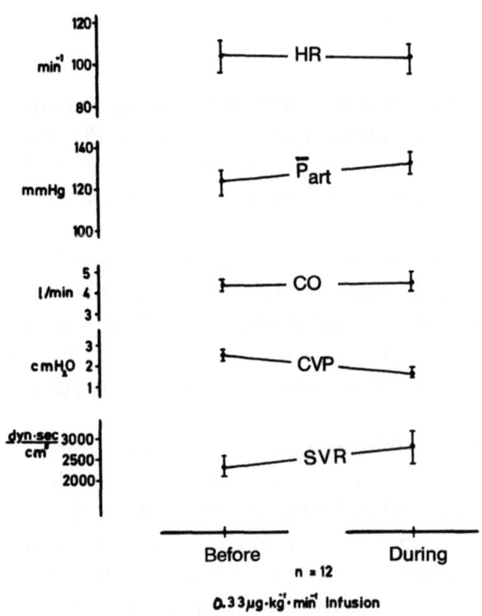

Fig. 3. Hemodynamic parameters

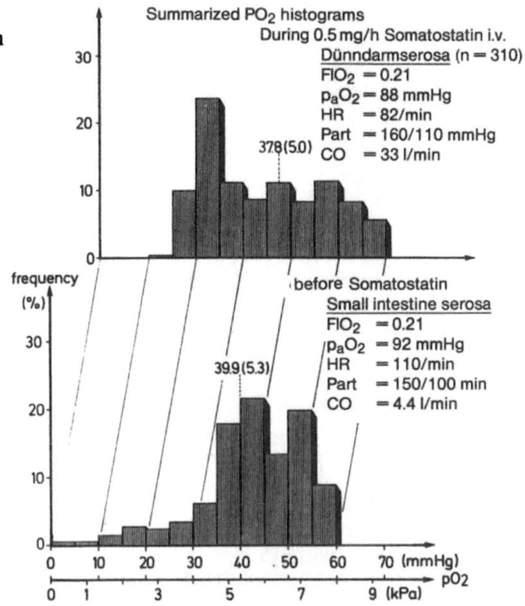

Fig. 4. Summarized PO_2 histograms of the small intestine with and before somatostatin

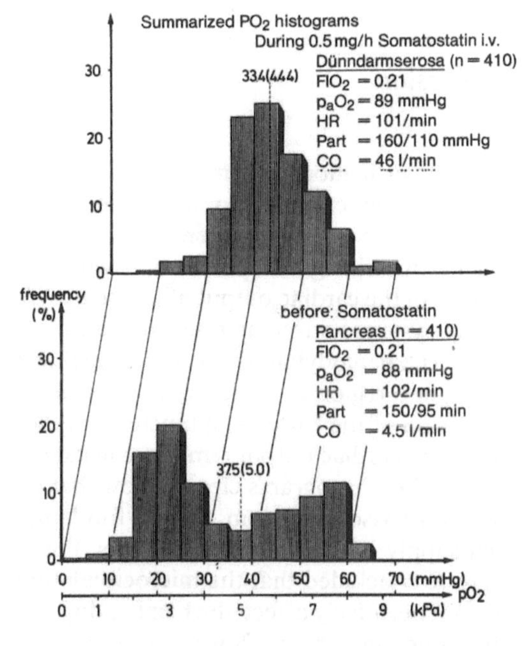

Fig. 5. Summarized PO_2 histograms of the pancreas

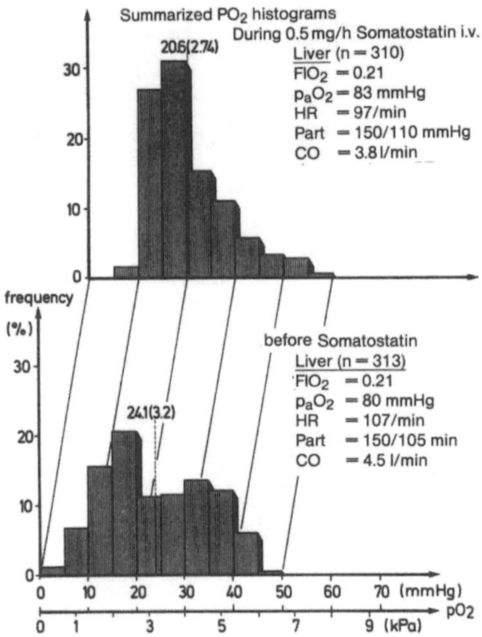

Fig. 6. Summarized PO_2 histograms of the liver

Discussion

In the investigation on hemodynamic changes under the influence of somatostatin, a marked reduction in blood flow in the visceral arteries was generally observed. Special somatostatin receptors in these vessels might constitute the mechanism producing this change. In correlation with other studies [1], we did not register any change in the cardiac output of the heart. We did, however, observe a slight increase in the systemic vascular resistance and, along with this, a slight increase in $\overline{p}_{art}$. There is a hemodynamically logical explanation for this vessel resistance increase: when, in a region as large as the visceral system, the blood flow is reduced by the activation of the "arterial sphincters" in the larger vessels, the arterial pressure will rise if the cardiac output remains constant.

The PO_2 histograms clearly show that in spite of reduced blood flow in the vessels of the visceral system, there is fundamentally no deterioration of the local oxygen supply in the tissues of the liver, the small intestine and the pancreas. It can thus be concluded that the microcirculation in these organs of the visceral system are not negatively affected when the dog is receiving somatostatin infusion. Clearly, the reduction in blood flow is not concurrent with a deterioration of the microcirculation. A physiologic explanation of this phenomenon can surely be found in the numerous arteriovenous shunts and precapillary sphincters in the visceral system.

Summary

Under a continuous i.v. infusion of somatostatin 0.33 μg/min kg b.w., we found under our experimental conditions a reduction of the central venous pressure and an increase in the systemic vascular resistance, of the $\overline{p}_{art}$, and of the $avDO_2$. We did not, however, establish any negative effect on the local oxygen supply in the liver, pancreas, and small intestine. Somatostatin therefore does not seem to have an effect on the microcirculation in the organs investigated in the dog, even under a dose that is six times higher than is clinically used.

References

1. Becker, R., J. Scholtholt, O. Speth, B. Schölkens: Regional blood flow and hemodynamic in anesthetized dogs during Somatostatin - and Secretin-infusion. Springer-Verlag 1981
2. Bünte, H.: Die Mikrozirkulation - eine entscheidende Kreislaufkomponente für das Überleben kritisch Kranker. In: W. Schellerer, F. W. Schildberg: Aspekte mod. Chirurgie (Chirurgie aktuell 4). Straube-Verlag, Erlangen (1977) S. 32-44
3. Kessler, M., D. W. Lübbers: Aufbau und Anwendungsmöglichkeiten verschiedener pO_2-Elektroden. Pflüg. Arch. 291 (1966) 82
4. Lübbers, D. W.: Die Bedeutung des lokalen Gewebesauerstoffdrucks und des pO_2-Histogramms für die Beurteilung der O_2-Versorgung eines Organes. Prakt. Anaesth. 3 (1977) 184

The Effect of Muscle Exercise on Partial Oxygen Pressure in Dogs with Unilateral Chronically Ligated Leg Arteries

E. Böhm, K. Strein, G. Sponer, B. Müller-Beckmann, and N. Würtz

Introduction

One of the advantages of animal experimental models of diseases is that pathophysiological and pharmacological problems can be thoroughly investigated. For peripheral arterial occlusive disease, Sunder-Plassmann and Messmer developed a model on a dog with unilateral chronically ligated arteries of the leg [1, 2, 3]. The femoral artery is ligated at several places and all of its lateral branches, with the exception of the caudal femoral artery, are detached. This allows the stage of intermittent claudication to be simulated. In this phase, circulation in patients is still sufficient while they are resting, but during exercise the oxygen demand can no longer be adequately met. We therefore tried to discover differences in the reaction of the ligated and nonligated leg on muscle exercise with respect to the partial oxygen pressure as a measure of ischemia.

Materials and Methods

For our experiments we used eight mongrel dogs, in which the arteries of the right hindleg were ligated at the points illustrated and any lateral branches of the femoral artery were detached (Fig. 1). In addition, an approximately 5-6 cm long piece of the proximal part of the femoral artery was removed. Most of the animals recovered from this operation within 2 weeks, after which we allowed the dogs to walk for 10 min at 10 km/h without any ascent. For the partial oxygen pressure measurements, which took place at the earliest 3 weeks after the operation, the dogs were anesthetized with 550 mcg/kg/h pentobarbital following premedication with 2 ml Combelen®. As a muscle relaxant, an initial dose of 5 mg Alloferin® was given, which was repeated during the investigation as required. Temgesic® was administered in 0.3 mg amounts to keep the heart rate below 110 min/h. ECG recordings were taken continuously and the heart rate registered. The concentration of O_2 and CO_2 in the expiratory air was determined with a respiratory gas analyzer. To measure the pO_2 pressure with the multiwire platinum surface electrode developed by Kessler and Lübbers [5, 6, 8], we used the m. vastus medialis. This muscle can easily

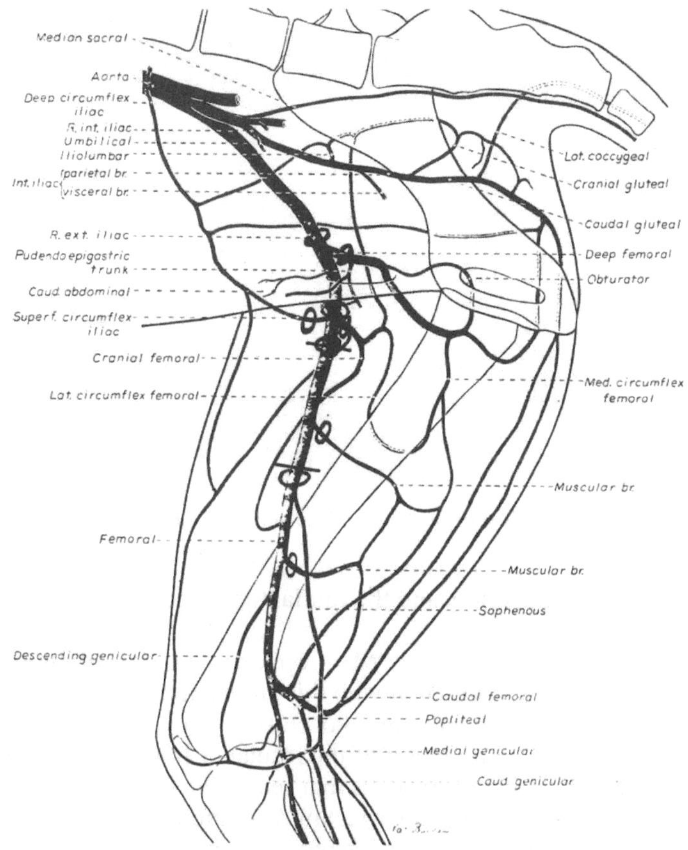

Fig. 1. Location of ligatures on the arteries of the right hindleg; proximal femoral resection between the marks (diagram after [4])

be detached from the fascia, even on the ligated leg, which is crucial for good measurements. Muscle stimulation was carried out with the Myotron stimulator (Datron Electronics, Mühltal). The stimulation impulse was a monodirectional rectangular impulse having a duration of 80 ms and a tactile ratio (=ratio of duration of periods to duration of impulse) of 7.5:1. The stimulation electrodes were placed in such a way that a maximum contraction occurred on the muscle being measured. Before and after each muscle stimulation at least one histogram was recorded. In two dogs, the arteries of the hindleg were represented angiographically.

Results

In the angiogram of one dog (Fig. 2), the formation of a collateral network typical for peripheral arterial occlusion can be clearly identified. Collaterals form between

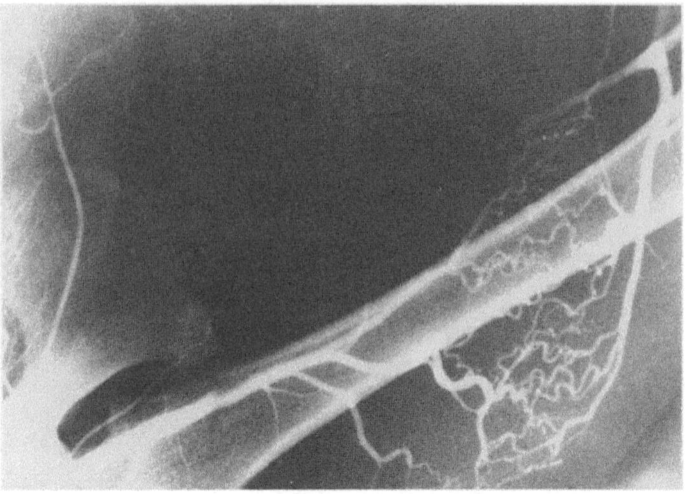

Fig. 2. Angiographic representation of the collateral circulation on the ligated right hindleg

the iliac internal artery and the caudal femoral artery. Their meandering is very typical.

All of our dogs limped. When walking on the treadmill they dragged their right hindleg. This indicated that with this animal model, the stage of intermittent claudication of the peripheral arterial occlusion can, in fact, be simulated.

Table 1 is a compilation of the results of pO_2 pressure measurements. Only the muscle stimulation of the dogs that reacted in the same way is shown (for the other two dogs see below). Under resting conditions, we can state the following: in one case, the initial value on the ligated leg is considerably higher than on the nonligated leg. In two other cases, there is no relevant difference between the mean values of the initial histograms. In the remaining dogs, the initial values on the ligated leg are lower than on the nonligated leg. Under the influence of muscle stimulation, all of the histograms show a shift to the left, even the ones taken on the nonligated (left) leg. Nevertheless on the ligated (right) leg, the pO_2 pressure decreases on average

Table 1. Results of pO_2-Measurements before and after muscle stimulation

No.	Weeks post op.	Initial values* before muscle stimulation		Decrease* due to muscle stimulation		Lowest pO_2-class in the histogram (mm Hg)			
						right		left	
		right	left	right	left	before	after	before	after
1	3	49.3	33.6	16	8	40–45	15–20	20–25	10–15
2	4	46.7	47.8	14	7	30–35	5–10	35–40	25–30
3	4	25.3	34.9	9	7	10–15	0– 5	15–20	10–15
4	8	26.3	56.1	11	5	10–15	0– 5	35–40	35–40
5	8	20.2	22.9	8	6	0– 5	0– 5	10–15	10–15
6	12	17.2	32.2	6	3	0– 5	0– 5	15–20	20–25

* mean values (mm Hg)

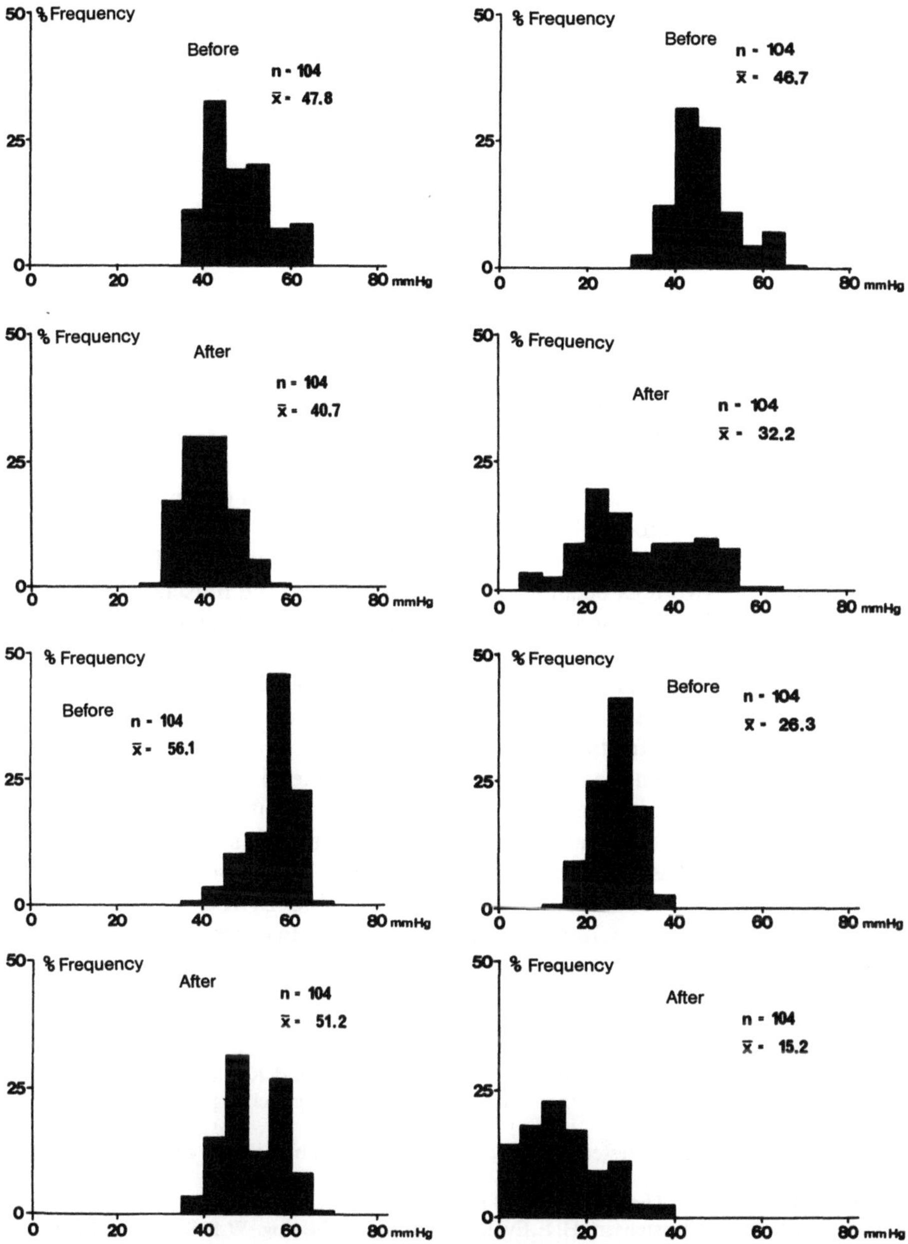

Fig. 3. pO_2 histograms before and after muscle exercise in two dogs (one above, one below). *Left,* the left, nonligated leg; *right,* the right, ligated hindleg, both before and after electrical muscle stimulation

more than on the nonligated leg. The difference becomes even more pronounced when the class with the lowest pO_2 pressure is regarded in the histogram: under muscle stimulation, this class in particular shifts to the left on the ligated leg much more than on the nonligated leg. Figure 3 shows examples of how the histograms of two dogs change as a result of muscle activity.

Discussion

The animal model with unilateral chronically ligated leg arteries is suited for simulating peripheral arterial occlusive diseases, as intermittent claudication can be induced. In angiography, the collateral network typical for this disease can be recognized. Partial oxygen pressure measurements show that there are animals which, when resting, show higher partial oxygen pressure in the muscle on the ligated leg than on the nonligated leg. This phenomenon is also found in humans. Wrabetz et al. [7] obtained the same results from their measurements using the multiwire surface electrode performed on patients with occlusive disease. The difference in pressure between the two legs becomes more evident when the muscle which is being measured is stimulated. Whereas the perfusion of the ligated leg in a resting state is maintained fairly well, there is a clear impairment on the ligated leg during exercise.

In two of our eight cases, however, this did not hold true: the partial oxygen pressure in the ligated leg in fact increased under muscle stimulation. We have yet to find the explanation for this phenomenon.

The analogy between the animal model and peripheral arterial occlusive disease stage II in humans allows conclusions to be drawn about the situation in human patients. The assessment of the microcirculation in a resting state does not necessarily provide conclusions about the degree of severity of the disease: the information obtained under conditions of stress is far more reliable. Investigations on ischemic muscle should therefore include the influence of muscle exercise.

References

1. Sunder-Plassmann, L.: Verbesserung der Mikrozirkulation und Kollateralstrombahn bei arterieller Verschlußkrankheit: Experimente und klinische Untersuchungen. Habil.-Schrift, Univ. München (1981)
2. Sunder-Plassmann, L., F. Hester, B. Endrich, K. Messmer: (Bibl. haemat. 47), Karger-Verlag, Basel (1981) S. 43–53)
3. Sunder-Plassmann, L.: Therapiewoche 33 (1983) 93–100
4. Miller, M. E., G. C. Christensen, H. E. Evans: Anatomy of the dog. W. B. Saunders, Philadelphia (1967) p. 370
5. Kessler, M.: Possibilities of measuring oxygen pressure fields in tissue by multiwire platinum electrodes. Progr. Resp. Res. 3 (1969) 147
6. Lübbers, D. W.: Principle of construction and application of various platinum electrodes. Progr. Resp. Res. 3 (1969) 136
7. Wrabetz, W., A. Creutzig, K. Alexander: Gewebesauerstoffdrucke bei arteriellen Verschlußkrankheiten. (Verh. Dt. Ges. Innere Med. Bd. 88) Bergmann-Verlag, München (1982) S. 844
8. J. Hauss, K. Schönleben, H. U. Spiegel: Therapiekontrolle durch Überwachung des Gewebe-pO_2. (Akt. Probl. Angiol. Bd. 41), Huber-Verlag, Bern (1982)

Comparative Hemodynamic Measurements Including Tissue PO_2 Measurements on the Liver under Vasopressin and Triglycyllysine Vasopressin

H. M. Kohaus, A. Holzgreve, U. Spiegel, and J. Hauss

Introduction

In 1979, our attention was drawn to the vasopressin derivative triglycyllysine vasopressin (TGLVP), which seemed to be well suited for the treatment of acute hemorrhage from esophageal piles. The drug Glycylpressin (Ferring Pharmaceuticals, Kiel) was introduced as a vasopressin derivative which has a selective effect on the visceral system lasting for more than several hours and without the side effects of vasopressin. Arguments against vasopressin treatment of acute esophageal pile hemorrhage included, first, cardiotoxic side effects, and, secondly, negative effects on the liver parenchyma, which is already damaged as a result of diminished arterial and venous perfusion. Korsback [3], for instance, reported an approximately 40% reduction of the partial oxygen pressure in the liver under intravenous vasopressin infusion: 37 mm Hg before vasopressin infusion and 20.5 mm Hg afterwards.

With the platinum multiwire surface electrode developed by Kessler and Lübbers [1, 2], a method has become available with which tissue PO_2 can be measured directly on the surface of the organ. The objective of our investigation was to control the partial oxygen pressure on the surface of the liver during lysine vasopressin application and under the administration of TGLVP.

Materials and Methods

Fourteen mongrel dogs (weight 18–23 kg) under high dosage piritramide basic anesthesia and controlled ventilation with a FIO_2 of 0.21 were investigated. The surface of the liver was exposed by an approximately 15-cm-long upper abdominal laparotomy. With the multiwire surface electrode, the tissue PO_2 was registered continuously. Before, during, and after the administration of lysine vasopressin and TGLVP, PO_2 histograms were taken.

Results

Under lysine vasopressin (Fig. 1), the mean value in the histogram dropped from 19.5 to 11.5 mm Hg. There was a marked shift to the left of the histogram with a noticeable threefold increase in the values in the lowest class between 0 and 2.5 mm Hg. The hypoxic values and those close to hypoxia thus rose distinctly under lysine vasopressin, while the arterial PO_2 hardly changed.

Under the administration of TGLVP (25 μg per kg body wt.), there was only a slight shift to the left of the PO_2 histogram with a drop in the mean value from 20.0 to 17.5 mm Hg (Fig. 2). In particular, the number of values in the smallest class increased by only about 2%. These changes could barely be registered during the continuous recordings.

The hemodynamic reactions after vasopressin and TGLVP were almost identical, though their duration varied, since TGLVP has a prolonged effect. Even after an hour following TGLVP application, the hemodynamic changes were still pronounced, whereas 30 min after the infusion of lysine vasopressin almost initial values had returned.

In detail (Figs. 3, 4) it was found that:

- The arterial pressure increased by about 20%
- The peripheral resistance rose considerably
- There was hardly any change in pulmonary pressure
- The pulmonary resistance of the vessel increased slightly

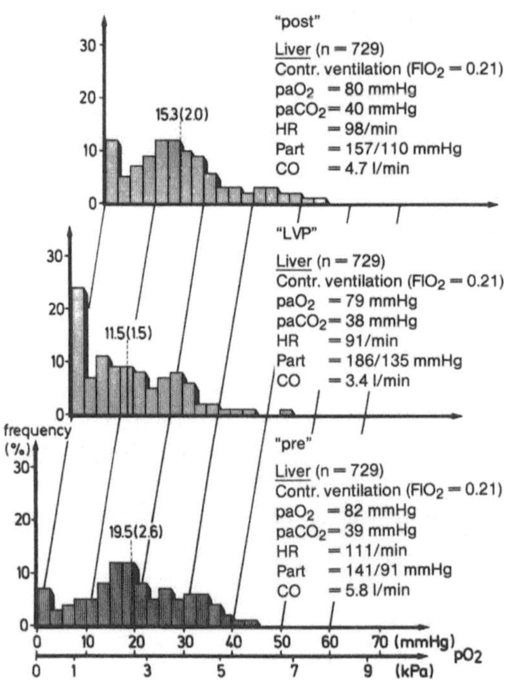

Fig. 1. Summarized PO_2 histograms before, during, and after lysine vasopressin infusion. Marked increase in hypoxic values and of those close to hypoxia in the middle PO_2 histogram during the infusion

- The heart rate dropped by about 10%
- The cardiac output declined significantly
- The stroke volume dropped slightly was a result of the reduced heart rate
- The central venous pressure remained constant

To clarify the hemodynamic changes, flow measurements were carried out under TGLVP on several visceral vessels, on the v. portae, and on the a. femoralis of experimental animals. Measurements were done with the electromagnetic flowmeter (Carolina Med.). A diminished flow was observed in the a. gastrica sinistra and in the truncus (t.) jejunalis, while there was a marked increase in flow in the a. hepatica. In the t. celiacus only a slight increase was found. A pronounced flow reduction of about 50% in the v. portae resulted from the diminished flow in the mesenteric region (Fig. 5). This flow reduction can be partly attributed to the reduced cardiac output. There must, however, also be a local reason for this reduction, since in the a. femoralis for example, only a slight flow change is registered. These flow changes are responsible for the hemostyptic effect during acute hemorrhage from esophageal piles and in the upper gastrointestinal tract.

Clinical Results

The results obtained in the animal experiment were confirmed indirectly in humans. A reversal of the ratio of arterial to portal venous flow was established by means of perfusion scintigraphy on healthy volunteers (Fig. 6). The end of the arte-

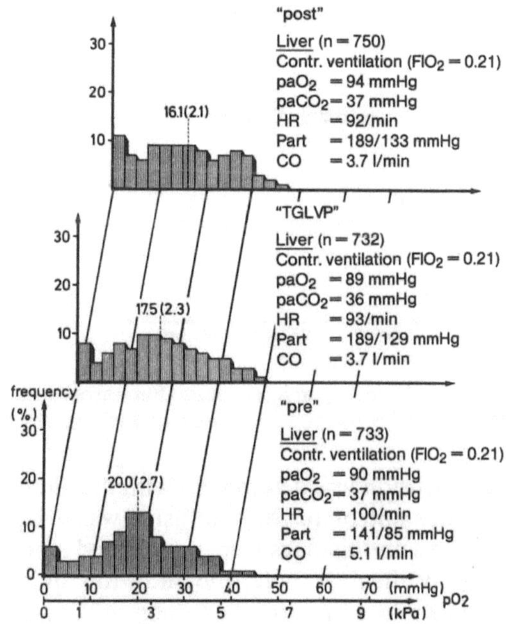

Fig. 2. Summarized PO_2 histograms before, 30 min after, and 60 min after i. v. bolus injection of TGLVP. Only a slight shift to the left of the histogram

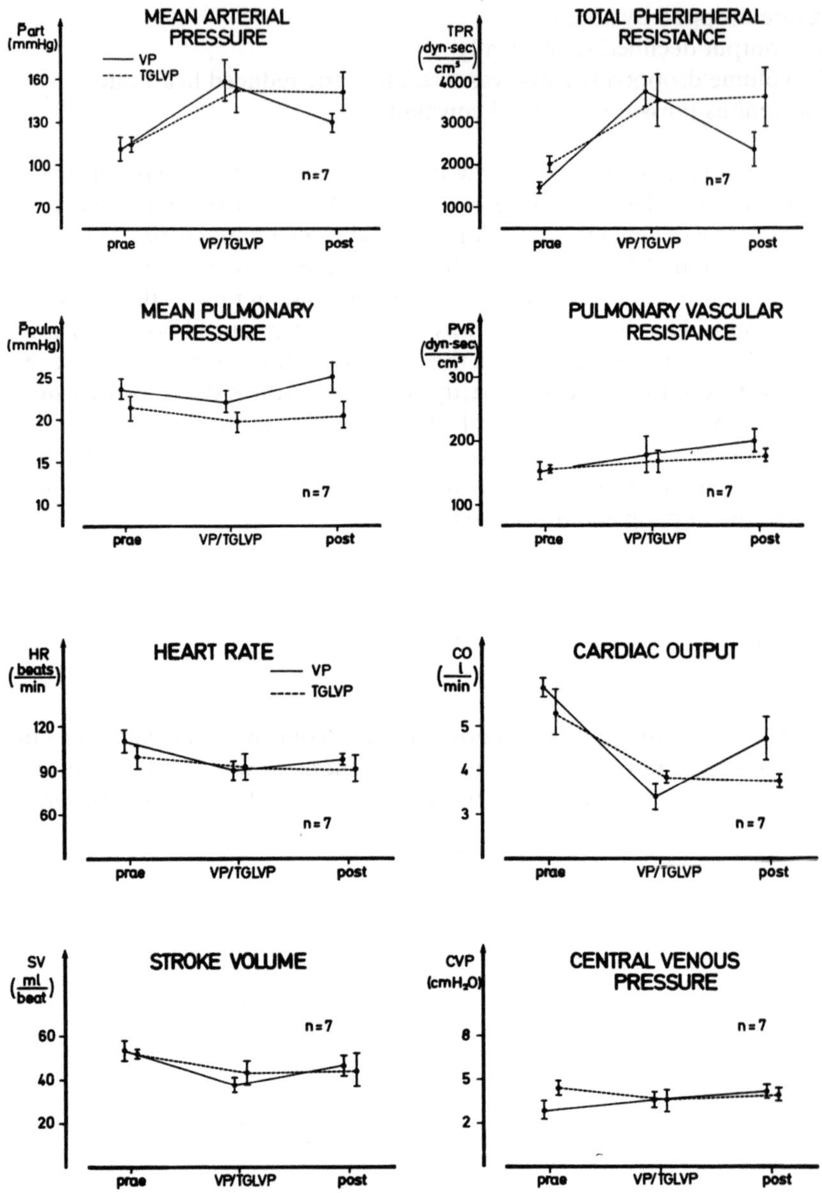

Figs. 3, 4. Hemodynamic changes after vasopressin and TGLVP. Effects of both drugs on the hemodynamics are the same, but there is a longer duration of action for TGLVP

rial perfusion was determined with the aid of kidney perfusion. After TGLVP infusion, the ratio of the liver perfusion was reversed: 85% portal venous to 15% arterial.

The same changes were established among patients with various degrees of cirrhosis of the liver. Independent of the initial value of the portal venous perfusion, the portal venous perfusion decreased consistently after TGLVP to about 10% of

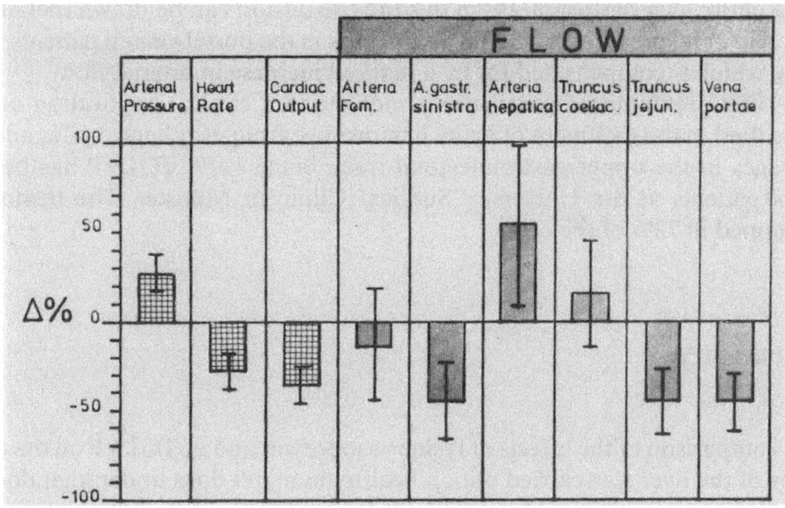

Fig. 5. Flow changes after TGLVP in t. celiacus, t. jejunalis, v. portae, a. hepatica, a. gastrica sinistra, and a. femoralis. Marked flow increase in the a. hepatica and flow reduction in the t. jejunalis, a. gastrica sinistra, and v. portae (bolus injection 20 μg/kg, $n = 12$, changes after 30 min)

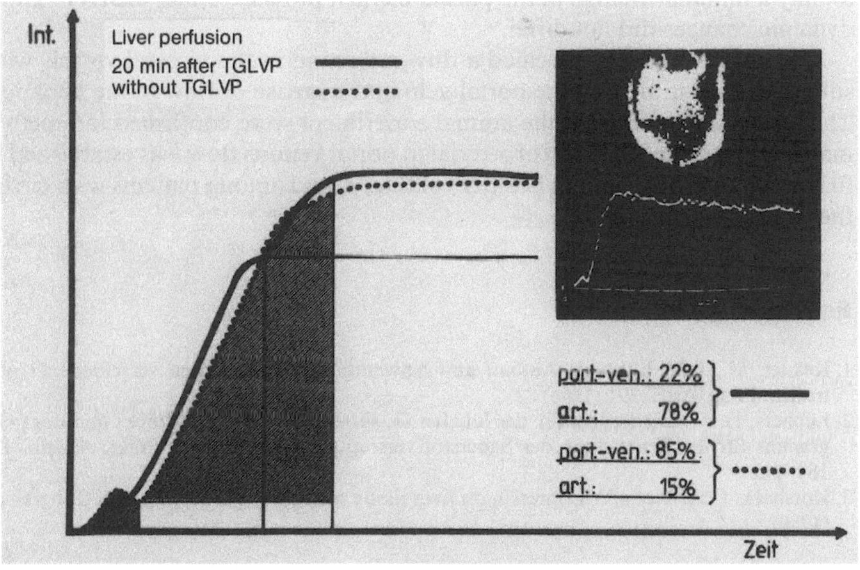

Fig. 6. Liver perfusion scintigram of a healthy volunteer; (···) normal perfusion scintigram; (—) perfusion scintigram after TGLVP; 22% increase in portal venous flow (mathematically calculated) to 85% after TGLVP

the entire liver perfusion. From this the conclusion can be drawn that after TGLVP application, pressure reduction also occurs in the portal vein in patients with cirrhosis, which is compensated for by a marked increase in arterial flow.

Our preliminary investigations indicate that clinical application of TGLVP is justified in the treatment of acute hemorrhage from esophageal piles and of hemorrhages in the upper gastrointestinal tract. Since 1979, TGLVP has been used on 160 patients at the University Surgical Clinic in Münster. The hemorrhage was stopped in 73% of the cases.

Summary

A comparison of the effects of lysine vasopressin and of TGLVP on the oxygen supply of the liver was carried out on healthy mongrel dogs under high dosage piritramide basic anesthesia. Measurements were performed with the platinum multiwire surface electrode developed by Kessler and Lübbers. The effect of TGVLP lasts for several hours, as opposed to that of vasopressin, which lasts only a few minutes. A marked reduction in tissue PO_2 was observed under lysine vasopressin, as opposed to only a minimal change in the partial oxygen pressure under TGLVP. The hemodynamic changes did not differ.

Flow measurements revealed a flow reduction in the visceral vessels with a resulting decline in flow in the portal vein and increase of flow in the hepatic artery. The results obtained from the animal experiment were confirmed indirectly in humans. A reversal of the ratio of arterial to portal venous flow was established by perfusion scintigraphy, both in healthy volunteers and among patients with cirrhosis of the liver.

References

1. Kessler, M., D. W. Lübbers: Aufbau und Anwendungsmöglichkeiten verschiedener pO_2-Elektroden. Pflüg. Arch. 291 (1966) 82
2. Lübbers, D. W.: Die Bedeutung des lokalen Gewebesauerstoffpartialdrucks und des pO_2-Histogramms für die Beurteilung der Sauerstoffversorgung eines Organs. Prakt. Anästh. 12 (1977) 184–193
3. Korsback, C.: Effect of Vasopressin on liver tissue pO_2. Abstr. E. S. S. R., 15th Congress, Brüssel (1980)

Measurement of Oxygen Pressure for Control of Isolated Muscle Grafts

H. P. Keller

Introduction

As was first reported in 1976, the transcutaneous (tc) measurement of partial oxygen pressure (PO_2) is a suitable method for controlling skin grafts. Using this technique in animal experiments, the healing of skin autografts and the rejection of allografts could be followed. On pedicle skin flaps, an improvement in the oxygen supply as a result of revascularization was observed between the 4th and 8th postoperative days (Figs. 1, 2). $tcPO_2$ measurements showed that the ischemia which results from early gradual cutting off of the circulation of so-called axial pattern flaps seems to effect a more rapid revascularization from the graft bed into the skin flap. This is clinically important because the flap's pedicle can then be safely detached at least 1 week earlier than usual. The $tcPO_2$ measurement also facilitates the postoperative control of free, microvascularly transplanted skin and skin muscle grafts, which in the presence of good venous drainage often make a valied clinical assessment difficult.

Fig. 1. PO_2 measuring electrode on a pedicle skin flap

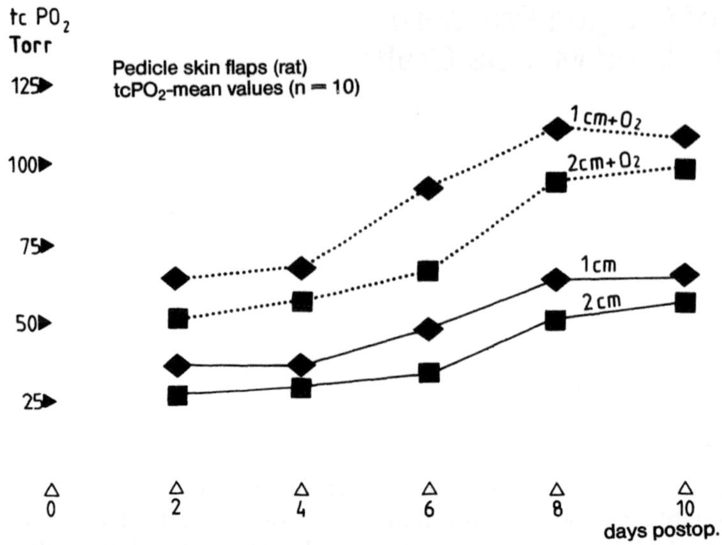

Fig. 2. $tcPO_2$ values on pedicle skin flaps at distances of 1 cm and 2 cm from the base of the flap; under ambient air *(bottom)* and at 40% O_2 *(top)*

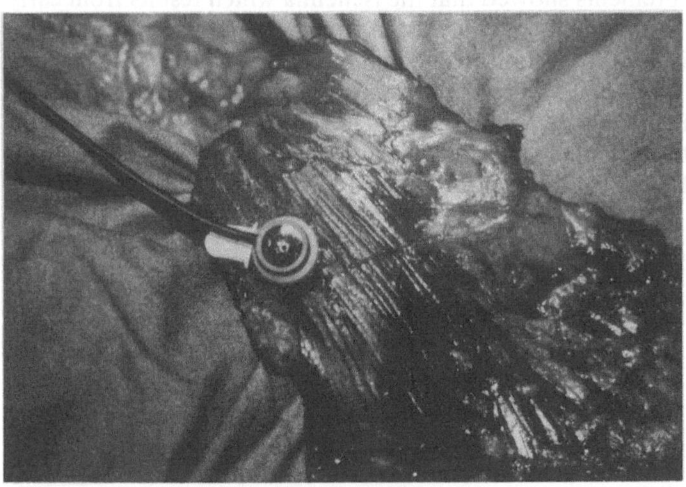

Fig. 3. PO_2 measuring electrode fixed with tissue glue on the m. latissimus dorsi of a dog

The rapid increase in muscle grafting, in the form either of vascular pedicles for the reconstruction of skin defects or of free microvascular grafting, and the sensitivity of the muscle to lack of oxygen, induced us to carry out PO_2 measurements for the purpose of verify the circulation in the muscles.

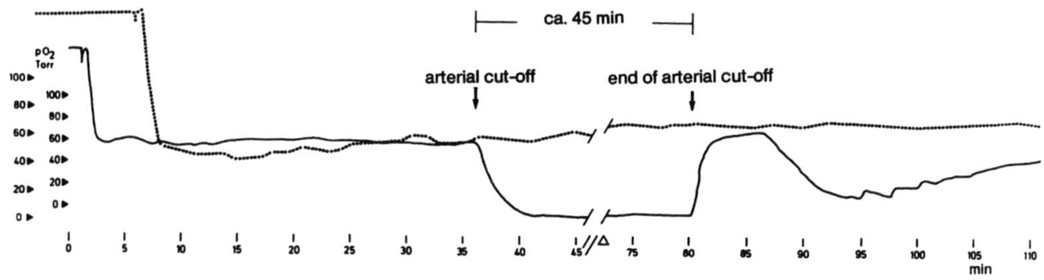

Fig. 4. PO_2 measuring curve on the m. latissimus dorsi with arterial flow cut off for approximately 45 min. —, PO_2 of isolated m. latissimus dorsi; · · · · ·, control PO_2

Methods

The PO_2 measurements were carried out with electrodes for $tcPO_2$ measurements [1]. The m. latissimus dorsi of the dog served as the experimental model. The muscle was isolated in such a way that the circulation of the muscle was maintained by only one artery and one vein. The contralateral m. latissimus dorsi was dissected by a small incision for control measurements.

Unlike measurements on the skin, the measurement of PO_2 on the muscle requires no heating of the area to be measured. On the skin, airtightness can be achieved simply by securing the measuring electrode with a self-adhesive ECG ring. For measurements on the muscle in situ, it is necessary to keep the electrode air- and moisture-tight, which can be done by adding tissue glue to the ECG ring (Fig. 3). After the electrode had been calibrated and secured to the muscle in the manner described, the muscle was transposed into the skin pocket that had formed. In this way, measurements could be easily continued for a number of hours. Arterial and venous flow were cut off by occluding the vessels concerned with a surgical clamp.

Results

After isolated clamping of the thoracodorsal artery feeding the m. latissimus dorsi, the PO_2 values dropped to 0 relatively fast. Up to approximately the 3rd h of ischemia, a typical increase in the PO_2 values to above 30–60 Torr was shown when the clamp was released. The PO_2 values then dropped to below these average values and finally climbed slowly back to them (Fig. 4). After about 3–4 h of ischemia, this physiologic PO_2 reaction failed to appear after the vessel block was released. Only a rapid increase back to the average PO_2 values could be observed, which then remained in this range (Fig. 5). In two cases, a slow decrease of the PO_2 value to below 10 Torr was observed after measurements had been carried out for a period of about 6 h.

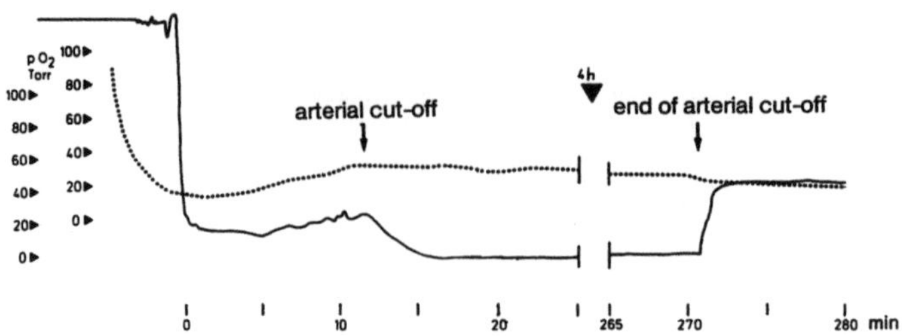

Fig. 5. PO_2 measuring curve with arterial cutoff of about 4 h. —, PO_2 of isolated m. latissimus dorsi; · · · · ·, control PO_2

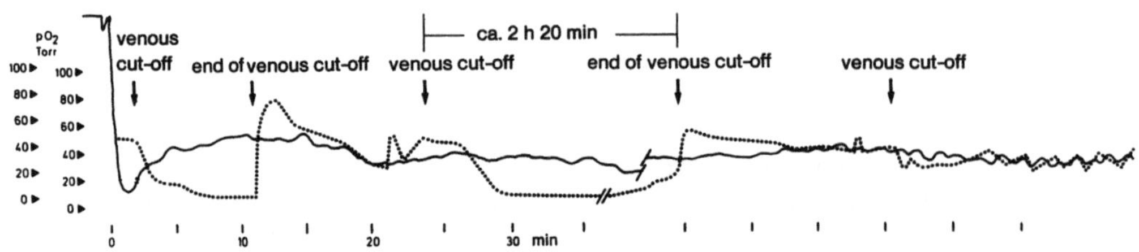

Fig. 6. PO_2 measuring curve with venous flow cut off for about 8 min and for about 2 h: PO_2 values increase again after venous flow is stopped. —, PO_2 of isolated m. latissimus dorsi; · · · · ·, control PO_2

Isolated clamping of the vein while the artery was kept open caused the PO_2 value to decline slowly - though faster than expected - to 0. After the vein was reopened, a short term average increase of the PO_2 to above the normal values was also observed. After about the 2nd h of ischemia, the PO_2 began to rise again slowly, without the clamp having been released, and then reached the initial values after release of the vessel block. When the vein was reclamped the PO_2 did not drop again.

Discussion

The $tcPO_2$ electrode, if secured properly, is well suited for PO_2 measurements on the muscle. Measurements can be made over a period of hours. On average, PO_2 values between 30 and 60 Torr were registered. The oxygen supply and thus the circulation of isolated muscle grafts can be monitored.

The experiments gave no clear indication of the muscle's tolerance of ischemia, nor did they give evidence as to at what particular point in time irreversible damage occurs. We only noticed that when the muscle was reperfused after about the 4th h of ischemia, there was no typical physiological reaction of an above-average increase in the PO_2 values and subsequent leveling off at normal values. The fact that the PO_2 reaches average values is a sign of reperfusion of the muscle vessels, but says nothing about possible irreversible damage of the muscle fibers.

Isolated clamping of the vein causes a surprisingly rapid decline in the PO_2 values on the muscle. This shows how threatening venous occlusion can be for the survival of a muscle. The repeated increase in the PO_2 values when the vein is blocked, consistently observed after an average of 2 h of ischemia, can be explained by an escape of plasma caused by ischemia through the damaged walls of the vein. The resulting venous flow makes arterial flow possible, so that when the venous block is released, reclamping the vein does not cause another drop in the PO_2 values. Moreover, it seems that the strong intramuscular pressure causes small vessels that had been previously coagulated to reopen. The considerable amount of serous fluid that is secreted from the isolated muscle after prolonged venous obstruction might be an indication that in the presence of ischemia of the muscle in compartments, damage to the muscle can be accelerated by additional tissue pressure (compartment syndrome).

Summary

Our investigations with PO_2 measurements on isolated muscle flaps show that the method is well suited for monitoring muscle grafts. The study provided information on the muscle's tolerance of ischemia.

References

1. Huch, R., D.W. Lübbers, A. Huch: Quantitative continuous measurement of partial oxygen pressure on the skin of adults and newborn babies. Pflüg. Arch. 337 (1972) 185–198
2. Keller, H.P., P. Klaue, D.W. Lübbers: Transcutaneous pO_2 measurement for evaluating the oxygen supply of skin allo- and autografts. Eur. Surg. Res. 10 (1978) 272–282
3. Keller, H.P., P. Klaue: Die transkutane pO_2-Messung zur objektiven Verlaufskontrolle von gestielten Hautlappen. Med. Welt 15 (1980) 556–557
4. Stroch, P.E., G. Majno: Vascular responses to experimental tourniquet ischemia. Surg. Gyn. Obstet. (1969) 309–318

The Effect of Local Hyperthermia on the Tissue Oxygen Tension of Melanoma in the Hamster

B. Endrich, J. Voges, and A. Lehmann

Introduction

The cytotoxic effect of hyperthermia in malignant tumors varies with the heat distribution in the tumor and in the surrounding tissue. Therefore, nutritive tumor circulation and morphologic changes in the microvasculature after the temperature increase are important variables in reaching a complete destruction of tumor cells under in vivo conditions [1–4, 11]. Hyperthermia not only induces environmental changes in the interstitial space, but also affects the thermosensitivity and the development of thermotolerance [11]. Nevertheless, since devices for noninvasive temperature measurements have not yet become available for routine use in clinical practice, hyperthermia is still being induced without knowledge of the actual temperature in the tumor. Moreover, to date, theoretical calculations of the heat distribution within a tumor have not been acceptable as an approximation of the temperature variations. If, however, a detailed quantitative analysis of tissue oxygenation and microhemodynamics under local hyperthermia of malignant tumors became available, a theoretical basis could be provided for mathematical models.

This study was carried out in an attempt to answer the questions (a) At what temperature is the maximum capillary perfusion observed? (b) What temperature is needed to induce stasis in the capillaries of the malignant melanoma? and (c) What is the mechanism of the microcirculatory disorder induced by local hyperthermia?

Methods

In 26 Syrian golden hamsters, a transparent chamber was implanted in the dorsal skin fold. Systemic pressures and heart rate were measured via two permanent indwelling catheters placed in the right carotid artery and the jugular vein [5]. In 17 animals, 4×10^4 cells of the amelanotic hamster melanoma A-Mel-3 were transplanted in the subcutaneous tissue within the chamber after 48 h. Five days later, with an average tumor diameter of 3 mm, the microcirculation of the tumor and the local tissue PO_2 were observed in conscious animals under the intravital microscope [9, 10] and the microvascular blood flow along different vessel segments was deter-

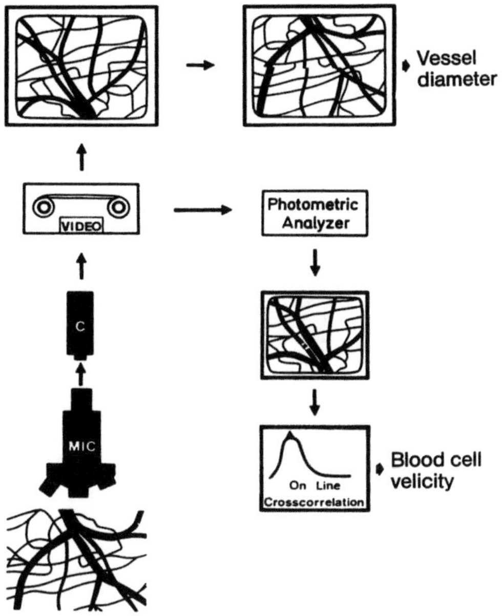

Fig. 1. Quantitative analysis of microcirculation by means of an intravital microscope *(MIC)*, a video camera *(C)*, and videotape recording with subsequent photometry and electronic television image analysis [cp. 6]

mined [6] by television methods (Fig. 1). Measurements were taken at 30 °C and 15 min after reaching a temperature of 35° and 42.5 °C.

Local hyperthermia was induced with a water-perfused heat exchanger placed beneath the skin fold chamber. In addition, the tumor was superfused with physiological Ringer solution, which was heated up to the temperature given by the protocol. A temperature probe was placed in the tumor for control and measurement of this parameter. A rectal thermometer served to register the body temperature of the animal.

Results

Upon heating the skin fold chamber, the apparent maximum of capillary perfusion in the tumor was reached 15 min after a tumor temperature of 35 °C was achieved. The volume flow of blood increased significantly from 2.0 ± 0.2 ml/min $\times 10^{-5}$ to 2.7 ± 0.2 ml/min $\times 10^{-5}$ ($P < 0.001$, Wilcoxon test, $n = 86$); the mean local PO_2 in the tumor tissue rose from 8.8 to 18.5 mm Hg (Fig. 2). When the heating was continued, a drop in capillary blood cell velocity was observed at a tumor temperature of 40.6°-41.3 °C. Fifteen minutes after 42.5 °C was reached, the capillary volume flow dropped to 0.5 ± 0.1 ml/min $\times 10^{-5}$ ($n = 86$). In the center of the tumor, the blood cell velocity was less than 0.1 mm/s in 50% of the capillaries. At the same time, the mean tissue PO_2 was reduced to 6.1 mm Hg. The majority of all the values registered (52.5%) were between 0 and 5 mm Hg. An interstitial edema developed as soon as a

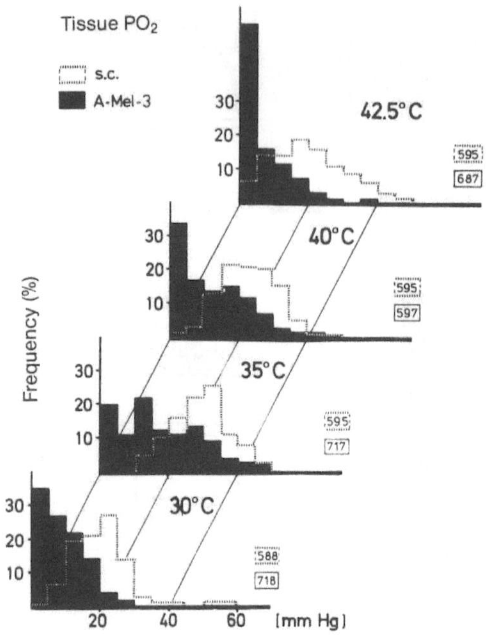

Fig. 2. Changes in local tissue PO_2 on the surface of the tumor *(A-Mel-3)* and in the normal microcirculation *(s.c.)* at various temperatures

disorder of microvascular perfusion, such as intermittent flow, occurred. There were multiple petechial hemorrhages, particularly in the center of the tumor. Despite the pronounced tissue hypoxia, a constriction of all arterioles feeding the tumor could be observed at the edge of the melanoma under study.

By contrast, in control preparations without tumors ($n = 5$), the tissue oxygenation did not show a marked decrease until the temperature of the tissue reached 45 °C. At this temperature, the mean local tissue PO_2 was of the order of 9.7 mm Hg ($n = 588$). As can be seen in Fig. 3, an intravascular pressure of 37.9 ± 1.5 mm Hg was measured in eight arterioles (five animals) with a diameter between 17.2 and 40.0 μm. While the pressure in precapillaries (36.8 ± 1.5 mm Hg) was almost identical, the pressure in the capillary lumen dropped to 24.7 ± 1.4 mm Hg under normothermia. Considerably lower values were measured in draining vessels of the tumor as compared with the feeding vessels. Heating the tumor to 35 °C caused no significant changes of intravascular pressure in arterioles and precapillaries. In draining vessels of the melanoma, however, significantly lower pressures ($P < 0.05$) were registered.

Local hyperthermia (15 min at 42.5 °C) resulted in a reduction of the arteriolar-venular pressure gradient from 26.9 mm Hg to 17.9 mm Hg. This was caused by the significantly increased pressure in postcapillaries ($P < 0.01$) and collecting venules ($P < 0.05$), and a result of the pronounced aggregation of erythrocytes.

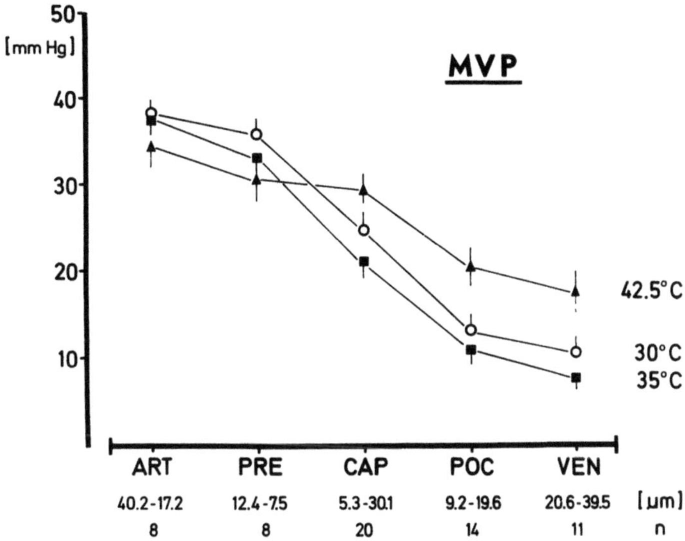

Fig. 3. Intravascular pressure in individual segments of the microcirculation of the amelanotic melanoma A-Mel-3 (*ART,* arterioles; *PRE,* precapillaries; *CAP,* capillaries; *POC,* postcapillaries; *VEN,* collecting venules) under local hyperthermia (*n,* number of single measurements; data derived from seven animals)

Discussion

By warming up an amelanotic melanoma from 30° to 35 °C, tissue oxygenation could be improved by almost 100% as a result of the increase in capillary circulation (volume flow). Heating the melanoma to more than 41 °C for 15 min caused a significant reduction of capillary perfusion, particularly in the center of the tumor, as well as a massive deterioration of tissue oxygenation. Since capillaries in the tumor have to be considered a network of rigid tubes, the increase in perfusion at 35 °C can only be explained by vasodilation of the host vessels. By contrast, at a tumor temperature of 41 °C, a disturbance of flow (aggregation of erythrocytes) is first observed in postcapillaries and collecting venules; later stasis in capillaries prevails. These changes are not found in the normal tissue which surrounds the tumor. This observation suggests that local hyperthermia might produce a selective blockade, particularly of the drainage vessels. When this blockade persists for more than 10 min, all of the arterioles feeding this tumor constrict, probably as a result of a myogenic contraction [8] (arteriolar constriction could occur when there is a 50% increase of vascular resistance in the draining vessels).

Since no revascularization and reperfusion of the terminal vascular bed destroyed by local hyperthermia was observed until 48-72 h after the treatment has been completed, one could assume that large portions of the tumor are excluded from the perfusion through the systemic circulation for this time interval. Consequently, drugs will not be able to reach the tumor cells via the microvascular network for at least 48 h upon hyperthermia. On the other hand, when using a combination of hyperthermia and radiation therapy, either both treatment modalities have to be applied simultaneously (at the time of maximum tissue oxygenation/maximum capillary volume flow at 35 °C), or hyperthermia should be applied immediately after radiation has been terminated.

References

1. Dickson, J.A., S.K.Calderwood: Thermosensitivity of neoplastic tissues in vivo. In: F.K.Storm, G.K.Hall (eds.): Hyperthermia in Cancer Therapy. Medical Publishers, Boston (1983) pp 63-140
2. Eddy, H.A.: Alterations in tumor microvasculature during hyperthermia. Radiol. 137 (1980) 515-521
3. Emami, B., G.H.Nussbaum, N.Hahn, A.Dritschilo, F.Quimby: Histopathological study on the effects of hyperthermia on microvasculature. Int. J. Rad. Onc. Biol. Phys. 7 (1981) 343-348
4. Endrich, B., B.W.Zweifach, H.S.Reinhold, M.Intaglietta: Quantitative studies of microcirculatory function in malignant tissue: Influence of temperature on microvascular hemodynamics during the early growth of the BA 1112 rat sarcoma. Int. J. Rad. Onc. Biol. Phys. 5 (1979) 2021-2030
5. Endrich, B., K.Asaishi, A.Götz, K.Meßmer: Technical Report - A new chamber technique for microvascular studies in unanesthetized hamsters. Res. Exp. Med. 177 (1980) 125-134
6. Intaglietta, M., N.R.Silverman, W.R.Tompkins: Capillary flow velocity measurements in vivo and in situ by television method. Microvasc. Res. 10 (1975) 165-179
7. Jain, R.K.: Bioheat transfer. Mathematical models of thermal systems. In: F.K.Storm, G.K.Hall (eds.): Hyperthermia in Cancer Therapy. Medical Publishers, Boston (1983) pp 9-46
8. Johnson, P.C.: The myogenic response. In: D.F.Bohr, A.P.Somlyo, H.V.Sparks (eds.): Handbook of Physiology (Vol II, Sect). Amer. Physiol. Soc., Bethesda (1980) pp 409-442
9. Kessler, M., W.A.Grunewald: Possibilities of measuring oxygen pressure fields in tissue by multiwire platinum electrodes. Prog. Respir. Res. 3 (1969) 147-152
10. Lübbers, D.W.: Principle of construction and application of various platinum electrodes. Prog. Respir. Res. 3 (1969) 136-146
11. Vaupel, P., W.Müller-Klieser, J.Otte, R.Manz, F.Kallinowski: Blood flow, tissue oxygenation, and pH-distribution in malignant tumors upon localized hyperthermia - Basic pathophysiological aspects and the role of various thermal doses. Strahlenth. 159 (1983) 73-81

Discussion

Summarized by **A. M. Ehrly**[1]

Introduction and Measuring Methods

Lübbers (Dortmund). Following this introductory presentation, questions were asked on the relationship of the hematocrit in the flowing blood to the transcutaneous (tc) PO_2 value. According to Lübbers, this constellation usually has no substantial influence on the $tcPO_2$ value, since the delivery of O_2 is compensated for by an increase in the cardiac output. Problems are posed, however, by several young (completely healthy) volunteers who have lower $tcPO_2$ values than normal, probably as a result of vasoconstrictor tonus being too high in the region of the candelabrum arteries. In these persons, local heat-induced hyperemia using the conventional measuring technique does not seem to be sufficient. This small group of "breakaways" deserves special attention, since they may provide new information.

Baumgärtl (Dortmund). Fleckenstein (Lübeck) commented that large needle electrodes involve considerable traumatization. This can be compensated for by reading the PO_2 values at very short intervals while the electrode is being advanced stepwise into the tissue. Fleckenstein believes that in this way an evaluable tissue PO_2 value can be obtained. Baumgärtl answered that this confirmed essential points of his report and emphasized that the tissue damage is time dependent. He did not want to say for certain, however, wheter this damage might cause a change in the tissue PO_2 values even within very short time periods.

Thermann (Kiel) asked how the needle electrode was sterilized. Baumgärtl answered that he had not dealt with that aspect. Fleckenstein uses ethylene oxide or hot water vapor. Ehrly said his electrodes are sterilized during the course of production with solvents. He claimed that in the course of several hundred measurements carried out in patients, there had never been a sign of contamination.

With regard to the problem of traumatization caused by needle electrodes, Ehrly explained that the thickness of the electrode as well as the geometry of the electrode

[1] This summary is taken from a tape recording of the discussions following the presentations held at the symposium. To keep the book from becoming too long, the questions and answers have been shortened and some have been omitted. Speakers who could be identified on the tape are named.

tip have an influence on the traumatization of the tissue being examined. Beyond that, the speed at which the electrode is introduced into the muscle tissue apparently is a contributing factor. The tissue needs time to get out of the way of the platinum needle electrode. In a histologic examination on an amputated leg, no puncture track from the microneedle electrode (diameter of tip: 1–3 µm) could be detected.

Spiegel (Münster). Baumgärtl (Dortmund) asked why KCl solutions were still being used for multiwire surface electrodes. Recent studies had shown that this is one of the worst electrolytes. Spiegel replied that several different electrolyte solutions had been tested and the buffered, bromide-free 0.3 molar KCl solution is still considered to be the best. If Baumgärtl had a better electrolyte, he would be happy to test it.

Fleckenstein (Lübeck) commented that in the histograms obtained with the multiwire surface electrodes, there are no classes of very high and very low PO_2 values, whereas Kunze's and Ehrly's results showed very low tissue PO_2 values. He asked whether this could be a result of the fact that the distance between the cathode and the tissue is relatively large when the surface electrode is used, which might cause extremely low and extremely high values to be shown. Lübbers (Dortmund) claimed that the explanation that the anesthesia or medication during intensive care treatment leads to an inhomogeneous blood flow is only partly valid. He had carried out simultaneous measurements on the brain of an experimental model, using both microplatinum needle electrodes and multiwire surface electrodes, and did not notice any major differences. This, however, could differ from organ to organ; the same was observed in measurements on the heart, where the anatomy of the vessel supply varies in the different myocardial layers.

Another question was how long it takes before the measured values can be evaluated after the electrode is placed on the musculature and whether the manipulation causes superficial damage to the tissue. Spiegel (Münster) answered that first, the electrode selected should be as small as possible and, secondly, the weight on the electrode must be kept light. The first values can be read 10–15 min after the electrode is first placed on the musculature. There is no criterion of damage to the tissue. Lübbers (Dortmund) added that he did not find any histologic changes in the tissue after careful application of the technique and putting very little weight on the electrode. Spiegel explained that during continuous measurements on the various organs over a 4-h period, the value obtained did not show any significant differences.

Ehrly (Frankfurt/Main) said that his group had carried out parallel measurements of the tissue PO_2 using platinum needle electrodes and transcutaneous PO_2 measurements in healthy volunteers and in patients. There was a discrepancy in the value obtained, i.e., relatively low $tcPO_2$ values on the skin with completely normal muscle tissue oxygen pressure values and, accordingly, normal distribution in the histograms. One reason for this poor correlation, Ehrly claims, could be the high oxygen consumption of the $tcPO_2$ electrode, which can occur when the flow rate in the skin is reduced. Another reason could be a steal phenomenon in the skin blood flow resulting from the local heating of the $tcPO_2$ electrode, which does not exist or should be prevented when the skin temperature is normal. Fagrell's diagrams have shown that there are also cross-connections in the region of uppermost capillary loops, which make a steal phenomenon feasible. When pharmacological investigations are being performed with so-called vasodilating substances, the question should be asked whether the increase in $tcPO_2$ is perhaps a result of a steal phenomenon in the electrode range (maldistribution).

Huch replied that she had never found any values of 35 Torr in measurements in healthy volunteers and that normal $tcPO_2$ values varied between 50 and 90 Torr. Moreover, additional pharmacological vasodilation could not intensify the hyperemia induced by local heat.

Lübbers (Dortmund) remarked that skin is not the same as muscle in terms of its hierarchical position within the circulatory system. The $tcPO_2$ value, which is measured while autoregulation is switched off by the application of local heat, is a value that pertains only to the skin. The regulation of the muscle lying beneath the skin is not affected, so that the tissue PO_2 values and the $tcPO_2$ values can be completely different. At 37 °C the situation is quite different, because the regulatory capacity of the skin can still be preserved when the electrode is heated only slightly. At this measuring temperature, questions on the regulation of the skin could be included. Careful consideration and measurement of the parameters at various temperature could allow further diagnostic conclusions to be made.

Huch (Zürich) stated that she sterilizes her complete, prepared measuring units in advance so that she can use them for newborn babies. She pointed out the significance of the amount of arterial PO_2 for the tissue oxygen pressure. Spiegel (Münster) indicated, however, that particularly under conditions of shock, the tissue PO_2 can be low even when the arterial PO_2 remains normal.

Kersting (Berlin) asked what ethical problems are involved when, for instance, the surface electrode is used in the intensive care unit. Spiegel replied that his project using the surface multiwire electrode in intensive care patients had been approved by the University Ethics Commission.

The discrepancy between the results of the microneedle electrode technique and the surface electrode technique is, in Ehrly's opinion, essentially caused by the angioarchitectonic differences on the surface of the muscle and also inside the muscle. Lübbers (Dortmund) added that compression could be done with needle electrodes. Ehrly confirmed the principle of this: by using suitable microneedle electrodes and, in particular, by introducing the electrode into the skin very gradually, the compression of the capillaries in the muscle tissue itself can be minimized to the point where it is no longer relevant. He further explained that, in his opinion, the histogram recorded with a multiwire surface electrode is, strictly speaking, only a "surface histogram," i.e., the results of recordings at different measuring sites on the surface of a tissue by repositioning of the electrode. With the microneedle electrode technique, however, the tissue is passed through so that the histogram achieved genuinely is a "histogram."

Huch (Zürich). Lübbers (Dortmund) said that he was pleased that what had been found in theory and in animal experiments was now being proved in practice. He was relieved that the parameters that could influence the $tcPO_2$ had been established. This did not mean, however, that in practice there were no deviations or that the applicability in patients was without limitations.

Kopp (Freiburg) referred to the restrictions with regard to the application of the $tcPO_2$ measurement in the field of anesthesia and especially wanted to know to what extent halothane anesthesia, which is often used, can affect the values. Huch replied that, in fact, excessively high $tcPO_2$ values had been obtained under halothane and that this was also the case for the blood gas determination of oxygen. At present, efforts were being made to find a membrane that was permeable for oxygen but not for halothane. Lübbers (Dortmund) added that Teflon membranes with these properties had already been developed, but were apparently not yet in use.

Tissue Oxygen Pressure

Landgraf and Ehrly (Frankfurt/Main). Funk (Heidelberg) wanted to know whether the Kolmogorov-Smirnov test, which was supposed to be suited for comparing the histograms, including comparison in terms of steepness, had been used for the statistical evaluation. Landgraf replied that the test was indeed one possibility of evaluation, but it did not completely solve the problem of statistically assessing and comparing histograms.

Intensive work with the statisticians at Frankfurt University had not provided any solution to the problem. It was not enough to simply feed any one test into the computer and consider the values thus obtained as being representative.

Lübbers (Dortmund) said that his former colleague Grunewald once theoretically calculated that a truly representative histogram would have to be made up of about 11000 points. To be more realistic, one would have to start from a minimum of 150 individual measuring points. Lübbers also thought that the median value alone was not sufficient and histograms should be compared with one another. This would cause problems which even a statistician could not solve.

Ehrly (Frankfurt/Main) reported that his group had dealt closely with this problem. Landgraf and the biostatisticians at the univeristy had reached the core of the questions on the comparability of the results from PO_2 measurements. Ehrly felt the only possibility was to compare the median values with one another and then to compare them statistically using the Wilcoxon test. Furthermore, the various groups of persons tested could be compared on the basis of a comparison of the pooled histograms.

Huch (Zürich) advocated revealing the changes in the histograms, after therapeutic influence, for example, by printing out a histogram every 10 min.

Hauss (Münster) asked the assembly how the changes in the tissue PO_2 should be described. Lübbers (Dortmund) said that descriptive changes, during the course of the therapy, for example, could be demonstrated with summarized histograms. A decisive factor, however, is the limited number of measuring sites and PO_2 signals, which makes analysis much more difficult, at least in terms of calculations and statistics.

Beerthuizen (Nijmegen). Lübbers inquired about the diameter of the needle electrode and whether the authors thought that the configuration of the histogram – as, indeed, the measured value itself - was affected by the traumatic damage to the tissue caused by the electrode. Beerthuizen said the electrode was 700 µm in diameter. During the discussion that followed, doubt was expressed as to whether the results of measurements performed with macroneedle electrodes could be compared with those of microneedle electrodes.

Ehrly et al. (Frankfurt/Main). The speakers were asked to what extent those patients who could not keep up the 4 min of exercise showed particularly low initial PO_2 values. Ehrly replied that the question could not be answered generally, but that the majority of patients had a low tissue PO_2. Partsch (Vienna) pointed out the heterogeneity of the tissue PO_2 behavior in patients after exercise and wished to know if the poststenotic blood pressure had been measured synchronously with these investigations. Ehrly answered that the main focus of his investigations was the behavior of the tissue oxygen pressure and not the development of arterial pressure. Because the method was so complicated, it was technically difficult to study additional parameters along with the tissue oxygen pressure. Although other authors had shown that in claudicants the postischemic pressure did change, the behavior of the flow in the periphery in comparison with the measured values of the tissue PO_2 was more significant than the behavior of the pressure.

Lübbers (Dortmund) referred to less recent investigations by Schroeder and associates, in which the infusion of acetylcholine brought about a shift to the left of the histogram, despite an increase in muscle blood flow. He asked whether such a distribution disturbance in the microcirculation took place in claudicants and whether stenosis or occlusion in the large artery concealed this phenomenon. Ehrly replied that the maldistribution in the smallest muscle vessels was, according to studies carried out so far, an additional factor responsible for the deterioration of the oxygen supply to the tissue distally from the stenoses and occlusions. Several signs indicated that inhomogeneous perfusion of the capillaries occurs, under stress as well as under resting conditions, which impairs the oxygen supply to the tissue. More information would not become available until better ergometers with defined amounts of exercise were used for studies on the tissue PO_2 in the musculature. Ehrly said it was likely that the local acidosis via a relaxation of the tone of the smallest vessels caused the inhomogeneity of the blood flow and thus the maldistribution. Hauss (Münster) asked whether there were diabetics among the claudicants examined. Ehrly said no.

Creutzig (Hanover). Wodick (Ulm) inquired about the time course of the PO_2 under compression treatment. Creutzig replied that for methodical reasons, a sequence of measurements could not be carried out and therefore he could not answer the question. Lübbers (Dortmund) pointed out that investigations on the $tcPO_2$ in the vicinity of the ulcus had been carried out earlier, by Partsch among others. Partsch (Vienna) added that even before his investigations, Krähenbühl had found low PO_2 values in the region of the ulcus cruris. He was surprised to hear from Creutzig that when the leg was hanging down the PO_2 values rose, although the ulcers were purely venous. Creutzig answered that in cases of purely venous ulcers, the PO_2 improved when the leg was elevated. When the causes of the ulcers were arterial and venous, the $tcPO_2$ increased when the leg was lowered to a hanging position.

Heinrich et al. (Tübingen/Lübeck). Ehrly asked if along with the PO_2 values, flow values had also been measured to clarify maldistribution. The answer was no, but that such measurements were being planned.

Schultheiß et al. (Dortmund). A question was asked regarding the length of response time of the electrode for the investigations presented. The answer was 5–10 s.

Other points of discussion were the clinical relevance of the findings, the role of cerebral edema, and whether corresponding changes in the tissue PO_2 were to be expected, even when the cranial cavity was still intact, i.e., not opened. Lübbers (Dortmund) added that the differences in the histograms from different brain tissues, measured at different sites, could also be explained by the different amounts of local oxygen consumption. This had been established in experiments with animals.

Hauss (Münster) commented that on the basis of other investigations in addition to tissue PO_2 measurement, blood flow could be assumed to be higher in these tumors. He asked why the tissue PO_2 was then so low. A lengthy debate followed with Endrich (Heidelberg), among others, on the question of the tissue PO_2 of tumors and their consumption of oxygen. No agreement was reached on these points.

Thermann et al. (Kiel). Kopp (Freiburg) wanted to know whether these investigations had confirmed earlier findings from animal experiments, according to which the PO_2 on the surface of the intestine approached zero shortly after the intestine had been eviscerated. He also asked whether this finding had played a role in the interpretation of the data. Thermann said he assumed that the dog intestine was not suited for such experiments and that, consequently, a comparison with his investigations was not possible. As to the question to what extent the anesthetic halothane had an effect when the membrane was used, Thermann conceded that this factor had not yet been sufficiently studied.

A discussion followed on whether the intraluminal pressure should be taken into account. It was decided that for ethical reasons comparative measurements of this parameter could not be carried out.

Wodick (Ulm) reported on his own investigations in small animals among which the PO_2 values on the intenstine were relatively low and asked whether the higher values measured in man could not be explained by better blood supply. Thermann replied that this was not known, but that different factors enter into his PO_2 results such as hyperemia of the intestine and controlled ventilation. Comparisons were therefore difficult.

Ehrly et al. (Frankfurt/Main). It was asked whether in multimorbid patients with heart insufficiency, another reason for the drop in oxygen pressure of the tissue could be that the cardiac output deteriorates, even though the flow properties improve. Ehrly said that this was basically plausible, but that he had excluded patients with cardiac insufficiency from his group for investigation.

Steinacker (Ulm) reported on measurements of the $tcPO_2$ after isovolemic hemodilution with human albumin solutions. He also found an inhomogeneous behavior and, in some cases, a drop in the $tcPO_2$ under the human albumin solution. Endrich (Heidelberg) asked whether similar investigations with dextran solutions had been carried out, as the drop of the tissue PO_2 in patients who had been infused with human albumin could be interpreted as an intolerance reaction. Ehrly replied that such investigations were presently going on.

A discussion followed among several participants on the so-called optimum hematocrit. Results from animal experiments performed by the Meßmer group and earlier studies on on the administration of albumin done by the Hauss group were discussed. Ehrly pointed out that a distinction had to be made between hypervolemic hemodilution in patients with limited cardiovascular compensation, the animal experiments, and the clinical preoperative isovolemic hemodilution, where the heart and circulatory conditions are not altered. Ehrly claims the optimum hematocrit according to his measurements in patients with cardiovascular diseases, particularly intermittent claudication, is not below 40%. Hauss (Münster) confirmed this: the improvements of the tissue PO_2 he measured after the infusion of human albumin solutions were a result of the fact that the patients were primarily hypovolemic and therefore urgently needed intravasal liquid to stabilize their circulation.

Kersting et al. (Berlin/Lübeck). Hauss (Münster) referred to his earlier studies with dopamine which he carried out using a multiwire surface electrode. He explained that using different doses of dopamine, one could quasi titrate the histograms. If too much dopamine is given, the α-adrenergic effect prevails and this results in a shift to the left of the histogram.

Kopp et al. (Freiburg). Lübbers (Dortmund) asked if when recording the signals of the surface electrode all eight electrodes react the same or differently. In this connection he referred to animal experiments on the brain, in which diverging modes of behavior had been observed. Kopp answered that he also had come across diverging courses of recordings of individual measuring sites. Hauss (Münster) quoted animal experiments in which different modes of behavior of the tissue PO_2 had been found without any relation to a deviating parameter having been established. Kopp replied that the diversity of results in patients was strongly dependent on the cardiac index.

Transcutaneous Oxygen Pressure

Abendroth and Sunder-Plassmann (Munich). Franzeck (Heidelberg) asked how the thermographic measurements were standardized. The answer was that this was the only standardizable instrument available so far, and that the investigations had taken place in a room with a regulated temperature. Partsch (Vienna) pointed out that both thermography as a method and the assessment of skin temperature as a criterion were subject to controversy and that one should therefore concentrate on the changes in the transcutaneous oxygen measurements.

Another question was whether there were any correlations between the $tcPO_2$ measurements and the postoperative and intraoperative flow measurement. Abendroth referred to studies by his co-author and said he hoped that the combination of the methods introduced would make flow measurements clinically unnecessary in the future.

A short discussion followed on the indication for sympathectomy and whether it could be checked. As was to be expected, the discussion did not yield sufficient answers to these questions.

Creutzig et al. (Hanover). Lübbers (Dortmund) suggested that the measuring parameter no longer be called "transcutaneous PO_2," since heating the skin puts the autoregulation out of commission. Creutzig replied that 37 °C was a compromise, since no PO_2 measuring signal could be picked up with the transcutaneous method at a normal skin temperature of 32°–33 °C.

To Partsch's (Vienna) question why higher $tcPO_2$ values were measured poststenotically in patients with diabetes stage IV, no satisfactory answer could be given at that time.

Ehrly (Frankfurt) brought up the discrepancy that the $tcPO_2$ drops during adenosine triphosphate infusions, although this therapy is known to increase poststenotic circulation in the skin. A similar mechanism could be assumed with the infusion of prostaglandin (pharmacological vasodilatation). It would be interesting, he said, to know what happens when a placebo is given. Prof. Huch (Zürich) agreed that the interpretation of results in arterial occlusive disease poses difficulties.

Steinacker et al. (Ulm). A brief discussion took place between Franzeck (Heidelberg) and Partsch (Vienna) regarding the correlation of the $tcPO_2$ values with the stage of the disease according to Fontaine. It was agreed that there was no correlation.

Steinacker stated again that the different levels of the oxygen pressure values measured on the same leg of one patients – whether with the multiwire surface electrode or with the transcutaneous electrode – were not necessarily in contradiction, as different parameters are determined with the two methods.

Franzeck et al. (Heidelberg) and Partsch (Vienna). Lübbers (Dortmund) pointed out the problem in the method of direct measurement of the capillary circulation. Most methods measure larger areas of skin, which results in the blood of the larger plexus as well as the feeding vessels being included in the measurement. It could be for this reason that the xenon clearance method shows evidence of hyperemia, even though the flow is reduced in the region of the capillary loop. Lübbers believed that when using photo-optic methods, the diameter of the capillary must be measured along with the blood flow velocity in order to obtain quantitative information on the blood flow. This received general agreement. Partsch claimed that the disadvantage of directly viewing the capillary vessel was that one could not see deep enough. Lübbers argued that 100 µm was a considerable depth for a platinum electrode, particularly when the electrode itself consumes oxygen.

Rühle et al. (Freiburg). It was asked whether the measuring site had been changed during the long period of measurement to prevent damage caused by heat. Rühle answered that different measuring sites were used, but that the instruments were recalibrated. Thermal reactions were already visible after the electrode had been left in place for 5 h.

Fallenstein et al. (Zürich/Dortmund/Ulm). A lively discussion took place regarding the size of the heated area of a transcutaneous electrode and the possibility of selecting various temperatures (stepwise increase in temperature from 37° to 44 °C) that might improve the information provided by this method and also, perhaps, its diagnostic capability. It became clear that the data obtained from athletes could not be applied to the low-flow situation in patients with circulation disturbances, since hyperemic reaction to heat stimulus is also dependent on how fast and how extensively the increased blood flow can occur. There was agreement that this was still an open field for investigation.

Animal Experiments

Hartel (Ulm). Lübbers (Dortmund) warned that one should be careful not to try to correlate flow changes with tissue supply values. Endrich (Heidelberg) was unable to follow Hartel's conclusion that human albumin offered more advantages than plasma substitutes, because of the different test times for the various substances. Hartel replied that the investigation was limited to 2.5 h. Ehrly (Frankfurt) stated that when he administered human albumin in a study in patients with intermittent claudication, the tissue PO_2 values were even worse, and he suggested that the data on the value of therapeutic measures always be evaluated in connection with the investigation model selected. A short discussion followed with other participants including von Ackern (Munich), which emphasized the comparability of the individual investigation models.

Lemke et al. (Dortmund). The speaker was asked to which forms of shock he was referring in his talk. Lemke answered that in each case it had been cardiogenic shock. It was also asked if the lactate level in the blood correlated with the behavior of the $tcPO_2$. Lemke replied that no such investigations had been carried out. Several participants then discussed the behavior of the $tcPO_2$ value in healthy volunteers when the extremity was lowered or elevated. Lowering the arm to below the level of the heart generally caused the $tcPO_2$ value to drop. One colleague maintained that cigarette smoking alone reduced the $tcPO_2$ value. Huch (Zürich) contested this. Lübbers (Dortmund) tried to coordinate the diverging opinions and stated that naturally the $tcPO_2$ was dependent on the blood flow. In the majority of the healthy volunteers there were no problems; in some healthy persons (Lemke, for example) the skin circulation was very low. For that reason it was difficult to distinguish between the diagnoses of disturbed circulation and healthy circulation.

Ewald and Rooth (Uppsala). The speakers were asked how long it takes, in the case of a child with newly detected diabetes with pathologic vessel reactions, for the situation to return to normal. Rooth replied that this usually happens very quickly – within days – in children with diabetes detected at an early stage; if the disease has already existed for some time, the normalizing process takes much longer. Abendroth (Munich) reported on his investigations on children with diabetes with skin transplants, which supported Ewald's and Rooth's findings. Endrich (Heidelberg) commented that in some diabetic patients, the anamnesis goes back quite some time and that therefore the active vessel response is already poor. According to Intaglietta's calculations, these patients are likely to have increased peripheral resistance due to their reduced vasomotion. Thus the question is whether the situation could be helped by improvement of the flow properties of the blood.

Ott and Stuttgen (Berlin). Abendroth (Munich) repeated his view that the level of amputation cannot be determined with the $tcPO_2$ method. He was supported by Schönleben (Ludwigshafen), who asserted that orthopedic, i.e., prosthetic, factors have the decisive say regarding the amputation level. Franzeck (Heidelberg) maintained, however, that skin blood flow could very well be decisive as to whether healing could be expected after amputation.

Funk et al. (Heidelberg). Heinrich (Würzburg) stated that in similar studies his group had not found any edema formation or flow inhomogeneity. He wanted to know whether the Heidelberg group could rule out inflammation in their chamber model. Funk answered that the animals had been carefully selected and that only those animals were accepted that had normal vasomotion in the tissue examined. Furthermore, the vessel status, the leukocyte adherence, and other parameters had been observed, including the permeability for high-molecular dextran.

Hauss (Münster) asked how high the mortality of the hamsters was and whether they might perhaps have a very sensitive reaction to the infusion of fluid. His question was whether this animal species was suitable for such experiments. Funk replied that there was no common cause of death among the animals; they were not treated any further after the experiment, however, which might be an explanation for the high mortality. Lübbers (Dortmund) claimed that although he had not had much experience with hamsters, he would assume that there could be a chronic wound presenting inflammatory changes. Funk and Endrich explained that when the chamber was inserted, there were signs of inflammation for a few days, but that histologically and functionally no sign of inflammation could be found for a period of several days. Not until later, i.e., after about 12 days. did signs of inflammation recur.

Fleckenstein et al. (Lübeck). Lunkenheimer (Münster) inquired as to whether body weight had been checked during the investigations in order to quantify water retention. The answer was no. Scheffler (Engelskirchen) wanted to know now the hematocrit values could return to normal so quickly after the infusion of saline solution. Fleckenstein answered that, in his opinion, the fluid left the blood stream very rapidly; signs of inhomogeneity after infusions of saline solutions did not show up until after 2 h. Lübbers (Dortmund) asked if there were any explanations for the discrepancy between the results presented and earlier findings obtained with the surface electrode. The answer was that possibly the weight of the surface electrode when the experimental systemic pressure values are low causes a reduction in flow. Lübbers maintained that the other way around, tissue damage could result from inserting a macroelectrode, which could after the regulation of microcirculation. Ehrly (Frankfurt) asked how one could be sure that, when making repeated measurements at the same site in the musculature, one is not measuring in a mixture of cell fragments and blood. Fleckenstein replied that because of the special features of his electrode, a different track could be used for the second insertion so that measurements did not have to be made in a used track. Endrich rejected Fleckenstein's claim that the discrepancy between his results and those of earlier investigations with the surface electrode was caused by compression of the microvessels on the surface. He cited studies by Hauss and others, who did not necessarily find a drop in the PO_2 measuring value when blood pressure was lowered with pharmacological substances.

Ruland et al. (Münster). It was asked whether the investigations had been carried out during normothermia as well as hypothermia. One could imagine that there could be a tissue-protecting effect during hypothermia. The answer was that there were plans to study this point. One member of the audience wanted to know if the investigators could yet state whether any transplant could be considered successful. The answer to this question was no; it was added, however, that studies in dogs in situ were being planned, in which the PO_2 measuring value was to be correlated with the function of lhe kidney.

Heinrich et al. (Tübingen/Lübeck). A participant asked if, in the case of larger test animals, it were possible to insert the electrode directly into the spinal cord. The answer was yes.

Hauss et al. (Münster). The speaker was asked whether it were possible that the volume situation differed among the various groups in the same way as the cardiac output and the oxygen consumption did. This could explain the different modes of behavior. Hauss replied that the control animals were also pregnant but did not recieve endotoxin. He said the cardiac output played a substantial role, but that it had not been measured in this particular series of measurements. In the pregnant animals, there was cardiac compensation by a markedly increased cardiac output compared with the norm. Slight disturbances, however, could cause cardiac decompensation. Another question was whether the postulated vasoconstriction could be corroborated by measuring the catecholamines. Hauss said that no such studies had yet been systematically carried out. As to the question whether there was a change in the potassium content of the plasma, the answer was that the values were within the normal range.

Ruland et al. (Münster). Upon request, the course of the experiment was described as follows: first, an oxygen histogram is made, and the PO_2 histogram is controlled after 20 min of continuous administration of somatostatin. A participant wished to know if direct blood flow measurements had been carried out on the pancreas, since it was known from the literature that the blood flow is considerably reduced under such pathologic conditions. Ruland replied that direct measurements of the blood flow had not been taken, but that his investigations had been prompted by the correlations known from the literature. Heinrich (Würzburg) commented that somatostatin had a vasoconstrictory effect, which could lead to reduced circulation. He suggested that prostaglandin be experimentally infused at the same time.

Asked if it were not so that the local oxygen supply and absolutely nothing to do with the circulation, Ruland answered that that was not necessarily true. Lübbers (Dortmund) supposed that one factor had to be changed for the balance to be right. That factor might be oxygen consumption; but more probably it was a redistribution that leads to hypoxia. Lübbers wanted to know if evidence had been found of inhomogeneous distribution. Ruland said he believed that those organs had a large complnsation range; measurements on the liver had yielded similar results.

Böhm et al. (Mannheim). Ehrly (Frankfurt) said that measurements made directly in the ischemic lower leg musculature of patients with intermittent claudication had never shown normal or even raised oxygen pressure values. On average, the tissue PO_2 in these patients was only half as high as in control persons with undisturbed vascular systems. He emphasized that his experience with tissue PO_2 measurements under exercise conditions showed that in claudicants, the method provided more information than the simple measurement of tissue PO_2 under rest conditions. He then asked if the electric stimulation on the dog leg caused a contraction against a resistance. Böhm responded that only a single contraction of the muscle had been generated, which meant that there were marked differences between Ehrly's investigations on patients and his using an animal model. Ehrly cited earlier studies by Schröder which showed that with nonligated arteries in the extremity and under exercise conditions, the tissue PO_2 in the musculature increased. Furthermore, it was pointed out that the two experiments were not comparable, because the experimental animal has to be anesthetized. Böhm explained that the type of stimulus was important; in his model, a nonphysiologic, rectangular type of stimulus was purposely used to quickly tire out the muscle. Lunkenheimer (Münster) pointed out that what is decisive is to determine not the type of stimulus, but rather the response to the stimulus, by somehow quantifying the exercise or performance. Böhm cited studies by Wrabetz et al., who found normal muscle PO_2 values using a multiwire electrode, even in the presence of arterial occlusive disease. Hauss (Münster) confirmed this, but explained that those investigations had been carried out on the thigh, while Ehrly's studies had been performed in the poststenotic area of the calf musculature.

Kohaus et al. (Münster). The speaker asked the audience if anyone had any ideas why there was a drop in the tissue PO_2 values under the administration of vasopressin, but not under triglycyl-lysil-vasopressin, even though the hemodynamic effects of both substances were comparable. Ehrly (Frankfurt) suspected that there was maldistribution in the microcirculation; Kohaus himself said that the effect of the two substances set in at different times, which could explain the differences.

Keller (Würzburg). It was asked if during long-term manipulations the microcirculation could be disturbed by vaporization on the surface. Keller replied that his model had special features that allowed the muscle to be returned to its original place after the electrode was positioned and the skin above it could be closed again. Lübbers (Dortmund) suggested the PO_2 be determined per minute, as one could derive information on the breathing activity grom the steepness of that parameter. Keller added that his studies should have shown that changes in the tissue PO_2 could be detected with the method. Lübbers said that such investigations could also be done with the multiwire surface electrode; the electrode could be applied with a special adapter by means of adhesive tape.

A member of the audience asked the speaker how long the response time of the electrode was when placed on pedicle skin flaps and how soon changes in the PO_2 values allowed a prognosis of beginning necrosis. Keller replied that the response time was about 20 min; however, this was not so important in a long-term measurement. He further explained that in one patient in whom the PO_2 value was low, there was superficial necrosis. During the further course of the discussion, questions on the revascularization of skin transplants were dealt with. Another point of discussion was what methods could best be used to compare the advantages and disadvantages of the individual procedures.

Endrich et al. (Heidelberg). Schönleben (Ludwigshafen) asked if histologic and electron-optic investigations had been carried out parallel to the PO_2 investigations. Endrich answered that histologic examinations had been performed on two types of preparations, though no long-term studies had been made. The histologic examinations were able to prove the formation of pseudoendothelia as a result of aggregates of thrombocytes. Further, extensive damage to the tumor cells and the cells in the neighboring healthy muscle tissue could be shown.

Finally, there was a lengthy discussion between the speaker, Lübbers, Heinrich, Hauss, and others on local regulation of the resistance vessels in the tumors. Of particular interest - and which remained unclear - was the mechanism of vasoregulation. To date there had only been hypotheses concerning the mechanism, which had been published by various authors, including Folkow. Both biogenetic regulation mechanisms and rheologic factors were suggested as bringing about a change in the microflow. Also striking were the findings of a very decided threshold temperature, above which the tissue PO_2 changes considerably. The intensity of the discussion not only demonstrated the amount of interest there was in that field of research, but also how many questions were still to be answered.

In his concluding remarks Ehrly (Frankfurt) thanked all of the speakers and participants and particularly emphasized the enthusiasm shown during the discussion. It should be the purpose of such a specialized symposium, he said, to offer a chance for thorough discussion of the latest findings, even if the time allotted for it is exceeded. Ehrly expressed his gratitude to the chairmen, the two organizers, Professor Huch and Dr. Hauss, and especially to his colleagues for their contribution to the success of the symposium.

MIX
Papier aus verantwortungsvollen Quellen
Paper from responsible sources
FSC® C105338

If you have any concerns about our products,
you can contact us on
ProductSafety@springernature.com

In case Publisher is established outside the EU,
the EU authorized representative is:
Springer Nature Customer Service Center GmbH
Europaplatz 3, 69115 Heidelberg, Germany

Printed by Libri Plureos GmbH
in Hamburg, Germany